Global Finance and Development

The question of money, how to provide it, and how to acquire it where needed, is axiomatic to development. The realities of global poverty and the inequalities between the 'haves' and the 'have-nots' are clear and well documented, and the gaps between the world's richest and the world's poorest are ever-increasing. But, even though funding development is assumed to be key, the relationship between finance and development is contested and complex.

This book explores the variety of relationships between finance and development, offering a broad and critical understanding of these connections and perspectives. It breaks finance down into its various aspects, with separate chapters on aid, debt, equity, microfinance and remittances. Throughout the text, finance is presented as a double-edged sword: while it is a vital tool towards poverty reduction, helping to fund development, more critical approaches remind us of the ways in which finance can hinder development. It contains a range of case studies throughout to illustrate finance in practice, including, UK aid to India, debt in Zambia, Apple's investment in China, microfinance in Mexico, government bond issues in Chile and financial crisis in East Asia. The text develops and explores a number of themes throughout, such as the relationship between public and private sources of finance and debates about direct funding versus the allocation of credit through commercial financial markets. The book also explores finance and development interactions at various levels, from the global structure of finance through to local and everyday practices.

Global Finance and Development offers a critical understanding of the nature of finance and development. This book encourages the reader to see financial processes as embedded within the broader structure of social relationships. Finance is defined and demonstrated to be money and credit, but also, crucially, the social relationships and institutions that enable the creation and distribution of credit and the consequences thereof. This valuable text is essential reading for all those concerned with poverty, inequality and development.

David Hudson is a Senior Lecturer in Political Economy. His principal research interests lie broadly in the political economy of development. More specifically he is interested in public engagement with global poverty, the international political economy of development finance and the politics of development.

Routledge Perspectives on Development

Series Editor: Professor Tony Binns, *University of Otago*

Since it was established in 2000, the same year as the Millennium Development Goals were set by the United Nations, the *Routledge Perspectives on Development* series has become the pre-eminent international textbook series on key development issues. Written by leading authors in their fields, the books have been popular with academics and students working in disciplines such as anthropology, economics, geography, international relations, politics and sociology. The series has also proved to be of particular interest to those working in interdisciplinary fields, such as area studies (African, Asian and Latin American studies), development studies, environmental studies, peace and conflict studies, rural and urban studies, travel and tourism.

If you would like to submit a book proposal for the series, please contact the Series Editor, Tony Binns, on: jab@geography.otago.ac.nz

Published:

Third World Cities, 2nd edition
David W. Drakakis-Smith
HB 0415-19881X, PB 0415-198828

Rural-Urban Interactions in the Developing World
Kenneth Lynch
HB 0415-258707, PB 0415-258715

Environmental Management & Development
Chris Barrow
HB 0415-280834, PB 0415-280842

Tourism and Development
Richard Sharpley & David J. Telfer
HB 978-0-415-37144-5 PB
978-0-415-37151-3

Southeast Asian Development
Andrew McGregor
HB 978-0-415-38416-2 PB
978-0-415-38152-9

Population and Development
W.T.S. Gould

Postcolonialism and Development
Cheryl McEwan

Conflict and Development
Andrew Williams & Roger MacGinty

Disaster and Development
Andrew Collins

Non-Governmental Organisations and Development
David Lewis and Nazneen Kanji

Cities and Development
Jo Beall

Gender and Development, 2nd edition
Janet Momsen

Economics and Development Studies
Michael Tribe, Frederick Nixson and Andrew Sumner

Water Resources and Development
Clive Agnew and Philip Woodhouse

Theories and Practices of Development, 2nd edition
Katie Willis

Food and Development
E. M. Young

An Introduction to Sustainable Development, 4th edition
Jennifer Elliott

Latin American Development
Julie Cupples

Global Finance and Development

David Hudson

LONDON AND NEW YORK

First published 2015
by Routledge
2 Park Square, Milton Park, Abingdon, Oxon OX14 4RN

and by Routledge
711 Third Avenue, New York, NY 10017

Routledge is an imprint of the Taylor & Francis Group, an informa business

British Library Cataloguing in Publication Data
A catalogue record for this book is available from the British Library

Library of Congress Cataloging in Publication Data
Global finance and development / David Hudson.
pages cm. — (Routledge perspectives on development)
1. Economic development.   2. Economic development—Finance.
3. International finance.   I. Title.
HD82.H783 2014
332□.042—dc23
2013022772

ISBN: 978-0-415-43634-2 (hbk)
ISBN: 978-0-415-43635-9 (pbk)
ISBN: 978-0-203-38125-0 (ebk)

Typeset in Times New Roman and Franklin Gothic
by RefineCatch Limited, Bungay, Suffolk

Printed and bound by CPI Group (UK) Ltd, Croydon, CR0 4YY

Contents

Tables

Figures

Boxes

List of abbreviations

ACU	Asian Currency Unit
ADB	Asian Development Bank
AfDB	African Development Bank
AfT	Aid-for-Trade
ASEAN	Association of Southeast Asian Nations
BAR	Bilateral Aid Review
BCBS	Basel Committee for Banking Supervision
BIS	Bank of International Settlements
BIT	bilateral investment treaty
CCT	conditional cash transfer
CGAP	Consultative Group to Assist the Poor
CPIA	Country Policy and Institutional Assessment
CPR	Country Performance Rating
CRS	Creditor Reporting System
DAC	Development Assistance Committee
DCD	Development Co-operation Directorate
DEC	Disasters Emergency Committee
DFID	Department for International Development
DFI	development finance institution
DRS	Debtor Reporting System
DSF	Debt Sustainability Framework
DTTs	double taxation treaties
EC	European Commission
ECAs	export credit agencies
EPZs	export processing zones
FAFT	Financial Action Task Force
FDI	foreign direct investment

FfD	financing for development
FIBV	World Federation of Stock Exchanges
GATS	General Agreement on Trade in Services
GE	grant element
GNH	gross national happiness
GNI	gross national income
GNP	gross national product
GRWG	Global Remittances Working Group
HDI	Human Development Index
HIC	high-income country
HIPC	Heavily Indebted Poor Countries
IADB	Inter-American Development Bank
IATI	International Aid Transparency Intitative
IBRD	International Bank for Reconstruction and Development
ICC	International Chamber of Commerce
ICP	International Comparison Program
ICSA	International Council of Securities Associations
ICSID	International Centre for Settlement of International Disputes
ICU	International Clearing Union
IDA	International Development Association
IFAD	International Fund for Agricultural Development
IFC	International Finance Corporation
IFI	international financial institution
IIA	international investment agreement
IMCU	International Money Clearing Unit
IMF	International Monetary Fund
IOSCO	International Organisation of Securities Commissions
IPOs	initial public offering
ISAB	International Accounting Standards Board
ISMA	International Securities Market Association
LDC	least developed country
LIC	low-income country
LLDC	landlocked developing country
LMIC	lower middle-income country
M&A	mergers and acquisitions
MDG	Millennium Development Goal
MDRI	Multilateral Debt Relief Initiative
MFI	microfinance institution
MIC	middle-income country
MIF	Multilateral Investment Fund
MIGA	Multilateral Investment Guarantee Agency
MIV	microfinance investment vehicle
MPI	Multidimensional Poverty Index
MTO	money transfer organisation
NGO	non-governmental organisation

NIC	newly-industrialised country
ODA	official development assistance
OOF	other official flows
PEF	political economy of finance
PPP	purchasing power parity
PQLI	Physical Quality of Life Index
PRI	Pakistan Remittance Initiative
PRSPs	Poverty Reduction Strategy Papers
RCT	randomised control trial
ROSCA	rotating savings and credit association
RSP	remittance service provider
SADC	Southern African Development Community
SAL	structural adjustment loan
SAP	structural adjustment programme
SBS	sector budget support
SDR	special drawing right
SDRM	Sovereign Debt Restructuring Mechanism
SEZ	special economic zone
SHGs	self-help groups
SIDS	small island developing states
SIMSDI	Survey of the Implementation of Methodological Standards for Direct Investment
SME	small and medium size enterprise
SWAp	sector-wide approach
TARP	Troubled Asset Recovery Plan
TI	transnationality index
TNC	transnational corporation
TPS	temporary protected status
UMIC	upper middle-income country
UNCITRAL	UN Commission on International Trade Law
UNCTAD	United Nations Conference on Trade and Development
UNDP	UN Population Division
USAID	United States Agency for International Development
WHO	World Health Organization
WTO	World Trade Organization

⬤ Acknowledgments

The original idea for this book came from teaching several brilliant cohorts of MSc students at UCL. The Political Economy of Development module covered many of the topics that became the chapters of the book. It was always a source of frustration and surprise that there wasn't an obvious text that covered the breadth of development finance in enough depth to be interesting, and yet accessible enough to ease in those with little prior knowledge of finance. I may now understand why such a book didn't exist. The task was much harder than I'd anticipated. But I hope the final text provides breadth and depth and finds the balance between setting out fundamentals as well as providing direction for the more critical or curious to travel. That was always the intention, at least.

Many people have helped in writing this book. Some through their support and love, some through their patience and others through their inspiration and encouragement that such a book needed to be written in the first place. First, many thanks go to all the students who made teaching the Political Economy of Development module such a joy. Second, a deep debt of gratitude goes to the editorial team at Routledge. They have given St Monica a run for her money. Thank you especially to Andrew Mould, Michael Jones, Faye Leerink and Sarah Gilkes for their carrots, sticks, advice and help along the way. It wouldn't have happened without them. Third, thanks to Rigas, Popi, Yannis, and Carmen for letting me hide away in Skyros one summer. Next time will be different. Likewise to my parents for the caravan in Durdle Door. Fourth, I also owe considerable thanks to Jon Ingoldby and Kyriaki Papachristoforou for immense help with the copy-editing and bibliography, respectively. Finally, the biggest and deepest thanks go to my family – most especially to Jennifer: thank you, for everything. I think she will be even happier than I that the book got finished.

1 Introduction

Learning outcomes

At the end of this chapter you should:

- Appreciate the importance attached to external development financing, especially in relation to meeting international development targets.
- Understand the logic of the financial system in mobilising capital for investment.
- Know the international context that development finance operates in, in terms of poverty, inequality and welfare indicators.
- Understand that there are different ways of viewing the relationship between finance and development: finance as a resource for development and as a social structure within which development happens.
- Develop your own position on what 'finance' is, understanding that it is an essentially contested concept.

Key concepts

Financing for development (FfD), political economy of finance (PEF), Millennium Development Goals (MDGs), vision, development, global finance, power

Introduction

The realities of global poverty and the inequalities between the 'haves' and the 'have-nots' are clear and well documented. The gaps between the world's richest countries and the poorest have grown larger, despite the sustained growth of a sub-group of large industrialising countries such as China, Brazil and India. And inequalities between individuals, across countries, have grown larger. The richest 5 per cent of people receive one-third of total global income, the same amount as the poorest 80 per cent (Milanovic 2005). While the average American's income doubled between 1963–2000, over that same period there were 180 million Africans living in countries that ended up being poorer by the end of that period than they were nearly 40 years earlier.

Nevertheless, good change is also in evidence. For example, Charles Kenny (2011) has been at pains to point out that things are getting better and global development can succeed; so much so that he calls for a 'realistic optimism'. He argues that almost every other indicator of quality of life (i.e., apart from income) points towards rapid and universal improvements in life chances. Plus, health, education and civil rights observance have all improved *even* in countries that have witnessed per capita income declines over the past 30 years.

The latest poverty figures from the World Bank report that in 2008 there were 801 million people living below $1 a day, which is 14 per cent of the developing world's population. The dollar a day metric is the target enshrined in the first Millennium Development Goal (MDG). The aim is 'to halve, between 1990 and 2015, the proportion of people whose income is less than $1 a day' (MDG1) (see Table 1.1 for a full list of the eight MDGs and accompanying targets). In March 2012, the World Bank announced that the goal had been met as the proportion of people in the developing world living on less than a dollar a day has fallen from 31 per cent in 1990 (and 42 per cent in 1981) (Chen and Ravallion 2012). Yet, despite this apparent success, critics point out that the MDG target is a very meagre measure of poverty (see Box 2.2). Moreover, much of this success can be explained by economic growth in China, but much less so elsewhere.

Table 1.1 *The Millennium Development Goals (MDGS)*

Goals and targets (from the Millennium Declaration)	Indicators for monitoring progress
Goal 1: Eradicate extreme poverty and hunger	
Target 1.A: Halve, between 1990 and 2015, the proportion of people whose income is less than one dollar a day	1.1 Proportion of population below $1 (PPP) per day 1.2 Poverty gap ratio 1.3 Share of poorest quintile in national consumption
Target 1.B: Achieve full and productive employment and decent work for all, including women and young people	1.4 Growth rate of GDP per person employed 1.5 Employment-to-population ratio 1.6 Proportion of employed people living below $1 (PPP) per day 1.7 Proportion of own-account and contributing family workers in total employment
Target 1.C: Halve, between 1990 and 2015, the proportion of people who suffer from hunger	1.8 Prevalence of underweight children under five years of age 1.9 Proportion of population below minimum level of dietary energy consumption
Goal 2: Achieve universal primary education	
Target 2.A: Ensure that, by 2015, children everywhere, boys and girls alike, will be able to complete a full course of primary schooling	2.1 Net enrolment ratio in primary education 2.2 Proportion of pupils starting Grade 1 who reach last grade of primary 2.3 Literacy rate of 15–24-year-olds, women and men
Goal 3: Promote gender equality and empower women	
Target 3.A: Eliminate gender disparity in primary and secondary education, preferably by 2005, and in all levels of education no later than 2015	3.1 Ratios of girls to boys in primary, secondary and tertiary education 3.2 Share of women in wage employment in the non-agricultural sector 3.3 Proportion of seats held by women in national parliament
Goal 4: Reduce child mortality	
Target 4.A: Reduce by two-thirds, between 1990 and 2015, the under-5 mortality rate	4.1 Under-5 mortality rate 4.2 Infant mortality rate 4.3 Proportion of 1-year-old children immunised against measles

Goal 5: Improve maternal health		
Target 5.A: Reduce by three-quarters, between 1990 and 2015, the maternal mortality ratio	5.1 5.2	Maternal mortality ratio Proportion of births attended by skilled health personnel
Target 5.B: Achieve, by 2015, universal access to reproductive health	5.3 5.4 5.5 5.6	Contraceptive prevalence rate Adolescent birth rate Antenatal care coverage (at least one visit and at least four visits) Unmet need for family planning
Goal 6: Combat HIV/AIDS, malaria and other diseases		
Target 6.A: Have halted by 2015 and begun to reverse the spread of HIV/AIDS	6.1 6.2 6.3 6.4	HIV prevalence among population aged 15–24 years Condom use at last high-risk sex Proportion of population aged 15–24 years with comprehensive correct knowledge of HIV/AIDS Ratio of school attendance of orphans to school attendance of non-orphans aged 10–14 years
Target 6.B: Achieve, by 2010, universal access to treatment for HIV/AIDS for all those who need it	6.5	Proportion of population with advanced HIV infection with access to antiretroviral drugs
Target 6.C: Have halted by 2015 and begun to reverse the incidence of malaria and other major diseases	6.6 6.7 6.8 6.9 6.10	Incidence and death rates associated with malaria Proportion of children under 5 sleeping under insecticide-treated bednets Proportion of children under 5 with fever who are treated with appropriate anti-malarial drugs Incidence, prevalence and death rates associated with tuberculosis Proportion of tuberculosis cases detected and cured under directly observed treatment short course
Goal 7: Ensure environmental sustainability		
Target 7.A: Integrate the principles of sustainable development into country policies and programmes and reverse the loss of environmental resources	7.1	Proportion of land area covered by forest
Target 7.B: Reduce biodiversity loss, achieving, by 2010, a significant reduction in the rate of loss	7.2 7.3 7.4 7.5 7.6 7.7	CO_2 emissions, total, per capita and per \$1 GDP (PPP) Consumption of ozone-depleting substances Proportion of fish stocks within safe biological limits Proportion of total water resources used Proportion of terrestrial and marine areas protected Proportion of species threatened with extinction
Target 7.C: Halve, by 2015, the proportion of people without sustainable access to safe drinking water and basic sanitation	7.8 7.9	Proportion of population using an improved drinking water source Proportion of population using an improved sanitation facility
Target 7.D: By 2020, to have achieved a significant improvement in the lives of at least 100 million slum dwellers	7.10	Proportion of urban population living in slums
Goal 8: Develop a global partnership for development		
Target 8.A: Develop further an open, rule-based, predictable, non-discriminatory trading and financial system		*Some of the indicators listed below are monitored separately for the least-developed countries, Africa, landlocked developing countries and small island developing states.*
Includes a commitment to good governance, development and poverty reduction – both nationally and internationally	*ODA* 8.1 8.2	Net ODA, total and to the least developed countries, as percentage of OECD/DAC donors' gross national income Proportion of total bilateral, sector-allocable ODA of OECD/DAC donors to basic social services (basic education, primary health care, nutrition, safe water and sanitation)
Target 8.B: Address the special needs of the least developed countries		
Includes: tariff and quota free access for the least developed countries' exports; enhanced programme of debt relief for heavily indebted poor countries (HIPCs) and cancellation of official bilateral debt; and more generous ODA for countries committed to poverty reduction	8.3 8.4	Proportion of bilateral official development assistance of OECD/DAC donors that is untied ODA received in landlocked developing countries as a proportion of their gross national incomes

Table 1.1 *Continued*

Goals and targets (from the Millennium Declaration)	Indicators for monitoring progress
Target 8.C: Address the special needs of landlocked developing countries and small island developing States (through the Programme of Action for the Sustainable Development of Small Island Developing States and the outcome of the twenty-second special session of the General Assembly) Target 8.D: Deal comprehensively with the debt problems of developing countries through national and international measures in order to make debt sustainable in the long term	8.5 ODA received in small island developing States as a proportion of their gross national incomes **Market access** 8.6 Proportion of total developed country imports (by value and excluding arms) from developing countries and least developed countries, admitted free of duty 8.7 Average tariffs imposed by developed countries on agricultural products and textiles and clothing from developing countries 8.8 Agricultural support estimate for OECD countries as a percentage of their gross domestic product 8.9 Proportion of ODA provided to help build trade capacity **Debt sustainability** 8.10 Total number of countries that have reached their HIPC decision points and number that have reached their HIPC completion points (cumulative) 8.11 Debt relief committed under HIPC and MDRI Initiatives 8.12 Debt service as a percentage of exports of goods and services
Target 8.E: In cooperation with pharmaceutical companies, provide access to affordable essential drugs in developing countries	8.13 Proportion of population with access to affordable essential drugs on a sustainable basis
Target 8.F: In cooperation with the private sector, make available the benefits of new technologies, especially information and communications	8.14 Fixed telephone lines per 100 inhabitants 8.15 Mobile cellular subscriptions per 100 inhabitants 8.16 Internet users per 100 inhabitants

The Millennium Development Goals and targets come from the Millennium Declaration, signed by 189 countries, including 147 heads of State and Government, in September 2000 (http://www.un.org/millennium/declaration/ares552e.htm) and from further agreement by member states at the 2005 World Summit (Resolution adopted by the General Assembly – A/RES/60/1, http://www.un.org/Docs/journal/asp/ws.asp?m=A/RES/60/1). The goals and targets are interrelated and should be seen as a whole. They represent a partnership between the developed countries and the developing countries 'to create an environment – at the national and global levels alike – which is conducive to development and the elimination of poverty'.

Source: Official list of MDG indicators, effective 15 January 2008. Online. Available: http://mdgs.un.org/unsd/mdg/Host.aspx?Content=Indicators/OfficialList.htm (accessed 21 December 2009).

In addition, economic growth and even reductions in income poverty do not necessarily increase people's freedoms nor reduce everyday deprivations such as access to clean water, schooling, or a stable and nutritious diet. Between December 2011 and January 2012, just before the World Bank announced its latest poverty figures, a set of surveys were being carried out in India, Pakistan, Nigeria, Peru and Bangladesh (GlobeScan 2012). The results highlighted the interaction of poverty and rising food prices; a third of households in four of the countries reported periods when they ate the same staple food for a week at a time. In Nigeria, only one in four can often afford nutritious food such as meat, milk or vegetables for their family. Worse still, 27 per cent of parents in Nigeria and 24 per cent in India report that their children go without food for an entire day. This matters. Maternal and child undernutrition is responsible for about 35 per cent of child deaths and 11 per cent of the total global disease burden (Black *et al.* 2008).

The charity Save the Children (2012) reports that 300 children die every hour of every day due to malnutrition, which totals 2.6 million child deaths per year (Black *et al.*

2008). And for those who survive, the consequences of malnutrition are life-long. It is estimated that some 48 per cent of children in India are stunted. Malnourished children cannot develop fully, physically or mentally, suffering from reduced IQs. And when they become adults, malnourished children can expect, on average, to earn at least 20 per cent less than those who weren't malnourished.

Yet there are solutions: food fortification, vitamin supplements, promoting healthy practices such as handwashing, cash transfers, agricultural investment in fertilisers and land reform are all proffered as potential fixes. But, of course, they have to be paid for. For example, a package of 13 direct interventions, which includes some of the fixes above, was recently costed by the World Bank. The Bank estimated that:

> An additional US$10.3 billion a year is required from public resources to successfully mount an attack against undernutrition on a worldwide scale. This would benefit more than 360 million children in the 36 countries with the highest burden of undernutrition – home to 90 percent of the stunted children worldwide – and prevent more than 1.1 million child deaths.
>
> (Horton *et al.* 2010: ix)

This estimated financing gap of US$10.3 billion, the report concluded, would be borne by households, national governments, aid donors, private sector corporations and international initiatives (Horton *et al.* 2010).

So what does it all add up to? The extraordinary growth of China and India, encouraging improvements in the quality of life, large-but-falling global poverty figures, obnoxious and increasing inequality, outrageous child malnutrition and morbidity figures, and relatively modest financing costs to address the worst deprivations all come together to paint a dynamic and complex picture. At the risk of sounding trite, it is a picture that suggests that much is to be done, *but* that things *can* be done; both in the sense that they can make a difference and they are affordable.

So what is to be done? The problem of *how* to successfully address poverty and foster sustainable and equitable development across and within countries is less clear (to indulge in a somewhat dramatic understatement). This is not, of course, to say that there hasn't been, and continues to be, a proliferation of attempts, plans, projects and ideas which have been proposed and tried.

Some have succeeded – for example, the improvements in literacy rates driven by increased access to education such as in Ethiopia, where primary enrolment had risen to 15.5 million by 2008/09, an increase of over 500 per cent from 1994/95 (Engel 2011), or the global eradication of smallpox by the end of the 1970s after a global campaign (WHO 2001), or improvements in access to safe water such as in Lao PDR where access to basic sanitation in rural areas rose from an estimated 10 per cent in 1995 to 38 per cent in 2008 (O'Meally 2011), or economic growth and industrialisation in East Asia (World Bank 1993), and investments in social protection leading to falling poverty and inequality such as in Brazil, where the Gini coefficient fell by 5.2 points and the percentage of households living below the poverty line halved between the early 1990s and 2008 (Holmes *et al.* 2011).

Yet many have failed – supply-led aid has resulted in 'empty schools and hospitals, crumbling highways and silted dams' (Browne 2007), the $2 billion Bataan nuclear power plant built on an earthquake fault line and never used (Hertz 2005), donated malaria nets ending up being sold, used as fishing nets or wedding veils (Easterly 2006), US food aid bankrupting local famers in Egypt and Haiti, sending them out of business (Bovard 1986), and the fact that sub-Saharan Africa's per capita income ended the 1980s lower than it was at the beginning of the 1960s (Lancaster 1983).

Geography, climate, culture, bad policies, weak institutions, are all explanations for poverty. Likewise, bad programme design, confounding politics, clientelism and corruption – domestic and international – poor monitoring, evaluation and not learning are all viable explanations for failing to reduce poverty, even if we can assume good intentions.

There is truth to all of these, but the premise of the current book is – whether a development intervention is large or small, whether it is to promote health outcomes or boost industrial productivity, whether it is an external initiative driven by wealthy countries or a home-grown programme, and even whether it succeeds or fails – that development requires resources, and more specifically financial resources: money and credit. This is true for a government, a firm, or households and individuals. Roads, infrastructure and factories need to be built, and populated with employees and machines. Individuals want to buy bicycles, or a shop or a plot of land, pay for school fees, books and uniforms, or the costs of attending the health clinic (see Box 1.1). Development, in the sense of increasing economic productivity or human freedoms, requires investment. Financial resources are a necessary, *but absolutely not sufficient*, element of engendering development. Somebody somewhere needs to pay for the investment; this is often the beneficiary, but not always. So, given that almost by definition, low- and many middle-income countries have historically lacked the savings and financial resources necessary for investment in development, this – from a financing for development perspective – is the problem for development.

Box 1.1

Finance and the poor

Isn't finance for the well off? Isn't borrowing and debt dangerous for the poor? Assumptions such as these are common, but the truth of the matter is that poor individuals and households rely on a great deal of financial activity – often highly complex and way in excess of their income. In their wonderful book, *Portfolios of the Poor*, Daryl Collins, Jonathan Morduch, Stuart Rutherford and Orlanda Ruthven (2009) document the financial portfolios of 250 poor households in Bangladesh, India, and South Africa. The researchers got the households in their sample to keep financial diaries to track how they managed their money. The findings are important and fascinating, revealing clearly the extent to which poor households rely upon borrowing and saving in order to live day-to-day.

The authors give the example of Hamid and Khadeja, a young Bangladeshi couple who moved from their coastal home town to the capital Dhaka. They were poor, living in a slum, sharing a toilet and kitchen with eight other families. He worked as a rickshaw driver and she ran the household and earned a little from sewing. Their monthly income was the equivalent of US$70 (taking into account the cost of living in Bangladesh). This puts them in the bottom two-fifths of the world's income and means that they're just below the international $2 per day poverty line. Yet, despite being poor, partially literate, with unsteady jobs and income, Hamid and Khadeja were active financial participants – they had money saved in a number of different places, from a life insurance policy, to a savings account with a microfinance institution and another with a moneyguard, as well as remitting savings home to Hamid's parents, a loan made to a relative and a small pot of $2 at home for emergencies. Plus they were actively borrowing money with a microfinance loan, interest-free loans from family, neighbours, employer, and local grocery store, as well as looking after money for neighbours who were trying to save. Collins *et al.* (2009) show that Hamid and Khadeja managed their money extremely well. Overall they were in debt, but by a small amount. Moreover, over the course of a year they actively saved, borrowed, repaid and refinanced their savings such that they moved more money than they brought in in income. Why is their financial life so intense? The simple answer is because they're poor.

Life goes on. Costs come along, both expected and unexpected. Weddings, funerals, school, doctors' fees, a business opportunity, a household repair. Things have to be saved for or paid in haste. In such circumstances you either go without, are forced to sell assets (not always possible, or you can be forced to accept a bad price if selling in desperation), or turn to finance – i.e. use past income (savings) or future income (borrow) to pay for present expenses. Individuals with small, uncertain, irregular incomes need to use financial services more than the non-poor with more financial security and reserves. In reality the financial services the poor make use of are a mixture of formal, semi-formal and informal; from microfinance institutions to savings clubs and moneylenders to friends and family. In sum, using financial services is both necessary and a daily reality for the world's poor.

The question that motivates this book is: what happens when external capital flows are required for development? This is a broader issue than just how can we best fund the investment needed for growth, poverty reduction and human development. The initial question leads to a whole series of further questions: does access to external sources of finance support positive development outcomes? Does it produce sustainable, equitable development? Do different types of financial flows – aid, foreign direct investment (FDI) or remittances, say – work better or less well than others? Under what conditions? Do they improve economic growth, or health outcomes, or gender equality? What are the social and political implications of increased external financial penetration into developing countries? For, while a lack of financial resources may well be a primary constraint on growth – preventing poor countries from developing or limiting individuals' freedoms – capitalist finance is not a neutral resource. Increased embeddedness within the circuits of financial capital can equally reinforce economic and political hierarchies and cement exploitative social relations. We must supplement the initial development financing question (Who pays and how?) with the subsequent political economy questions listed above. The rest of this chapter is dedicated to outlining a series of 'visions' – as ways of looking at – of finance and development.

Box 1.2

The political economy of development

This box could also have been called: 'Do I need to be an economist to talk about the economy?' The answer is no, but neither can you ignore economics. This book is written from a political economy perspective. That is, it is focused on questions of well-being and the economy, but it is also explicitly driven by moral questions about this distribution as well as an empirical and theoretical concern with the broader political and social context in which economic outcomes come about.

Political economy is an older tradition than modern economics and one which stretches back to Karl Marx, Thomas Malthus, Adam Smith, François Quesnay and Aristotle. Political economy is concerned with the interaction between wealth and power, between markets and states. This is because it is an approach driven by a belief that the distribution of wealth and life opportunities are driven as much by questions of power and institutional conflict as a narrower concern with understanding the efficient, welfare maximising allocation of resources which tends to define the study of mainstream economics. It was only with the work of Jevons, Menger and Walras at the end of the nineteenth century that economics became a separate science and stopped asking questions about politics and morality. In contrast, political economy retained a separate identity from the narrow concern with the methodological assumptions of rationality and utility maximisation of economics, and incorporates an explicit emphasis on issues of power, political institutions, distributional conflicts, ideas and moral agency. Similar to classical political economy, it retains an identity as a branch of moral philosophy; a concern with the justness of outcomes as well as a more hard-headed analysis of 'who gets what, when, and how', which is the very stuff of politics and political science (Lasswell 1936; Strange 1994).

The relationship between economics and politics is a complex and contested one. For the purposes of this book we can broadly identify two ways of thinking about the role of politics. First, there is a narrower, more agency-centred focus on politics as policy; that is, attention to the public policy implications and solutions. Second, there is a broader, more institutional or structural focus on how the capitalist economic system serves to structure outcomes. This draws on a long-tradition of viewing the economy as a social system which is embedded in wider political and cultural structures (Marx 1857–58; Polanyi 1944; Granovetter 1985). Each chapter will address both the broader international context which structures outcomes as well as the governance and policy responses, proposals, initiatives and regulation of the different capital flows.

This also means that as an approach, political economy tends to be far more interdisciplinary, drawing on insights from sociology, law, philosophy and geography. In many ways similar to development studies itself. With an eye on the words of the French statesman, Georges Clemenceau, Richard Higgott (2002: 91) suggests that 'If war is too important to be left to the generals, then globalization is too important to be left to economists'. Similarly, Noreena Hertz, the economist most famous for her work popularising the issue of international debt, has stated in an interview that (Betts 2004):

> I realised that economics is not about models, graphs and curves, but about people, politics and society, about history and culture; that these are all legitimate things to be concerned about as an economist. In fact, you would be a much better economist if you did understand these things.

This is sound advice and is all very liberating. But, there is also a danger of sliding too far the other way. To legitimately and constructively engage with questions of finance and development it is absolutely necessary to know something about the economy and indeed economics too (Tribe *et al.* 2010). Oftentimes popular or casual critics of neoliberalism – not to mention its advocates – are guilty of crude generalisations and ill-informed analysis. While such caricatures tend to be popular they are rarely constructive. This book tries to strike a useful balance between research and analysis from economists and insights from other disciplines about the economy.

Vision

Joseph Schumpeter (1954), in his *History of Economic Analysis*, argues that all analysis involves a 'preanalytical cognitive act', something he famously described as 'vision'. Robert Heilbroner (1990: 1111) describes visions as 'complex narratives combining many prognoses' that can be narrow and technical or extend to complex socio-economic dramas such as Marx's. Schumpeter's point was that 'vision' was a *necessary* prelude to scientific analysis, noting that any 'analytic effort is of necessity preceded by a preanalytic cognitive act' (Schumpeter 1954: 41). Thus, any analyst's theoretical framework flows from this, admittedly potentially ideological and certainly subjective and value-laden, starting point. In short, our vision about the way things are is inevitably coloured by what we wish to see.

Schumpeter's view was that the subsequent scientific processes of theory-building would depose false visions (or more properly, as Blaug (1962) suggests, it is the empirical results of the analysis that will depose false visions). But Schumpeter also conceded that this may take time and, in the meantime, new ideologies would always emerge. Therefore economic analysis will never be entirely free from the ideological effects of vision. What we analyse and how we analyse it reflects a set of prejudices and interests we bring to scholarly enquiry. In contrast to Schumpeter's (naïve?) hope of the purging potential of scientific analysis, Heilbroner (1993: 93) suggests that we should embrace 'the unexpungeably political character of economic analysis itself'. And, because vision and ideology are the necessary preconditions for economic analysis, we should accept and indeed valorise vision!

As such, this introductory chapter invites the reader not just to consider the 'good' and the 'bad' of global finance and development, but also to reflect on the fact that there are fundamentally different ways of thinking about their nature. At first glance, examining the relationship between global finance and development seems a simple enough question, even if the answers may be less straightforward. Are capital flows beneficial, what are the costs, how do these balance out, how can benefits be maximised and the costs minimised? But the question also assumes a common ground. That is, we know what we mean, and we agree about what we mean, when we talk about global finance. And while we're at it, the same point goes for development; that is to say, we know what we mean, and we agree about what we mean, when we talk about development.

Yet both remain contested – although, I would argue, the debates surrounding the contested nature of development are often more explicit and well-documented than the contested nature of finance. And so, in general, most people are much more sensitive to the conceptual and theoretical disagreements about the nature of development than they are about finance and money. Before introducing the theories in Chapter 2 (moving from Schumpeter's preanalytical to the analytical sphere) this chapter documents two common-but-different ways of approaching the issues of global finance and development.

The contention is that public and academic debates about global finance and development tend to begin from one of two different visions. They can be summarised easily. First, that *finance is a necessary and liberating resource used to boost savings, increase investment and drive the capital accumulation process necessary for economic growth and development. The more financial resources that are available, the more potential there is to release a positive developmental dynamic.* The second vision of finance is far more negative, or sceptical at the very least. *Financial markets are speculative casinos largely divorced from the needs of everyday economic activity and investment. The financial system is a source of indenture, a mechanism of social control, which is essential to the reproduction of capitalist inequalities, exploitation and the commodification of life.* The first vision we call financing for development (FfD), and the second vision the political economy of finance (PEF).

Visions of global finance and development

Take 1: Financing for development

20–22 September 2010, New York, United States of America. World leaders convened for a three-day summit at the United Nations to address the MDGs, the eight global anti-poverty goals agreed back in 2001 (see Box 2.5, p. 35). The purpose of the summit was to discuss the progress made, lessons learned, and to get the international community to recommit – in terms of energy, focus, resources and words – to the MDGs and to reenergise the process towards meeting the goals for the 2015 deadline. The outcomes of the summit were summarised in the document *Keeping the Promise: United to Achieve the Millennium Development Goals* along with a 'global action plan'.

One of the headline successes of the summit was a big push on women's and children's health, specifically a pledge of over $40 billion for the next five years. In a telling phrase, Ban Ki-moon, the UN Secretary General, was confident that 'We know what works to save women's and children's lives'. The implicit message here is that, without the political will or financial resources to help realise and apply it, this knowledge is useless. The purpose of the summit, in the words of the UN, was to provide the 'leadership'; that is, a public statement of commitment and the pledging of financial resources.

A UN (2010c) press release lists examples of 'leadership' which relate to the different MDGs:

● The World Bank will increase its support to agriculture to between $6 billion and $8 billion a year over the next three years, up from $4.1 billion annually before 2008, under its Agriculture Action Plan to help boost incomes, employment and food security in many low-income areas . . .
● Japan will provide $3.5 billion over five years for education in developing countries, beginning in 2011 . . .
● UPS International pledged $2 million to the World Association of Girl Guides and Girl Scouts to empower women through leadership and environmental sustainability programmes in 145 countries . . .

- The United Kingdom announced a tripling in its financial contributions to fight malaria, increasing its funds for malaria from £150 million a year to £500 million by 2014 . . .
- The United States announced a commitment of $50.82 million over the next five years for a Global Alliance for Clean Cookstoves, a public-private partnership led by the United Nations Foundation seeking to install 100 million clean-burning stoves in kitchens around the world . . .
- Belgium pledged €400,000 for the UN Conference on Least Developed Countries, to take place in Turkey in 2011 . . .

When it comes to discussions of political will, especially with respect to the international community, it is easy to be cynical. As is so often the way in politics and international diplomacy, fine words and good intentions are all too easy. And all too easily broken. Nevertheless, nothing here goes against the view that the relationship between development and finance is one of resources. The debate is how best to channel global financial flows into the developing world, so that the international community might fulfil what one scholar has called the 'world's biggest promise' (Hulme 2009). Success is measured in dollars and failure to meet pledges and perceived backsliding are the focus for criticism. Hence the subsequent debates are about the provision of these resources, the lack of these resources, the reasons for shortfalls, and to whom and how money should be given to maximise their effectiveness. In sum, the FfD paradigm emphasises the role of capital as a resource and the importance of money in meeting the identified gaps.

This is the classic and mainstream view of global finance and development and has a well-established logic (see Chapter 2 for an extended discussion). Financial resources are a necessary catalyst to kick-start the process of development. This injection of capital will allow poor countries to address the causes of economic backwardness and develop. Jeffrey Sachs, a well-known proponent of a 'big push' for development aid, argues that developing countries are caught in a 'poverty trap' (Sachs et al. 2004). A poverty trap is a situation where high levels of initial poverty combine with low household savings rates (because money is taken up by necessities/survival), low levels of infrastructure capital, and population growth. Together these interact to make it almost impossible for countries or households to exit this self-reinforcing cycle. An external injection of capital, it is argued, can break the cycle and allow a positive developmental dynamic of increased savings and investment.

So how much will it cost to break the worst poverty traps and where should the resources come from? The authoritative mainstream statement of how to fund development is the Monterrey Consensus. In March 2002, heads of state and government adopted the Monterrey Consensus of the International Conference on Financing for Development. This was a 'landmark' moment and a 'watershed' for development financing as it provided an international framework for cooperation on these issues (UN Millennium Project 2005). The Monterrey Consensus identifies, in its different chapters, six broad areas of work. These are: (1) mobilising domestic resources for development; (2) increasing and harnessing FDI and other private capital flows; (3) harnessing international trade as an engine for development; (4) increasing the effectiveness of official development financing; (5) ensuring sustainable external debt financing; and finally (6) addressing systemic issues to do with the international monetary, financial and trading systems (Bouab 2004).

In a well-cited report by the High-Level Panel on Financing for Development, which was prepared to feed into the International Conference on Financing for Development, it was estimated that an additional $50 billion per year on top of existing aid would be necessary to help reach the MDGs (Zedillo *et al.* 2001). More recently the Millennium Development Project led by Jeffrey Sachs estimated that the total costs to support the low-income and middle-income countries, as well as the costs of capacity building and debt relief, would start at $121 billion per year in 2006 and reach $189 billion by 2015. And as such, official development assistance (ODA) should be $135 billion in 2006 and rise to $195 billion in 2015, which would be equivalent to 0.54 per cent of donor gross national product (GNP). As you might imagine, all of these kind of exercises are inevitably rough and ready estimates, which make a number of 'heroic assumptions' (Clemens *et al.* 2007: 737), not least an improved policy environment within developing countries and the resolution of a number of bottlenecks in the effective delivery of such aid. Table 1.2 summarises a range of the key costing exercises and what they cover (Clemens *et al.* 2007).

This all underlines the FfD view of finance as resources-to-fill-a-gap. Indeed the UN even has an annual report, prepared by its Millennium Development Goal Gap Task Force, which outlines how much aid donor countries have provided and how much they have fallen short of their commitments and what is necessary to reach the MDGs (MDG Gap Task Force 2011). This is the most common way of approaching global finance and development.

Take 2: The political economy of finance

20 September 2008, Washington DC, United States of America. Exactly two years earlier than the three-day summit on the MDGs, to the very day, the US Treasury Secretary, Henry Paulson, was announcing a rescue plan for the US banking system to a press conference at the White House. The banking system had been hit by extraordinary turmoil over the previous week after the decision to allow the investment bank, Lehman Brothers, to collapse. It was the biggest bankruptcy in US history. The collapse had been preceded by months of instability in the financial markets as the extent of bad mortgage loans was revealed. The bad loans were built on the foundations of cheap credit, lax regulation and financial innovation based around the process of securitisation (the pooling of individual loans into a security which can then be sliced up into tranches which can be bought and sold like any other financial asset). The rise in interest rates and the fall in house prices had triggered widespread foreclosures, leaving many of the securities worthless. This led to liquidity concerns in the banking system and the inter-bank money markets drying up – i.e. as almost all the confidence

Table 1.2 *MDG costings*

Study	Covering	Annual new money
Brossard and Gacougnolle (2001)	Africa, primary education	$2.9–3.4 billion
Delamonica *et al.* (2001)	Global, primary education	$9.1 billion
Zedillo *et al.* (2001)	Global, MDGs	$50 billion
African Development Bank (2002)	30 African countries, MDGs	$20–25 billion
Devarajan *et al.* (2002)	Global, poverty goal	$54–62 billion
Devarajan *et al.* (2002)	Global, social and environmental goals	$35–75 billion
Devarajan *et al.* (2002)	Global, primary education	$10–15 billion
Filmer (2002)	Global, primary education	$30 billion
Greenhill (2002)	Global, poverty goal	$15–46 billion, plus 100% debt cancellation
Greenhill (2002)	Global, other goals	$16.5 billion, plus 100% debt cancellation
Mingat *et al.* (2002)	33 African countries, primary education	$2.1 billion
Naschold (2002)	Global, primary education	$9 billion
Oxfam (2002)	Global, MDGs	$100 billion
Vandemoortele (2002)	Global, MDGs	$50–80 billion
World Bank (2002)	47 IDA countries, primary education	$2.5–5 billion
World Bank (2002)	Africa, primary education	7x current aid
Bruns *et al.* (2003)	Low-income countries, primary education	$5–7 billion
World Bank (2003a)	Asia and South Asia, MDGs	2–3x current aid
World Bank (2003a)	Africa and Central Asia, MDGs	60% increase
UN Millennium Project (2005)	Global, MDGs	$73 billion in 2006, $135 billion by 2015

Source: Clemens *et al.* (2007: 737, Table 1).

had drained out of the financial system, banks refused to lend to other banks. Credit default swaps – an insurance policy against bankruptcy – were trading at 30 times their normal prices, indicating the levels of fear and uncertainty.

Later that day, the Treasury Department issued a fact sheet on its plan to save the US banking system from collapse; it noted that: 'Removing troubled assets will begin to restore the strength of our financial system so it can again finance economic growth'. To do so required huge financial resources – i.e. a pledge to inject $700 billion into the financial system (compare this figure with the estimates to halve global poverty in Table 1.2). The financial markets reacted positively to the statement. The Dow Jones industrial average climbed by 3.3 per cent and in London the FTSE registered its biggest one-day gains finishing 8.8 per cent higher. The Troubled Asset Recovery Plan (TARP) – as it was called – was passed by the Senate and the House in early October, at the second attempt, and essentially provided a series of enormous loans to banks. The TARP bailed out the financial sector through an enormous transfer of wealth from the public to the private sector, in effect socialising the banks' losses. It was understandably unpopular with the US public. However, the TARP has also been credited with saving the US financial system and preventing a second Great Depression. Meanwhile, the fall-out from the financial crisis was global.

At the very beginning there were hopes that the developing world had decoupled from the developed world. That is to say, that economic growth in parts of Latin America,

Asia and Africa was now independent from US economic performance. In other words, the old chestnut about the rest of the world catching a cold when America sneezed no longer held. However, as it turned out, in the short term this was absolutely not true; negative financial spillovers spread around the globe.

Credit markets froze up and capital flows to developing countries dropped precipitously, with flows turning negative (World Bank 2010). Somewhat ironically – given it was the source of the crisis – capital flooded into the US. This is because investors still saw it as a safe haven. International trade, an important source of foreign exchange earnings, also fell, again precipitously. The WTO (2010) estimates that between 80 and 90 per cent of world trade relies on short-term trade credit, which had dried up as a result of the crisis. Trade financing to the developing countries had fallen by around 6 per cent year-on-year. Meanwhile, Western governments, the major aid donors, facing recession and increasing budget deficits, started to wobble on their aid commitments, with individual countries such as Spain and Ireland announcing a reduction in overseas aid and the G8 collectively missing their Gleneagles commitments to double aid to Africa. Meanwhile, international remittances – the money sent home by migrant workers – were also hit, with the International Monetary Fund (IMF) predicting the impact on relatively remittance dependent countries to be in the region of 2 per cent of their GDP for 2009, though the impact has been, as expected, relatively short lived (Barajas *et al.* 2010).

Unsurprisingly, the impacts on development outcomes were negative with falling growth rates, increased poverty and hunger. The World Bank estimated that an extra 64 million people slipped into extreme poverty in 2010; and in 2009 an extra 40 million more people went hungry. The financial crisis, which originated in the financial markets of the rich countries, was also responsible for leaving a huge fiscal hole of $65 billion in the 56 low-income countries' budgets as their governments sought to cope with these consequences (Kyrili and Martin 2010). The budget deficits were a result of developing country governments having to increase expenditure in health, infrastructure and agriculture to combat the downturn. But, with international grants and loans only covering about one-third of the expenditure gap, the developing country governments found the extra spending was unsustainable. As a result, by 2009, the low-income countries ended up in a 'fiscal hole' (even with spending cuts) of almost 10 per cent.

These are the costs of the global financial crisis for the developing world. Yet the solution is not necessarily a simple restoration of financial flows and liquidity. From a PEF perspective it is less a question of what developing countries are lacking, but rather to enquire how they are *already inserted* into a politically and socially constituted economic system. To explain why, it's useful to return to Susan Strange's (1994) famous question '*cui bono*?', i.e. to whose benefit? Because:

> The power to create credit implies the power to allow or to deny other people the possibility of spending today and paying back tomorrow, the power to let them exercise purchasing power and thus influence markets for production, and also the power to manage or mismanage the currency in which credit is denominated, thus affecting rates of exchange with credit denominated in other currencies.
>
> (Strange 1994: 90)

Taking power seriously means being sensitive to how, for example, greater integration into international financial markets, instead of increasing the investment available for governments, can actually reduce governments' ability to spend, because of increasing borrowing costs and the likelihood and severity of debt crises (Hardie 2012). And being sensitive to the power and authority private actors such as the rating agencies – such as Standard & Poor's and Moody's – have over government policy (Sinclair 2005, 2008). And being sensitive to the way in which states use currencies as an instrument of coercive power (Kirshner 1995; Andrews 2006). And being sensitive to the ideological power and influence of some economists and the IMF in promoting neoliberal ideals of greater liberalisation of capital flows (Soederberg 2004; Chwieroth 2007).

Box 1.3

Finance and power

Finance is power. This can be positive and enabling – such as the example of Hamid and Khadeja where access to financial services provided them with the ability to survive and have some control over their lives (see Box 1.1). But this power can also be negative and dominating. For example, Haiti is the poorest nation in the Western hemisphere, but it was 'born' poor and indebted. The country's independence from France was won in 1804 after a 12-year slave revolt. The new nation was subject to economic sanctions by the former colonial power which were eventually lifted in 1825 in return for Haiti paying France for its loss of property. The amount was 150 million gold francs, some $22 billion in today's dollars. This was about five times Haiti's annual export revenue at the time. In order to repay the money Haiti had to borrow money from France at a very high interest rate. The loan taken out to repay the original amount was only paid off in 1947. This 'double debt of independence' was one that has hung over Haiti ever since, acting as drag on economic development, but also significantly undermining its official independence. The history of Haiti is a long sad tale of corruption, misrule (of Papa Doc and Baby Doc Duvalier), invasions and foreign interference. It shows the significance and power of finance to enslave.

It is important to recognise that this kind of coercive power is also institutionalised and legitimated within the international economy. The World Bank – the largest multilateral donor – wields significant power by virtue of its financial resources. When countries borrow from the World Bank or the IMF it is usually in return for commitments to policy change or changing the underlying organisation of the economy and society (structural adjustment). These demands and commitments are referred to as 'conditionality'. Conditionality became increasingly important and controversial from the early 1980s when structural adjustment loans (SALs) became the norm. Conditionality is not new – countries have long granted aid in return for alliances, voting, using it to buy back products from the donor (tied aid), and demanding that money for projects was spent as intended and that certain performance targets were reached. The difference with structural adjustment was that recipient governments were increasingly having their economic policy set by institutions in Washington and in increasingly liberal and austere ways (Peet 2003, 2007). The demanded reductions in government spending, the liberalisation of markets and tightening of the money supply severely circumvented governments' sovereignty and subjected their populations to hardship and violations of their human rights (Abouharb and Cingranelli 2007). Let's be clear, finance is a source of power and conditionality, is a form of coercion, analogous to the use of economic sanctions (Killick 1998).

Such overt expressions of power and coercion generate identifiable moral problems and debates. But financial power can also be much less obvious. For example, the fact that the US dollar is used as the de facto world currency provides the US government with a huge source of power. Because of the demand for US dollars, because of world trade and the depth and openness of US financial markets, as well as the ability of the US to influence the rules of the international monetary order have meant that it has built a dollar empire and the ability to shape the behaviour of other countries and the rules of the game (Strange 1994; Kirshner 1995; Cohen 1998). In addition, technically known as 'seigniorage', the issuer of money can profit from the difference between the cost of issuing (printing) money and the market value of that money. As we will see in Chapter 3, the US, as the issuer of the international currency, used its power to fund domestic reforms and international interventions, such as in Vietnam.

The collective conclusion of more critical political economists is that finance is deeply woven and implicated into not just the economy, but also society and politics too. This underlines the importance of political questions for the study of finance. Finance is itself a structure of power not just a resource for the powerful. And the increasing power of finance is a consequence of financialisation: the increasing importance of the financial sphere – the financial markets especially – over the economy, production and the organisation of everyday life. Critical observers of financialisation note how the 'handmaiden' role of finance has been 'turned on its head' (Grahl and Teague 2000: 170). So much so that finance is less a tool – a source of credit that can be drawn upon in order to produce, invest or consume – than a structure of control.

For example – to stay with the global financial crisis – consider the way in which sub-prime mortgages, which were initially presented as the democratisation of house-ownership, ended up being an extension of indentureship. People who were seeking access to loans and credit who were usually turned down, were offered teaser rates on mortgages with the seller's full knowledge that they probably couldn't pay. But the sub-prime market offered the potential for high profits and the lenders could just repackage the mortgages and sell them on through securitisation and therefore not have to worry about whether it was a good loan or not. This occurred within a larger context of increasing economic inequalities, so much so that US households were increasingly reliant on credit because of stagnant real wages and personal savings falling below zero. In essence, then, poor individuals were directly exploited through the extraction of equity from their homes, personal incomes and savings (Lapavitsas 2008). Likewise, consider the way in which the management of developing country indebtedness during the 1980s resulted in debt restructuring in return for structural adjustment (Chapter 5). Likewise, consider the way in which the joint-liability model of microfinance has the potential to reinforce social inequalities, indebtedness and result in household or communal violence (Chapter 8). See also the discussion in Box 1.3. This, as should be evident, is a very different way of looking at global finance and development.

Conclusion

This chapter has introduced two broad visions of the relationship between finance and development. These are the FfD and the PEF. The FfD perspective is the mainstream problem-solving approach which views finance as a monetary and functional resource, and the aim of policy is to ensure more of it gets where it is needed. Advocates can and do disagree how best to achieve this, but the framing of the problem is a shared one. On the other hand a more critical tradition views finance as a social relation, one that defines and shapes the relationships of power between states, classes and people. In contrast to the resource-instrumentalist view, the PEF approach treats global finance as something which is deeply implicated in the everyday operation of developing societies and their populations – structuring societal values, opportunities and power relations in favour of capitalist actors and norms, and perpetuating indebtedness and costly crises. Proponents of a critical political economy approach tend to stand back from the world and ask how the current state of affairs has come about (Cox 1981; Payne 2005; Payne and Phillips 2010). It's a quite different form of analysis.

Chapter 2 defines and disaggregates the notions of finance and development, before Chapter 3 introduces the theoretical frameworks: the neoliberal, liberal institutionalist, critical reformist and radical schools of thought.

Discussion questions

1   Make a list of the most important 'binding constraints' on the development of poor countries and people. How high or low does money and finance come on this list?

2   What, in your view, poses the greater threat to development: financial crises or a lack of financing from donors and private investors?

3   Is global finance a resource which provides the power to develop, or a structure of power over the poor?

4   To what extent are the two visions of finance and development (FfD and PEF) mutually exclusive? Can they be usefully combined?

5   Take a developing country of your choice and assess the role of external finance in its development since 1945.

Further reading

● It's always worth taking a look at the World Bank's (2000) *Voices of the Poor*, a quite remarkable piece of participatory research based on conversations with 64,000 people from around the world which 'chronicles the struggles and aspirations of poor people for a life of dignity'.

● For a definitive statement of the FfD perspective take a look at *Investing in Development* (2005), the final report of the UN Millennium Project which sets out the amounts, the strategies and investment needed to meet the MDGs. The solution is in the hands of the Western donors!

● For a critical and wonderful class-based analysis of the role of the US and the international financial institutions in promoting free capital mobility and the Monterrey Consensus see Susan Soederberg's (2004) *The Politics of the New International Financial Architecture*.

Useful websites

- For a topical, well-written, well-read, daily blog discussing the latest issues in and around development I would go no further than Duncan Green's blog *From Poverty to Power*. It covers a huge range of issues and provides plenty of links and thoughts on the issue *du jour*. Green offers well informed commentary which is usefully sceptical but always practical, and places a good deal of weight on the politics of developmental change. Available at: www.oxfamblogs.org/fp2p/.

2 Global finance and development

Learning outcomes

At the end of this chapter you should:

- Appreciate that money is a necessary but insufficient part of the solution in addressing economic and human development and poverty reduction.
- Develop your own position on what 'development' is, understanding that it is an essentially contested concept.
- Know the key measures and indicators of development, including economic growth, poverty, human development and inequality.
- Identify the key types of financial flows to developing countries and appreciate their varying volumes (in total and for individual countries), their different drivers and impacts on development.

Key concepts

Development, poverty, inequality, finance, money, capital, investment, Millennium Development Goals (MDGs), Monterrey Consensus, global financial crisis, global finance

Introduction: unpacking global finance

This chapter builds on the visions presented in Chapter 1 and the proposition of what happens when external capital flows are required for development. The chapter introduces the notions of finance as capital and finance as a system, and development as a goal and development as a strategy. Finally, in reviewing the key debates on conceptualising and measuring development it sets out the evaluative framework for the rest of the book, one which incorporates economic growth, investment, a multidimensional view of poverty, inequality and autonomous development.

The chemical power of society

> If money is the bond which ties me to human life and society to me, which links me to nature and to man, is money not the bond of all bonds? Can it not bind and loose all bonds? Is it therefore not the universal means of separation? It is the true agent of separation and the true cementing agent, it is the chemical power of society.
>
> (Marx 1844)

Money and finance are literally the 'lifeblood of the international economic system' (Strange 1994; Baker *et al.* 2005). Finance accelerates economic growth and development because it provides the money and credit needed to facilitate economic exchange and investment in production (Strange 1994). Money has a number of different functions (see Box 2.1 for more on the nature of money). Money's function as a means of exchange is the most intuitive, i.e. the use of coins as tokens in lieu of barter. However, the real properties of money come into play with the emergence of a credit-based monetary system – i.e. it is not just about physical tokens, but a way of accounting for who owes whom, and the ability to literally create money through promises to pay.

The emergence of money as a system of accounting has many histories, but financial historians often point to the use of tallies in the eleventh century. Tallies were initially used for accounting purposes but eventually came to be used as a form of receipt. The thirteenth and fourteenth centuries saw the emergence of bills of exchange. Bills of exchange – written orders promising to pay a set amount of money – were used to address the problems of physically transporting expensive, heavy and vulnerable commodity money long distances to fund commercial trade, as well as the problems of converting foreign exchange (Davies 1994; Leyshon and Thrift 1997; Graeber 2011).

Money as a unit of account seriously changes economic possibilities. As the sociologist Anthony Giddens (1990: 24) puts it, 'money provides for the enactment of transactions between agents widely separated in time and space'. Thus, the characteristics of finance, allowing for the 'storage' of purchasing power, the accumulation of capital, and thus the financing of productive enterprise, make it incredibly important for development. As well as its centrality to economic development, there is an inescapable political dimension to finance as it 'confers power on those able to accumulate capital or with access to credit' (Strange 1994: 97). The premise of this book is that money is a key part of explaining development, both its success and its failure. It is truly the chemical power of society; both binding and separating, enriching and setting free, cutting loose, denying and indenturing in equal parts. Finance is central to economic development and the global financial system now dwarves the rest of the economy.

If money makes the world go round, then it is the financial system that makes money go round. Two measures of the size of cross-border financial flows underline the scale of the contemporary financial system. First, the foreign exchange markets – which ostensibly facilitate the operation of trade in goods and services – now dwarf trade volumes. The Bank of International Settlements' (BIS) latest triennial survey reports that average *daily* turnover in the global foreign exchange market was $4.0 trillion in April 2010 (BIS 2010). This is compared to global merchandise exports in the whole of 2010 of $14.9 trillion (WTO 2011). Second, capital flows have been steadily growing as a proportion of global income. By 2007, international capital flows had risen to one-fifth of world GDP from about 5 per cent in the mid-1990s; this is a rate of increase about three times faster than that of world trade flows (OECD 2011c).

Box 2.1

What is money?

The nature of money is often neglected in discussions of financing for development. In contrast to the debates about the amount of money needed to reach development goals and the extent to which financial markets are pro- or anti-developmental, the question 'What is money?' can, at first glance, appear a distinctively academic debate, and quite an esoteric one at that. But it is crucial. Money is not just a thing, a commodity or a token demarcating value, but also a complex set of social relations of power, authority and promises to pay which both bind and set free. To use Keith Hart's (1986) expression, we need to look at 'both sides of the coin', the role of money as a commodity or resource and how that impacts on development outcomes, but also the role of money as a token of authority and the social relations of credit on the economy and societies of the developing world.

Over the past couple of decades a fascinating and productive debate has developed among economists and sociologists over the nature of money. The common implication is that money is a resource and the financial system is a kind of machine by which money is moved from savers to borrowers; it's a functionalist account of finance based on an instrumental view of money. Money is treated as a neutral instrument; 'Money is what money does' (Ingham 1999: 103). This is a commodity theory of money which is the mainstream or neoclassical view. However, there is an alternative theory of money which views money as a social relation (Innes 1913; Smithin 2000b). This view is more common amongst heterodox economists, economic historians, sociologists and anthropologists.

Classic textbook accounts identify three functions of money: it acts as a medium of exchange, a unit of account, and a store of value. Neoclassical accounts of money understand the nature of money primarily through its role as a medium of exchange. This is a view which is derived from Carl Menger's (1892) classic account. Menger argues that money spontaneously emerges as a technical solution to the problems associated with a barter economy. Key to this is what's known as the double coincidence of wants. That is, if you want to buy your neighbour's cow you are relying on them also wanting, say, the salt which you have for exchange. The problem of the double coincidence of wants has long been recognised as an inefficiency of a barter-based economy and works against specialisation (Smith 1776). To solve this problem, money was invented. Not to mention the additional problems of cows being cumbersome to carry around and difficult to use as change without damaging their life expectancy.

This view of money is significant insofar as it relegates money to a supporting role, a water-carrier. The neoclassical textbook orthodoxy is that 'money is a veil', and behind this veil is the proper stuff of economics: the production and exchange of real goods and services. As a medium of exchange money is no different to other commodities, apart from the fact that it is more easily divisible and transportable, and is a token which symbolises 'real' commodities (Ingham 2000). It is a lubricant, allowing real economic activity to occur; it is necessary but doesn't make a difference in and of itself.

This barter-exchange theory of money is a nice logical story, but it's a myth. Anthropologists have failed to identify a single example of the barter economies envisaged by Carl Meger and Adam Smith (Graeber 2011). But there is a perfectly logically explanation as to why this is so and why it is better to focus on the 'unit of account' role of money to understand its nature. The problem of a double coincidence of wants only arises if all transactions are 'spot' transactions. That is, the buyer and seller settle the transaction immediately. Instead what happens is, to return to the cow-salt example above, that the buyer takes the cow and owes the seller (Graeber 2011). So debt actually precedes money, not the other way around. This is hugely significant. The problem with the neoclassical view, the commodity theory of money, is that it ignores everything except exchange. What has happened here is that the commodity form of money, 'money-stuff' as Geoffrey Ingham

(2000) refers to it, has become conflated with money of account. Instead of thinking about money as a thing, the alternative view asks how that thing exists – what are the social relations which make it possible. The question therefore becomes, how does a system of debt and credit turn into money in the senses of a system of measurement (unit of account) and as a coinage and currency (medium of exchange)? The historical records showing the use of money and credit can be traced back to Mesopotamia in 3,200 BC where money emerges as a unit of account to allocate resources within temple complexes (Graeber 2011). Debt, for Graeber, is a promise which can be quantified in the form of money.

The alternative view of money is a social construct – constructed through moral, social and political relations. Ingham (1996, 2000) argues that money is a social relation in a number of senses. Unlike regular commodities, goods and services, only certain agencies can create money. This is because it is a social institution which requires legitimate sanctioning, without which money's 'promise to pay' does not exist. It is often the state which performs the crucial role of providing confidence in a particular money through fiat (Wray 1998). Money's formal validity is partly established through fiat, in the first instance, but the ongoing validity is a function of a communal belief in its exchange value and validity as a means of payment (Ingham 2000). This belief is manifest as a willingness to hold money which is itself a function of trust in others and confidence in the future. It is not a coincidence that the word 'credit' comes from *credere* (Ganssman 1988; Ingham 2000). This is why sociologists of money hold that *all* money is credit, that it has a fiduciary character (Dodd 1994; Simmel 1907; Ingham 2000). Marxists also note the social relations which are codified into money – notably the exploitation of workers to produce profit which is then objectified into money (see Chapter 2 for a discussion). Because this system is a social one, relying on trust and confidence, it is inherently precarious – it needs to be constantly regulated and managed. As such, the social reproduction of money is an ongoing negotiation between monetary authorities, the banking system and economic agents (Ingham 2000). Thus, studying money and finance is not just about economics, but about power, morality and social relations (Dodd 1994; Zelizer 1994; Leyshon and Thrift 1997; Fine and Lapavitsas 2000; Ingham 2001). There is much diversity within this school of thought; but there is a basic agreement that money 'is a system of social relations based on power relations and social norms' (Ingham 2000: 19).

What is finance?

At this stage it is worth setting out two preliminary definitions of finance. First, finance as an object; that is to say, 'money stuff' or financial capital. This can refer to money or credit in general or to its particular forms or instruments such as bank loans, bonds and so forth. Finance-as-an-object tends to equate finance with numbers on a balance sheet and as a resource capable of funding investments and other economic activity. A second definition of finance is as a system. The financial system refers to the organisation of money and credit, its constituent institutions and markets, which function to reallocate capital. In theory, capital is reallocated to where it is most demanded; that is to say, from lenders (those with surplus capital) to borrowers (those with a capital deficit).

Global finance is the extension of these principles to a world scale, where capital flows across borders not just within a nation state. The problems of cross-border financing and the extent to which this capital allocation is efficient, fair or effective and how it works are all highly contested, as we will see. But these functional understandings of finance – as a resource and as a system – offer a useful place to begin.

The financial system: savings, investment, borrowers, savers, banks and markets

The financial system is comprised of the markets, institutions, instruments and transactions by which financial capital (money and credit) is transferred between savers and investors. Financial systems are typically made up of two components, a banking system (or intermediated credit system – see below) and open financial markets. As well as the private market actors which operate within the financial system, such as commercial banks, hedge funds and institutional investors, there are also public institutions such as governments, central banks and the IMF which are often simultaneously operators within the system as well as playing a regulatory role.

Time is a key aspect of finance. Money and the financial system allow economic actors to bring future consumption and investment decisions into the present by accessing capital which they wouldn't otherwise have access to. This is done on the presumption that temporary indebtedness will be profitable over the longer term, i.e. that future income levels will be such that the borrower can repay the loan with interest and be better off. This may well not be the case, but it is nevertheless the logic behind financial transactions. While the bringing forward of economic time is directly beneficial for those actors who choose to engage in this, proponents have also argued that society benefits as a whole, since a more efficient allocation of resources raises overall productivity. As such, the financial system is designed to efficiently transfer money, risk and information to maximise the potential return on investments for individuals and the economy overall (Goodhart 1975). The extent to which a financial system does this effectively and rationally is referred to as its 'allocative efficiency'.

The simplest way of thinking about the allocation of capital is through a bank. Banks are economic actors that specialise in financial intermediation; that is to say, they mediate between lenders (those who have a surplus of capital) and borrowers (those who have capital deficits). Banks play their intermediation role by taking deposits from lenders, pooling these deposits and then using this pool of savings to make loans to other economic agents who wish to borrow to fund their consumption or investment (or indeed other borrowing). The bank itself earns its profits in the difference between the return it earns on its investments, i.e. from the loans made to borrowers, and the money paid in interest payments to its depositors, i.e. the initial savers.

It is common, at a personal level, to think of depositing capital in a bank as saving, but from a larger perspective it should be clear that savers are effectively lenders and investors – lending their excess capital which they do not have another immediate need for. The simplest form of lending is having a savings account, but any time an actor buys a financial product – a bond or a stock – they are effectively lending to the issuer of the asset. The interest on the savings is the return (earnings) on the investment which reflects the relative riskiness of the investment.

Borrowers can be individuals or households, companies or governments, all of whom need or want to fill the gap between their demand for financial capital and their current stock of capital. See the discussion of Chile in Box 6.1 (p. 171). So, on the one hand, lenders are making a financial investment, earning a return on their capital, while on the other hand borrowers are making an investment in, say, new machinery for a business venture, or funding a new highway or hospitals if a government, or a school uniform if an individual. Borrowers don't always invest, or even invest wisely, but this is the logic.

Although the banking analogy is useful, it's worth noting that in the industrialised world, bank operations are now dwarfed by, but also operate within, open financial markets. The move towards using financial markets to directly lend and borrow, essentially cutting out the banking middle-man, is called 'disintermediation' (Sinclair 1994, 2008). Some developing countries are also beginning to access the open financial markets through issuing bonds or via inflows of portfolio equity. Chapters 6 and 7 provide further details on the extent of this and its consequences.

Why external finance for development?

It was as far back as 1911 that Joseph Schumpeter suggested that financial institutions play a crucial role in economic development. In his *The Theory of Economic Development* he identified the key role of financial intermediaries, such as banks, in mobilising savings, managing risk, and facilitating and monitoring investment. But developing countries lack money. Hence when viewed from a financing perspective, the problem of development is the relative scarcity of capital. Indeed, wealthy countries are wealthy by the very virtue of having access to capital, and poor countries are poor by virtue, and often by definition, of lacking capital and access to it. In short, money and finance are crucial to capital accumulation and thus the possibility of development. In order to fund investment or consumption, actors – whether they be individuals, states or firms – have to borrow money. And because financial markets tend to be small and incomplete within developing countries, borrowing needs to be external. This can be from international banks, governments or the international capital markets.

There are a number of different financial markets, such as the capital markets, derivatives markets, foreign exchange markets, commodities markets and the money markets. These markets allow traders to do a number of different things: the capital markets are where actors raise capital to finance projects, so these can be equity (Chapter 7) or bond markets (Chapter 6). Meanwhile, the derivatives markets allow agents to transfer (or create) risk; foreign exchange or currency markets facilitate international trade; commodities markets cover coffee, wheat, copper, gold and so forth; and the money markets allow agents to borrow very short term to ensure sufficient liquidity.

For developing countries looking to finance development there are two main forms of capital available to fund their investment: equity and debt. Equity financing is where investors provide money in exchange for an ownership share in the enterprise. Borrowers are not, therefore, obligated to repay the money, but instead sacrifice a portion of future profits as well as some control over economic decisions. Debt financing takes the form of loans that must be repaid over time, usually in regular instalments with interest. While debt usually limits the total repayment to the investor it can leave borrowers vulnerable to changes in economic conditions while repayments are still due. As such, there is less risk-sharing than with equity. Most of the following chapters focus on one of these two major forms of capital.

- Chapter 5: ODA. Aid is comprised of concessional or soft loans and grants, so it's public and often subsidised debt.
- Chapter 6: Bank loans and bonds. These are the main flows of mostly private debt.
- Chapter 7: FDI and portfolio equity. These are the main flows of private equity.
- Chapter 8: Remittances. These are workers' earnings which are sent home, so strictly speaking it's not really capital at all, just a source of funds.
- Chapter 9: Microfinance. This is mainly debt, but also includes insurance and savings products.

Unpacking development

Next, we need to outline what is meant by development. As already suggested, development is a contested concept. Nevertheless, similar to finance, there are two ways of thinking about development: development as an object and as a strategy. Development, at its most generic, can be thought of as 'good change' (Chambers 1997). But as has been pointed out elsewhere this says nothing about what constitutes 'good', which aspects of change are good, and how to bring about (good) change (Thomas 2000). The answer to these questions depends on one's social values and theory of political and economic change. Hence, thinking about development as a goal requires us to think about it as an 'intended goal of somebody or something', an 'object of strategy' (Payne and Phillips 2010: 5). Hence development is purposive, positive change, often towards an ideal or objective, which implies agents – individuals, governments, non-governmental organisations (NGOs) and international organisations – in bringing about this change.

Development as strategy: political economy

A useful way of thinking about development as strategy is through its relationship with capitalism (Thomas 2000). As Cowen and Shenton (1996) argue, the modern idea of development emerged during the era of industrial capitalism, and is best understood in this historical context. Alongside the great progress of industrialisation was the mass immiseration of large swathes of the population, so much so that social order was threatened. It was in this context that development as 'intentional constructive activity' emerged. 'To develop, then, was to ameliorate the social misery which arose out of the immanent process of capitalist growth' (Cowen and Shenton 1996: 116). Hence what emerges is an issue of political economy: how governments and other social and political actors interact with capitalism and economic markets.

As Thomas (2000) outlines, answering the question of how development and capitalism fit together – why poor countries are poor and why rich countries are rich, and how poor countries can best develop – suggests three visions: the development of capitalism, development alongside capitalism and development against capitalism (Thomas 2000). This threefold distinction is useful in identifying general dispositions towards development and the market, and in Chapter 3 we adapt it slightly to derive a fourfold distinction that is more useful in analysing the relationship between finance and development.

Development as object: from growth to freedom

If the strategy is to develop, to bring about good change, how do we know what and who is developed? Trying to conceptualise and define development – who is and who isn't 'developed', the rich world and the poor world – is fraught with difficulties. Old labels such as the industrialised world (versus the non-industrialised world?) or the Third World (as opposed to the capitalist First World and communist Second World) seem outmoded. Newer labels such as the Global North and South hide as much as they reveal, given the enormous diversity of countries and life experiences in those two categories. Moreover, the language of the developed and developing world somewhat unfortunately suggests that the rich countries have somehow 'arrived' at their destination, with the rest of the world on their way. It seems that few labels satisfactorily capture what is needed. Nevertheless, both intuitively and as students of development, we know what they are *meant* to mean. The history of development studies is in part a history of how to define and measure deveopment.

Development is one of those words which falls into W.B. Gallie's (1956) category of an 'essentially contested concept'. That is to say, concepts such as morality, democracy, art or terrorism are polysemantic because they are evaluative, internally complex and subject to historical change. There are whole library shelves devoted to debating the nature of development. And these debates will continue. It is fair to say that we can't resolve the problem here. As such, in this book, we take a pragmatic and pluralist approach and work with a composite concept of development – incorporating both the economic and human perspective – which is defined and operationalised through a series of measurable outcomes.

Each chapter will assess the potential for finance to foster growth and poverty reduction, increase welfare indicators such as education and health, as well as further gender equality, but also the negative impacts financial flows have on these measures of development. But, in addition, the chapters will consider broader, critical issues such as the effects of financial structures on the possibilities for autonomous and sustainable development, the perpetuation of underdevelopment and the operation of power in the international system. Autonomous, locally-owned, locally-appropriate and sustainable development is something which can be thought of as operating at the national level – especially the way in which dependency theorists use the term – or at the community or individual level, which is closer to Amartya Sen's notion of capabilities. The reader is free to take an equally catholic view of development or choose which indicators they feel are more or less important in evaluating development outcomes in the following chapters.

Measuring development

National measures of income and growth

Economic development has long been equated with economic growth (Lewis 1955; Dollar and Kraay 2002). The World Bank has a fourfold classification which is the most well known and widely used measure of national development. Countries are classified according to their gross national income (GNI) per capita (i.e. total country income divided by its population size). The decision to focus on income per capita reflects the World Bank's desire for a single broad measure that captures development progress.

The World Bank reports national income in US dollars for all countries through the Atlas conversion factor, which is the average exchange rate of the country in question for that year and the two previous years adjusted for domestic and international inflation (which is based on inflation rates in the Euro zone, the UK, US and Japan). This figure is then divided by the country's population at the middle of that year to provide the GNI per capita figure which is so widely used. The income classifications are set each year on 1 July. In 2013, the income groups were divided as follows:

- low income (LIC), $1,025 or less
- lower middle income (LMIC), $1,026–$4,035
- upper middle income (UMIC), $4,036–$12,475
- high income (HIC), $12,476 or more

As of January 2013 there were 36 LICs (e.g. Afghanistan, Kenya, Uganda), 54 LMICs (e.g. Indonesia, Morocco, Vietnam), 54 UMICs (e.g. Costa Rica, Lebanon, Thailand) and 70 HICs (e.g. Andorra, Japan, Sweden). For a full list of countries and their classifications go to: http://data.worldbank.org/about/country-classifications/country-and-lending-groups. And, to clarify, when the World Bank refers to 'developing economies' it is referring to all middle- and low-income countries. The usefulness of this schema is that it allows for comparisons across countries and over time using a common indicator. For example, the historical record immediately tells us that the number of LICs has fallen sharply – from 63 in 2000 to 36 in 2013. This, of course, then raises important and interesting policy implications for development donors (see Box 2.2).

Box 2.2

The falling number of LICs

As noted, the number of LICs – those that depend the most on official aid from donors – has been falling sharply. There were 63 LICs in 2000, but only 13 years later this had fallen to 36. Commentators have noted that this raises a number of issues. First, Kenny and Sumner (2011) suggest that it should be a cause for optimism. It shows that development can and does work. But second, moving forwards, there are bigger issues. It's worth noting that the classification scheme is used for operational not just analytical reasons, that is to say it determines what type of lending countries are eligible for from the World Bank. And here's the crucial point. Given the smaller number of LICs but stubbornly high levels of poor people, is the World Bank's classification still fit for purpose? For example, now only a quarter of the world's poor live in LICs, and mostly in sub-Saharan Africa. However, back in 1990, 93 per cent of poor people lived in LICs (Sumner 2010). So, given this, should aid be given to poor countries or to (countries with) poor people? And if so, what other categories might be more appropriate? Box 2.3 discusses measurements based on the population under the poverty line. Plus, the UN provides some additional categories for the most vulnerable and poorest countries in the international system based on the nature of their challenges, namely: least developed countries (LDCs), landlocked developing countries (LLDCs), and small island developing states (SIDS). For further details see UN Office of the High Representative for the Least Developed Countries, Landlocked Developing Countries and Small Island Developing States (UN-OHRLLS 2010).

The other schema the World Bank uses is a geographical one, placing all countries into regions, namely East Asia and Pacific, Europe and Central Asia, Latin America and the Caribbean, Middle East and North Africa, South Asia and sub-Saharan Africa. See Figure 2.1, which compares the growth rates of the developing world regions over time.

Looking at Figure 2.1 on growth rates since the 1960s, the variations over time and across regions are quickly apparent. The 1970s were the last decade in which all regions experienced positive economic growth rates. The 1980s, the so-called 'lost decade', was marked by the debt crisis (see Chapter 6). East Asia, however, bucked these trends to successfully maintain positive growth rates. And of the 47 sub-Saharan African countries, only Botswana and Equatorial Guinea have had annual growth rates greater than 7 per cent (Clemens *et al.* 2007).

Figure 2.1 *Growth rates across the developing world 1960–2012.*

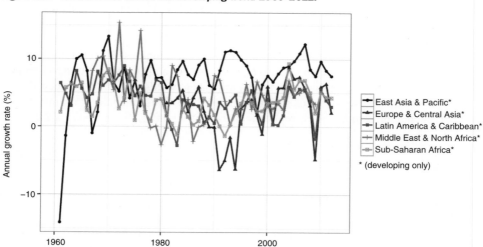

Data: WDI 2013.

However, economic development cannot be reduced to economic growth for five reasons. First, income per capita or growth rates don't directly measure qualitative changes in the structure or quality of the economy (Chenery 1960; Kuznets 1973). Examples of structural economic development include changes in the types of economic activities pursued, such as moving from dependence on the agricultural sector to diversifying into manufacturing and service activities; increases in levels of economic productivity; and improvements in the quality of jobs and employment conditions and opportunities. The difference between the growth and structural change is important. The latter underlines that it is not just an issue of increasing quantity (something growth rates capture well), but it is *qualitative* change in the nature of the economy that counts as development. Hence other measures are needed.

Second, just to focus on growth neglects improvements in health (e.g. infant mortality, life expectancy), education (e.g. literacy) and other welfare outcomes. As such, increases in, say, military expenditure, will register as an increase in national income but probably do little to improve the life chances of the population. The same is true for pollution. Expanding economic activities that have negative consequences (externalities) will register as 'development' since they increase GNI even while they reduce the quality of life for many.

Third, GNI per capita only tells us the average wealth in a country, it says nothing about how this wealth is distributed. If increases in GNI per capita are explained by large increases in the incomes of a small portion of the elite, or a sizable section of the population which lives above the poverty line, while a large number of people living in extreme poverty see no changes in their life chances, this isn't development. National growth rates say nothing about *who* benefits from increases in wealth, and nothing about what it can be used for. As an old joke goes, using average income to measure development is analogous to accepting that the temperature of a man with his feet in a freezer and his head in an oven is probably about right. The distribution of temperature may well be a cause for some alarm.

Fourth, GNI per capita only captures the value of goods and services that are traded on the market. This means that unwaged labour doesn't count. This ignores work done inside the household in the form of subsistence farming as well as looking after any family business or reproductive labour (caring for the family). Notably these are tasks often done by women (Boserup 1970; Benería 1999). Famously, an estimate which monetised the non-market work of women came out at $11 trillion per year, equivalent to 70 per cent of global output (UNDP 1995). Official GNI figures also ignore the informal sector – i.e. the work done that is done outside of the household but is not registered or recorded because the work is too small scale, one-off, or the workers and employers are avoiding taxes and other business regulation requirements. Estimates of the size of this activity place Africa's informal economy at 42 per cent of official GDP, 41 per cent for Latin and South America and 26 per cent for Asia. And individual countries are much higher, for example Zimbabwe (59.4 per cent), Bolivia (67.1 per cent) and Thaliand (52.6 per cent). By way of comparison, Greece and Italy have informal economies of 28.6 and 27.0 per cent of GDP, and the US and Switzerland are both at 8.8 per cent (Schneider and Enste 2000).

Poverty: income and multidimensional

From the 1960s other critics of the growth-centred view argued that we can only talk about economic development proper if we experience reductions in income poverty, inequality and unemployment (Seers 1969, 1979). This perspective, championed by the likes of Dudley Seers, put questions of basic needs and distribution back at the heart of what development means. There were two aspects of these critiques – both have strongly influenced current ways of measuring development.

First, instead of just economic growth, Seers (1979) argued that we can only properly talk about development if we witness a reduction in levels of poverty and malnutrition, falls in the degree of income inequality, and more and better employment. Growth can indeed support these things but they cannot be assumed to follow automatically nor universally; for example, there is the documented phenomenon of 'jobless growth' (UNDP 1996). Despite the growth, this should not, following Seers, count as development. The year 1990 marked the official refocus on poverty in the mainstream with publication of the World Bank's *World Development Report* on poverty. See Box 2.3 for details on measurement of international poverty lines.

Box 2.3

International poverty

The World Bank (2000) defines poverty as a position where individuals or households do not have command over enough resources to meet their needs. This is a multidimensional understanding of poverty. It could be income, food, access to health and education resources. Nevertheless, in terms of *measuring* poverty, the narrower (but catchier) 'dollar a day' international poverty line became the Bank's, and the world's, standard definition of absolute poverty. In addition to the more straightforward headcount (incidence) measure (i.e. the proportion of the population living below a poverty line) there are further ways of measuring poverty (Foster *et al.* 1984; Ravallion 1998). There is a depth of poverty measure which is a measure of the resources required to bring the poor up to the poverty line, and a poverty severity measure which places a higher weight on those who are further away from the poverty line. The headcount/incidence measure is the most commonly used, partly because it's easier to calculate, but the other two can be more useful for specific policy interventions and more telling impact evaluations.

The international poverty figure is calculated using purchasing power parity (PPP). The PPP data comes from price surveys from the International Comparison Program (ICP) which measures the costs of basic goods and services in different countries and converts them into international prices in a common currency (US$). In 1985 prices this was indeed $1 per day (World Bank 1990; Ravallion *et al.* 1991). However, a combination of inflation and new data meant that in 2009 the World Bank revised the absolute poverty line upwards to $1.25 for 2005 (Ravallion *et al.* 2008). Using this new poverty line it is estimated that 1.4 billion people live in absolute poverty, which is 25 per cent of the developing world's population (Chen and Ravallion 2008). The better news is that, according to these figures, 25 years earlier there were 1.9 billion poor people which was equivalent to half of the developing world's population. Some regions have faired much better than others, with East Asia performing particularly well and sub-Saharan Africa becoming poorer. Figure 2.2 tracks the poverty rates across the developing world.

Figure 2.2 Poverty rates across the developing world.

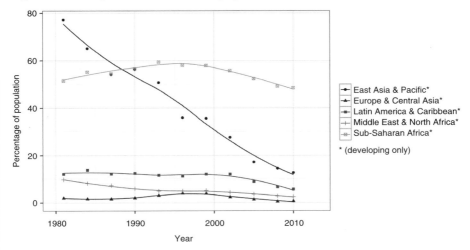

Source: Adapted from Chen and Ravallion 2008.

Critics have argued that the poverty line used by the World Bank is too arbitrary and doesn't reflect the real needs of human beings, the PPP equivalences still miss important cultural and nutritional cross-country differences, and the figures are extrapolated from limited data (Reddy and Pogge 2010; for a response see Ravallion 2008a).

The most recent data address some of these criticisms, but not all, and especially not the frugality of the international poverty line. In a wonderful piece, Lant Pritchett (2006) has proposed an 'upper bound' poverty line of around $10 a day. The implication of this? About 95 per cent of the developing world's population live below this line. As Chen and Ravallion (2012) note, $1.25 is the average of the national poverty lines found in the poorest 10–20 countries. It's a pretty frugal standard. And many countries have higher poverty lines than this.

But the latest data is much improved, based on over 850 household surveys for almost 130 countries; which represents 90 per cent of the population of the developing world. The 2005 and 2008 data are based on interviews with 1.23 million randomly sampled households.

The second key success of the 60s/70s growth critique came out of the observation that development shouldn't just be measured in economic terms, whether this is growth, poverty, income inequality or jobs. Development should be about the quality of people's lives more generally and should therefore utilise welfare indicators – for example, education, access to public services, mortality, health and nutrition, and political rights. There have been various attempts at measuring this – for example, David Morris's (1979, 1980) Physical Quality of Life Index (PQLI) which combined the basic literacy rate, infant mortality and life expectancy at age 1 to come up with a measure of well-being. Morris's aim was to paint a less fatalistic, pessimistic picture than that provided by stagnant GNP figures – a point recently reinforced by Kenny's work cited at the outset of this book. The point that development should be judged through measurable improvements in the quality of life for individual people paved the way for the perspective known as human development.

Human development

In 1990, in addition to the World Bank's *WDR* on poverty, the UNDP published a new measure of development, the Human Development Index (HDI). The HDI was an attempt to operationalise the human development perspective now so strongly associated with Amartya Sen. The human development perspective underlines the fact that development and poverty are multidimensional and that income is but a means to an end, not an end in itself.

Human development has a long tradition of thought – essentially back to classical discussions of the good life in Ancient Greece – but it is also a relatively recent challenger to the economic perspective that has tended to dominate development studies and practice. Human development refers to the eradication of deprivations and the expansion of human freedoms. The aim of development is to attack and eradicate 'unfreedoms' such as hunger, ignorance, prejudice, premature death and so forth. Famously, Amartya Sen defines development as freedom; i.e. the ability to 'live long, escape avoidable morbidity, be well nourished, be able to read, write and communicate, take part in literary and scientific pursuits and so forth' (Sen 1984: 497).

Crucially – given this book's focus on finance and development – Sen (1992) is clear that poverty is not poor well-being per se, but the inability to pursue well-being because of deprivations. This means that economic resources and income are relevant to achieving this, but not in an absolute way nor exclusively so. 'Not absolutely' because the relative wealth and power of others matters too: inequality can be as important to an individual's freedom as whether or not they live on more or less than $1.25 per day. 'Not exclusively' because the ability of individuals to participate in a community or go out in public without shame also depends on other social factors such as patriarchy, caste or racism, for example. As such, income becomes a means to development, but not the end in itself and it is important not to conflate the two. Sen (1990: 44) follows Aristotle on this when he states that 'wealth is evidently not the good we are seeking; for it is merely useful and for the sake of something else'. This goes some way towards underlining the claim made at the outset that financial investment in development is a necessary, but certainly not a sufficient, condition for achieving positive development outcomes. Critics of the human development perspective – and proponents of the income view – tend to assume that the expansion of these freedoms is a more or less automatic outcome of economic growth (Dollar and Kraay 2002); i.e. look after economic growth and the good life will follow.

Sen has done more than anyone else to develop and popularise the view that the link from economic growth to development is not an automatic process. But it was the

Figure 2.3 *HDI and income in Equitorial Guinea and Bhutan.*

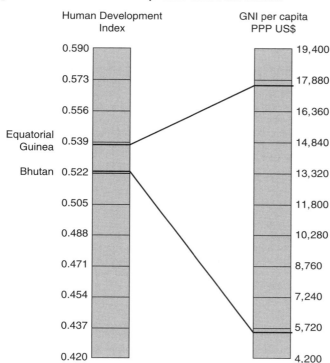

Data: http://hdrstats.undp.org/en/tables/

Pakistani economist Mahbub ul Haq who was responsible for developing the HDI. The HDI is a summary composite index which combines three elementary aspects of the ability to live a full and meaningful life to give a single score between 0 and 1. The elements of the index are education (adult literacy rates and combined gross enrolment ratio for primary, secondary and tertiary schooling), health (life expectancy) and wealth (income). The popularity of the HDI is that it gives a single broad-based comparison.

Comparing countries' relative positions on the HDI and the per capita income measure of development underlines the claim that income does not capture changes in human development (UNDP 2006). A comparison of the measures suggests that 'income variations tend to explain not much more than half the variation in life expectancy, or in infant and child mortality. And they explain an even smaller part of the differences in adult literacy rates' (UNDP 2010). Figure 2.3 shows a direct comparison of Bhutan and Equatorial Guinea based on data from the 2011 *Human Development Report* (UNDP 2011). It suggests that economic growth alone is insufficient, but it also starts to show that governments and countries can 'make' more or less of the same levels of GNI per capita. This requires access to finance, a plan, good investment, good leadership and smart, well-functioning politics (see Box 2.4 on Bhutan).

Box 2.4

Bhutan's investment in development

The Bhutanese government has intentionally and carefully engineered the outcome shown in Figure 2.3 via a clear and shared national vision and strong leadership. Since the 1960s, Bhutan, a traditional, rural and mountainous country, has emerged from self-imposed isolation, setting out to modernise and develop. Through a series of five-year plans it has followed a path of modernisation through planned and state-led development. Investments in infrastructure, universal primary health care and free schooling have seen gains in human development. These investments were initially financed by the Indian government. Then, as Bhutan joined the UN, new sources of financing emerged from the World Bank and the OECD donors.

In the years from 1984 to 2011, life expectancy at birth has gone from 47.4 to 67.2, and the adult literacy rate has gone from 23 per cent to 59 per cent (Royal Government of Bhutan 2000: UNDP 2011). Although Bhutan has a long way to go – given its starting point, its geographical circumstances and its GNI per capita – it is proudly punching above its weight.

Significantly, the Bhutanese way has been to emphasise the importance of maintaining a harmony between 'economic forces, spiritual and cultural values, and the environment' (Royal Government of Bhutan 2000: 6). The government has always maintained that development is about more than material progress and an accumulation of wealth. This is famously captured in the Bhutanese concept of maximising gross national happiness (GNH) rather than GDP.

Most recently, the 2010 UNDP *Human Development Report* has introduced a newer and, they argue, fuller measure of poverty: the Multidimensional Poverty Index (MPI) which measures three dimensions of poverty with 10 indicators. The advantage of the MPI is that it better captures extreme deprivation. Plus, the way in which the indicators are measured means that the index captures both the extent and average intensity of the multiple deprivations which poor people face (Alkire and Santos 2010). This measure looks set to become more important in coming years as more data emerges to support MPI analysis (see the debate between Sabina Alkire and Martin Ravallion in the 'Further reading' section at the end of the chapter, for more on this).

The MDGs

The MDGs have been referred to as 'the most important promise ever made to the world's most vulnerable people' (UN 2010a: 5). Table 1.1 (p. 2) details the eight goals and the more specific targets and indicators that pertain to each one. The advantage of the MDGs is that they offer a broad, multidimensional view of poverty and one which has garnered a great deal of consensus and use among the international development agencies such as the World Bank, UNDP and IMF. Plus they offer quantifiable targets that fulfil a dual purpose of measuring and evaluating progress as well acting as a means to hold the international community to account. See Box 2.5 for further detail on the MDGs.

Box 2.5

The Millennium Development Goals (MDGs)

The MDGs emerged out of the UN General Assembly's Millennium Declaration and provide eight goals to be reached by 2015. Progress on each goal is measured through a set of 18 targets and 48 technical indicators. For example, MDG1 is to eradicate extreme hunger and poverty, which leads to three targets, an income-, employment- and hunger-based target. There are then nine different indicators to assess whether these targets are being reached. The full list of goals, targets, and indicators is shown in Table 1.1.

The MDGs are far from perfect and reflect the compromises of international politics and diplomacy. David Hulme (2009) provides an excellent discussion of the emergence of the MDGs. Contrary to 'whig' views of history he conveys the mixture of serendipity and purpose well, and how the emergence of the MDGs was largely a piecemeal process of mudding though (Lindblom 1959). He argues that the critical factors which led to the MDGs were the rise of UN summitry post-Cold War, the role of the OECD's Development Assistance Committee (DAC) in pushing a set of performance-based international development goals into the international agenda, the efforts of individual policy entrepreneurs such as Clare Short, the UK Development Secretary, and Kofi Annan and the UN Secretariat. While the Millennium Summit of the UN General Assembly produced the Millennium Declaration it was only following a good deal of further negotiation between the World Bank, the DAC the IMF and UNDP that the MDGs came into being through blending the Declaration with the existing international development goals.

There is now, of course, ongoing debate around the post-MDG, post-2015 agenda. The discussion began in earnest some years back in academic circles (Sumner and Tiwari 2010), and in 2012 the UN appointed a High Level Panel, co-chaired by President Yudhoyono of Indonesia, President Johnson-Sirleaf of Liberia and Prime Minister David Cameron of the UK, to look at the issue of what should come after 2015.

There are two issues which arise at this point: (1) the issue of financing these goals, and (2) progress towards the MDGs. On the first, we have already seen a range of costings from Zedillo, the UN Millennium Project and others, as well as the Monterrey Consensus, on where the financing should come from (see Table 1.2, p. 13).

Second, progress on the MDGs has been mixed; the UN issues a progress report each year (UN 2010a, plus see the CGD Progress Index in 'Useful websites' below). Despite the recent setbacks of the financial and food crises, as we have seen, the world as a whole recently met the poverty target of MDG1; though individual countries and indeed entire regions such as sub-Saharan Africa and Eastern Europe are in all likelihood going to miss the goal. And MDG4, where the target is to reduce the maternal mortality ratio by three-quarters, is way off track. Unsurprising, considering that fewer than half of women in parts of the developing world receive sufficient care when giving birth (UN 2010a; Leo and Thuotte 2011). Nevertheless, the MDGs have had a significant political and economic effect in providing a focus for the efforts of national and international development agencies, a tool for campaigners to hold governments to account, and a common measure for progress on development.

Inequality

In addition to growth and poverty (whether income or multidimensional), the other crucial issue is the distribution of resources, i.e. the inequality of access to income or education. This can be measured as inequality between individuals, or more commonly households, or between countries or even globally (Milanovic 2005). One of the major criticisms of the MDGs is that, for all their good work in providing a multidimensional view of poverty, they provide little to nothing by way of focusing efforts and resources on addressing inequality (Fukuda-Parr 2010). Indeed, Jeffrey Sachs (2005: 289) is famously on record saying that:

> the goal is to end extreme poverty, not to end all poverty, and still less to equalize world incomes or to close the gap between the rich and the poor. This may eventually happen, but if so, the poor will have to get rich on their own effort.
>
> (see Dasandi 2009 for a critique)

At the international level we have a long-term divergence in income as the gap between the income of the wealthy and non-wealthy countries has grown. This has been because of the consistently positive growth rates of the HICs (until very recently) compared with the generally worse performance of the developing countries (emerging markets excepted) – indeed growth rates have been negative in some regions for recent periods. In fact, the current income gap between the richest countries and the poorest countries is the largest in history (Maddison 2007). In a famous article, Lant Pritchett (1997) estimated that between 1879 and 1990 the ratio of the richest countries' income and the poorest went from 9:1 to 45:1.

Crucially, the growing gap is not just because developing countries have failed to grow (fast). Milanovic (2011: 103) identifies an important aspect of the dynamics of divergence that holds even when poor countries grow faster.

> If the U.S. GDP per capita grows by 1 per cent, India's will need to grow by 17 per cent, an almost impossible rate, and China's by 8.6 per cent, just to keep absolute income differences from rising. As the saying goes, you have to run very, very fast just to stay in the same place. It is therefore not surprising that despite China's (and India's) remarkable success, the absolute income differences between the rich and poor countries have widened.

Nevertheless, if we switch back to a welfare-based measure instead of income-based measures, most indicators point to convergence – for example, in education (literacy levels, enrolment) and health (child mortality, life expectancy). Some have argued that it is less important to focus on income if what we are interested in is development more broadly (Jensen 2008). Kenny (2005) has shown that over 85 per cent of variables are converging. Moreover, it is becoming relatively easier to improve welfare indicators than it used to be: in order to reach the same life expectancy it takes only 10 per cent of the income today than it took in 1870 (Kenny 2005). This is all well and nice but, critics retort, income also allows individuals and countries to do things other than improve the quality of life – i.e. it also allows countries to work against a level playing field, to prevent all countries having a fair voice in global decision-making and to systematically distort trade rules (Pogge 2008; Wade 2010). As such, income inequalities remain very important.

To summarise, it is common to distinguish between growth or income-based measures and welfare indicators that capture the broader sense of development. It is also common for critics of neoliberal development and the World Bank to reject growth and income-based measures of development, both normatively and practically in the sense of guiding and assessing policy. For the purposes of this book we follow Amartya Sen in arguing that for development to occur improvements in both growth and welfare are necessary. Without autonomous growth, welfare gains cannot be sustained, and without welfare gains growth is utterly without purpose. Growth and income gains should be seen as means not ends, but no less necessary for that. Each chapter will, where the evidence is available, cover how the different types of financial flows impact upon economic growth and development, on reducing poverty and on promoting human development and various welfare indicators.

Box 2.6

Evaluating development, measuring impact

The practical questions of how to promote development and reach internationally-agreed goals are tricky. How much will it cost? How should it be financed? But, before we get to these, there is another fly in the ointment: even if we can agree that progress is being made, can we attribute it to one particular initiative or another? Because there are still billions of people in poverty, does this mean that development has failed? Impact evaluation is different from just monitoring outcomes, because it attempts to assess the causal effect of a particular intervention. The World Bank defines impact evaluations as 'the counterfactual analysis of the impact of an intervention on final welfare outcomes' (White 2006: 5).

Because the world is not a laboratory, there is no simple counter-factual: we can never know what would have happened in the absence of, say, aid flows. It is not implausible that growth rates could have been (more) negative. Relatedly, aid is not given in a neutral context, there are significant contextual and structural factors at both the international and domestic levels which influence the impact of financial inputs or policy changes. There are other complications too, such as the problem of lagged impacts: how long does it take for an intervention to impact on growth, employment, health indicators (Riddell 2007)? Take the example of international aid: there are different types of aid (modalities) and aid money may go into very different sectors or projects – for example, into hospitals, roads, schooling or institution-building. These will impact on different measures and have different lags. As such, sensitivity to the different modalities of aid matters. Moreover, correlation is not causation. Increases in aid sometimes follow deteriorations in the situation of developing countries, for example the large inflows of aid money into Haiti in January 2010 following an earthquake there. The coincidence or the correlation between aid and development would appear to be negative in this case, but it's clear that the increase in aid did not *cause* the earthquake. This means that we should be careful in reading off that aid is ineffective, or that aid is linked to the downturn in the country's fortunes; correlations may well be spurious.

Trying to disentangle the effects of development assistance on development outcomes is a tricky matter. There are often too many other confounding variables – these need to be controlled for. Because counterfactuals cannot be directly observed, researchers often make use of comparison groups which do not experience a change in aid inflows or adopt a new policy or change their institutions. The most common approaches to controlling counterfactuals are standard regression-based methods such as through the use of instrumental variables, propensity score matching and experimental techniques (Ravallion 2008b; Khandker *et al.* 2010).

Experiments, or randomised control trials (RCTs), where individuals are selected at random to receive a treatment (say a micro-loan) or not, are often held up as the gold standard of impact evaluations as they are the surest way to eliminate selection biases (Duflo *et al.* 2008). The control and treatment groups should be as similar as possible to each other – differentiated only by random allocation. A carefully designed evaluation has to stop contamination of the control or treatment groups such that the effect of the intervention is no longer isolated, or that there is sample selection bias whereby the groups chosen are not representative of the population as a whole, for example entrepreneurs who put themselves forward for micro-loans (White 2006).

However, experiments are not always appropriate for every type of problem – for example evaluating the growth effects of international aid does not lend itself to choosing a treatment and control group. Experiments can result in problem selection bias where researchers focus on topics which lend themselves to randomisation (see Copestake's intervention in Karlan *et al.* 2009). Plus other critics have ethical issues about including or denying individuals access to an intervention at random (Karlan *et al.* 2009; Ravallion 2009). Moreover, even if these issues of internal validity can be resolved, how representative is the evaluation? Does it focus on one project, in a particular locale, in a country with a distinct historical and cultural legacy (Karlan *et al.* 2009)? Could these findings be assumed to hold true if the intervention took place elsewhere, i.e. are the findings generalisable?

Nevertheless, the use of RCTs and the debates surrounding them have massively increased the quality of developmental impact evaluations in the past decade or so. Many of these issues are being worked through and the quality of what we know about certain types of development interventions on certain outcome indicators in certain environments and locales is much better than it used to be. Unfortunately (and amazingly) far too little is known about the impact of different development policies and interventions. Savedoff *et al.* (2006) note that only around 15 per cent of UNICEF reports include impact evaluations and argue there are too few incentives to conduct them. The situation slowly looks to be changing with the champioining efforts of the likes of the International Initiative for Impact Evaluation (3ie). See http://www.3ieimpact.org/ and the Abdul Latif Jameel Poverty Action Lab, http://www.povertyactionlab.org/. Both websites are well worth visiting.

Measuring finance

Having now introduced finance and development, what is the relationship between the two? In order to foster sustainable and equitable development, which type of financial resources should be encouraged? Who should be responsible for attracting and delivering them? How should they be delivered? What are the negative consequences – or harms – done by the current global organisation of finance?

Structure of the book

The next chapter introduces the key theoretical approaches for the study of global finance and development. The different theories help make sense of the two visions of global finance and development as well as reveal key disagreements within the visions, presented in Chapter 1. The key aim of Chapter 3 is to show why the positions of 'practical men' (or 'madmen in authority') are often distilled from the ideas of 'some academic scribbler of a few years back' (Keynes 1936: 383).

The theoretical approaches help make sense of whether and why capital flows are, on balance, a good or bad thing but more usefully the extent to which and how to intervene in terms of policies to liberalise or restrict capital flows. This is especially important as the mainstream agenda has moved away from crude liberalisation that ruled at the IMF and World Bank in the 1980s to late 1990s in terms of privatisation and capital account liberalisation. Indeed, state intervention and industrial policy is back in fashion at the heart of the World Bank (Lin 2012). But the intellectual foundations for interventions owe more to neoclassical theory than Keynesian approaches. Yet on the surface the policy conclusions can sound similar, so it is especially important to distinguish between orthodox and heterodox reformist agendas.

Before the book introduces a comparative assessment of each of the major types of capital flows to developing countries, Chapter 4 examines the international monetary system as the structure through which trade occurs and financial flows flow. Then Chapters 5–9 provide a comparative examination of the major sources of finance for development. The forms of finance all have their strengths and limitations; some are limited by their volume, some by the interests of the actors who are behind them, while others, because of their own logics, have proved downright harmful for poverty reduction.

Disaggregating finance: different flows, different drivers

Access to finance and the developmental impact of finance are key concerns for the book. It is assumed that access and impact take many different forms, work through many different actors, and operate and manifest themselves at different levels: global, national and local. For example, access to finance can be through international bond markets or aid flows for developing country governments to pursue national development strategies, or corporate financing through equity and bond markets for firms to conduct business, or at the level of the household and individuals accessing credit or savings through microfinance schemes and the flow of remittances from family members working abroad. The chapters move through the global and international, through the transnational to the local.

There are different drivers apparent across different types of capital. Financing development is partly a standard economic investment. Even leaving the attractiveness of natural resources such as oil, diamonds, rare earth minerals, gold and timber to one side there are many profitable investments in the developing world, especially over the long run. Note that liberal theory needs not assume any altruism on the behalf of banks and investors, a natural harmony of interests is assumed. But development financing is also an investment in reducing poverty. Thus, in addition to a narrow profit-motive sense of self-interest, there can be an enlarged, enlightened notion of self-interest in funding development. Many argue that this is a good basis for delivering development aid (USAID 2003; DFID 2011). There is also, of course, a moral driver of development financing – for example, charitable giving and arguably many donor aid programmes (Lumsdaine 1993). These different drivers are behind the different types of capital which, in turn, has important policy implications depending upon which drivers are in the driving seat.

Figure 2.4 *(a) and (b) Net capital flows to developing countries.*

(a)

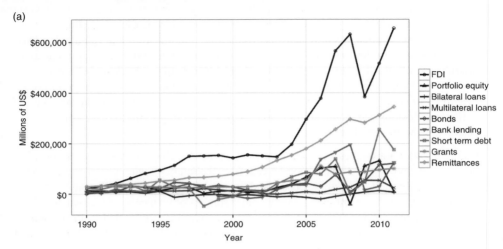

(b)

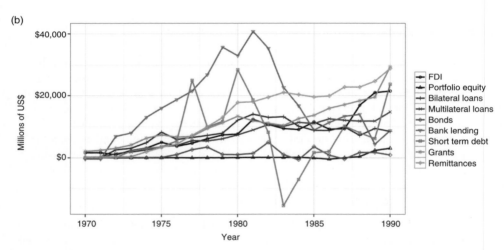

Data: WDI 2013.

Box 2.7

The capital flow indicators

The World Bank's Data Bank holds all of the World Development Indicators data as well as descriptions of the data. See the 'Useful websites' section at the end of the chapter for further details. The description of the indicators used in Figures 2.4 and 2.5 is provided below as well as their coding IDs.

FDI, net inflows (current US$)
FDI is the net inflows of investment to acquire a lasting management interest (10 per cent or more of voting stock) in an enterprise operating in an economy other than that of the investor. It is the sum of equity capital, reinvestment of earnings, other long-term capital and short-term capital as shown in the balance of payments. This series shows net inflows (new investment inflows less disinvestment) in the reporting economy from foreign investors. Data are in current US dollars.

BX.KLT.DINV.CD.WD

Portfolio equity, net inflows (current US$)
Portfolio equity includes net inflows from equity securities other than those recorded as direct investment and including shares, stocks, depository receipts (American or global) and direct purchases of shares in local stock markets by foreign investors. Data are in current US dollars.

BX.PEF.TOTL.CD.WD

Net financial flows, bilateral (current US$)
Bilateral debt includes loans from governments and their agencies (including central banks), loans from autonomous bodies and direct loans from official export credit agencies. Net flows (or net lending or net disbursements) received by the borrower during the year are disbursements minus principal repayments. Data are in current US dollars.

DT.NFL.BLAT.CD

Net financial flows, multilateral (current US$)
Public and publicly guaranteed multilateral loans include loans and credits from the World Bank, regional development banks and other multilateral and intergovernmental agencies. Excluded are loans from funds administered by an international organisation on behalf of a single donor government; these are classified as loans from governments. Net flows (or net lending or net disbursements) received by the borrower during the year are disbursements minus principal repayments. Data are in current US dollars.

DT.NFL.MLAT.CD

Portfolio investment, bonds (current US$)
Bonds are securities issued with a fixed rate of interest for a period of more than one year. They include net flows through cross-border public and publicly guaranteed and private non-guaranteed bond issues. Data are in current US dollars.

DT.NFL.BOND.CD

Commercial banks and other lending (current US$)
Commercial bank and other lending includes net commercial bank lending (public and publicly guaranteed and private non-guaranteed) and other private credits. Data are in current US dollars.

DT.NFL.PCBO.CD

Net flows on external debt, short-term (current US$)
Net flows (or net lending or net disbursements) received by the borrower during the year are disbursements minus principal repayments. Short-term external debt is defined as debt that has an original maturity of one year or less. Available data permit no distinction between public and private non-guaranteed short-term debt. Data are in current US dollars.

DT.NFL.DSTC.CD

Grants, excluding technical cooperation (current US$)

Grants are defined as legally binding commitments that obligate a specific value of funds available for disbursement for which there is no repayment requirement. Data are in current US dollars.

BX.GRT.EXTA.CD.WD

Personal remittances, received (current US$)

Personal transfers consist of all current transfers in cash or in kind made or received by resident households to or from non-resident households. Personal transfers thus include all current transfers between resident and non-resident individuals. Compensation of employees refers to the income of border, seasonal and other short-term workers who are employed in an economy where they are not resident and of residents employed by non-resident entities. Data are the sum of two items defined in the sixth edition of the IMF's *Balance of Payments Manual*: personal transfers and compensation of employees. Data are in current US dollars.

BX.TRF.PWKR.CD.DT

Net flows on external debt, short-term (current US$)

Net flows (or net lending or net disbursements) received by the borrower during the year are disbursements minus principal repayments. Short-term external debt is defined as debt that has an original maturity of one year or less. Available data permit no distinction between public and private non-guaranteed short-term debt. Data are in current US dollars.

DT.NFL.DSTC.CD

Source: descriptions taken from the World Development Indicators, available at: http://databank.worldbank.org/data/home.aspx

Types of financial flows: the chapters

Aid

Aid is the generic term for describing the transfer of resources from rich to poor countries – it can be grants, loans, technical assistance or food, and so forth. But more accurately aid counts as ODA if the grants and loans are provided at concessional terms and by public agencies to developing countries for the expressed aim of promoting economic development and welfare. Chapter 5 examines the world of official development assistance or aid. A lot of books start with ODA, i.e. aid, and there are many good reasons for this. It is what many people think of intuitively when they think of financing development. Furthermore, aid constituted the modern development project; that is to say, rich countries granting aid to poor countries created the norm of international paternalism. And for the poorest countries, aid remains the single most important source of development finance – see Figure 2.5(c) for the example of Rwanda. Nevertheless, aid-centricism can skew our view of the larger reality of finance and development and so we should always seek to see aid within the larger context of development and development financing.

Figure 2.5 *Net capital flows to (a) Jamaica, (b) Indonesia and (c) Rwanda.*

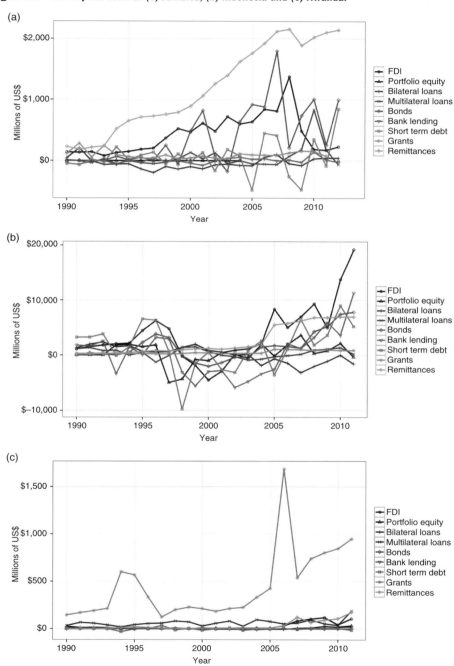

Private capital flows

International private capital flows cover the short- and long-term flows through or into banks, bond markets, stock markets or FDI by transnational corporations. International private capital flows have enjoyed two booms, one in the early to mid-1990s and a second in the 2000s which saw these capital flows rise to 5.1 per cent of developing countries' aggregate GDP (Word Bank 2006a). Figure 2.4 underlines just how important private financial flows are when considering total resources to the developing world. However, this also hides the important fact that these flows are incredibly concentrated – only a few countries receive them – so for the majority of the developing world the importance of official flows (aid) is more important than this picture suggests.

The organisation of the chapters means bank loans and bond issues are considered together and that equity and FDI are considered together. The logic for this is that they represent different forms of the same *type* of financial flows: debt in the case of bank lending and bond issues, and non-debt based investment in the case of portfolio equity and FDI. Equity-based financing is where a borrower generates investment by offering a share of the ownership in a project. On the other hand, debt-based financing is where a borrower promises to repay the lender via a series of fixed instalments via a bank loan or a bond. Chapter 6 deals with debt flows and Chapter 7 with equity flows. Many other books and reports have different ways of dividing private capital flows up – often lumping portfolio equity, bonds and bank loans together and giving FDI a separate chapter. All schemas have their advantages and disadvantages. The logic for organising by type rather than fashion seemed compelling. Too often we see generalised comments about private capital flows, referring to portfolio (equity, debt) and bank lending, but treating FDI differently. This understates the differences between the first three.

Remittances

Remittances are the portion of a migrant's earnings that they send home to their country of origin, often to their family. The latest estimates suggest that remittance flows to developing countries reached $351 billion in 2011, and as such dwarf aid flows (Mohapatra *et al.* 2011). Yet workers' remittances are not a capital flow, properly speaking. Instead they are private unilateral transfers, i.e. they are unrequited transfers between households which do not register as a debit and a credit on any accounts. However, the World Bank *does* report them in its *Global Development Finance* reports; crucially, as a 'memorandum item' rather than a capital flow. But for many developing countries they represent the largest source of external financing, larger than aid flows and larger than FDI (see Figure 2.5). As such, when considering the financial resources available to developing countries, we can't afford to ignore them. Chapter 8 reviews the drivers, impact and policy debates around remittances and how the development community is trying to harness their potential.

Microfinance

Microfinance is the provision of financial services – such as deposit accounts, loans, payment services, money transfers and insurance – to poor and low-income households and enterprises which do not have access to the formal financial sector. Chapter 9 reviews the debates and evidence around microfinance and development. Microfinance is slightly different from the other chapters since it doesn't strictly relate to an international flow. Instead it is a very local form of finance. However, it is impossible to ignore because microfinance has emerged as a major part of the global development architecture. It is a part of the push towards meeting the development goals, meaning that many international and global donors are working to either subsidise or invest in the industry. And with large increases in private capital flows into the sector the microfinance industry now operates on a global scale. Microfinance is a key plank in the efforts, for better or worse, to further global financial inclusion.

Conclusion

The purpose of this chapter has been to introduce and set out the subject matter of the book. It has introduced the notion of finance as an object – money stuff or capital – and as a system to move savings around effectively and enable investment for development. It has also introduced development as an object – growth, poverty reduction and so forth – and as a strategy based on an understanding of the political economy of capitalism, of whether to promote, guide or resist free markets.

The crucial role financial institutions play in facilitating economic development is not a new insight (Schumpeter 1911). And the emphasis placed on how much money it will take to reach the international development targets is not a neglected issue. So, it is somewhat surprising that – despite its centrality – there have been so few book-length treatments of finance and development. There are honourable exceptions to this: namely, Hulme and Mosely (1996), Rao (2003), Soederberg (2004) and Ocampo *et al.* (2007).

By making the case for a more sustained examination of the nexus between finance and development, at no point should it be assumed that money, credit and investment are the most important elements or only route to development, whether that be economic growth, poverty reduction or freedom. Yet, there is clearly a popular tradition of 'capital fundamentalism' in the international community and the public more generally. That is to say, the problem of development is often seen through the lens of money. Specifically, there is not enough of it. Poverty is a function of being capital poor. This is the standard place to begin in thinking about the relationship between global finance and development. As Heilbroner's discussion of vision made clear, we all have to start somewhere (Chapter 1). This book sets out from the issue of transfers of financial capital from rich countries to poor countries; in particular the differences between the various types of capital flows. But in doing so the book is also about the wider consequences of these flows; the vision is of finance capital as a socioeconomic and political structure that can equally entrap as enable. The following chapter expands upon this in reviewing the neoliberal, liberal institutionalist, critical reformist and radical schools of thought.

Discussion questions

1 Does the analogy between the domestic banking system and global finance make sense? If not, what doesn't work about the analogy?

2 Do you think the MDGs capture most of what we need to focus on when we think about development? What's missing? What's superfluous in them?

3 Is development about countries or people?

4 How do you think development should be measured and how should interventions be evaluated?

5 Which financial flow do you anticipate is going to be the most significant for sustainable development? Can you give an illustration to back up your answer?

Further reading

- Sen's classic statement of his views is *Development as Freedom* (1999), in which he advocates the view that freedom is both the primary end and the principal means of development. It is difficult to understate the impact of the Nobel Laureate's work on bringing individual agency and capabilities into the mainstream, though others lament the lingering methodological individualism and lack of political economy (Fine 2010).

- For a fascinating debate between two excellent proponents of whether a narrow or broad measure of poverty is best, see the debate between Martin Ravallion from the World Bank and Sabina Alkire from Oxford's Poverty and Human Development Initiative. See here for Duncan Green's introduction (http://www.oxfamblogs.org/fp2p/?p=3061), Ravallion's case (http://www.oxfamblogs.org/fp2p/?p=3070) and Alkire's response (http://www.oxfamblogs.org/fp2p/?p=3092).

- For two recent and really good books which present a sophisticated but accessible introduction to the new movement towards controlled experimental research, see *Poor Economics* (2011) by Abhijit Banerjee and Esther Duflo and *More Than Good Intentions* (2011) by Dean Karlan and Jacob Appel. The approach is not to everyone's taste, nor without its critics, but it's impossible to ignore – see Box 2.6.

Useful websites

- The MDG Progress Index. The Washington based Center for Global Development provides an interesting way of examining countries' progress towards the MDGs. The website offers an interactive MDG web tool with a graphical illustration of each country's progress towards the MDG targets. The tool helps to identify those countries which are trailblazers and the laggards. The methodology compares a country's performance against required achievement trajectories for each of the examined MDG indicators. The performance across the eight goals is then aggregated to give a single indexed score (Leo and Thuotte 2011). It is also possible to download the raw data. Available at: http://www.cgdev.org/section/topics/poverty/mdg_scorecards.

3 ▶ Theories of finance and development

Learning outcomes

At the end of this chapter you should:

- Understand the classical approach to the problem of development as a gap to be filled by investment and capital accumulation.
- Appreciate the role of savings and investment in this classical view and the need to turn to external sources of development financing.
- Describe the four main approaches to the study of global finance and development: neoliberal, liberal institutionalist, critical reformist and radical.
- Know the work of four key representatives of these schools (Anne Krueger, Justin Yifu Lin, Ha-Joon Chang and David Harvey).
- Critically engage with the four different approaches; understand the family resemblance in the two orthodox and the two heterodox approaches; as well as understanding what's different and what's similar about liberal institutionalism and critical reformism.

Key concepts

Investment, savings, capital accumulation, Harrod-Domar, neoliberalism, liberalisation, neoclassical growth theory, new institutionalist economics, Keynes's economics, the developmental state, Marxist political economy, theories of imperialism, dependency theory, underdevelopment.

Introduction

The previous chapter opened up as full a range of questions as possible by juxtaposing the 'FfD' and 'PEF' visions of global finance and development. This chapter moves from the preanalytical to analytical, introducing the key theoretical approaches to the study of global finance and development. The aim is to go beyond the ability to examine whether finance is 'a good or bad thing' for development and to understand the assumptions upon which different authors base such claims.

For instance, few observers of development would suggest that the processes of financial development and globalisation are cost-free. To take one example, banking crises cost developing countries in excess of a trillion dollars during the 1980s and 1990s (Caprio and Klingebiel 2002). Given that since the global financial crisis a trillion dollars ain't what it used to be, it's worth putting that into context: during the same period, the total amount of official development assistance given to all developing countries totalled $870 billion (in current US$). However, and here's the crucial point, there are fundamental disagreements over how to interpret this.

1   Are financial crises an inevitable part of development? Are crises in some sense the growing pains that all countries have to go through in their drive to maturity? And, as developing countries become increasingly developed, will crises become less common and less costly?

2   Or, are crises avoidable and the result of poor economic governance, policies and institutions in developing countries that, if improved, will reduce the frequency and mitigate the future effects of such crises?

3   Or, alternatively, are financial crises the consequence of highly liberalised and under-regulated financial markets that require strong regulation and state guidance to produce sustainable and pro-development capital flows? Without state management of financial markets they produce greater inequality and fail to generate productive investment in the economy.

4   Or, do aid and capital flows primarily represent political as well as economic ties and help reproduce global inequalities? As such, is the best solution for developing countries to delink from global financial capitalism since as long as there is capitalist financialisation there will be crises?

These differences represent not only differences in politics and vision but are explained and justified by competing theoretical, conceptual and methodological tools and assumptions. The rest of the chapter fleshes out the intellectual underpinnings for each one of these four positions: i.e. neoliberalism, liberal institutionalism, critical reformism and radicalism (see Box 3.1 for an outline of each).

Recall, from the previous chapter, the competing visions of development as an intentional constructive activity: development of capitalism versus development alongside capitalism, versus development against capitalism. A review of the literature makes it very apparent that there are a lot of approaches which fall into the 'development alongside capitalism' or the states *and* markets view. Too many to be useful. Therefore, we can use two political economy questions to subdivide the field better, as shown in Table 3.1. The first question is whether the view of the economy is an equilibrium or non-equilibrium view, i.e. whether markets are self-regulating. The second question relates to what governments can and should do; can they usefully intervene in a capitalist economy to improve economic and social outcomes? This provides us with four distinct intellectual positions, summarised in Box 3.1. The positions are used throughout the text to frame key debates, understand points of contestation and interpret policy responses. These different positions could be comfortably captured by the traditional right-left political spectrum: as we move from neoliberalism to radicalism the view of the market becomes increasingly malign. The aim of this chapter is to dig a little deeper and unpack the intellectual underpinnings for the different views of the necessary state-market balance.

Table 3.1 *Deriving the main positions on finance and development*

	Development of capitalism		Development alongside capitalism	Development against capitalism
Approach	Neoliberalism	Liberal institutionalism	Critical reformism	Radicalism
	Orthodox		Heterodox	
Does the economy and financial system tend towards equilibrium?	Yes	Yes, but only with the right institutions	No, uncertainty means that the state needs to manage markets	No, the economy systematically extracts wealth from the periphery
What is/should be the role of the state?	Meddling rent-seeker/minimal, nightwatchman	Market-supporting	Market-replacing	Instrument of the ruling class/delinking from circuits of global capital
View of money	Money as a commodity, functional view of money stuff, money as veil		Money is a social construct, fiduciary character, power relation	
Examples	Friedman, Krueger, McKinnon, Shaw, Lucas	Rodrik, Lin, Easterly, Rostow	Chang, Wade, Amsden, Domar	Harvey, Frank, Marx

The structure of the current chapter is as follows. The next section begins by introducing the original 'classical' approach to development which focused on capital accumulation, savings and investment. This is introduced to ground the rest of the chapter and indeed the book. It is useful to see the legacies of this perspective as well as where more recent approaches have distanced themselves from it. Sections two and three introduce the orthodox approach to finance and development. Section two looks at the free market perspective more commonly known as neoliberalism or the Washington Consensus; its main claims, logic, and rise and fall. Section three examines the shift away from the pure free market position to the so-called post-Washington Consensus with its greater emphasis on equity, institutions and aid. We label this liberal institutionalism given its orthodox heritage. Sections four and five flesh out the heterodox views of finance and development. The emphasis here is on the different ontological roots – i.e. their view of money and finance as well as the emphasis on the need for state-driven investment and the exploitative nature of finance, respectively. As such, there are important differences between liberal institutionalism and critical reformism.

The chapter reviews the economic theory behind each of the positions. This will highlight the evolving views on capital accumulation, investment and savings, the sources of economic growth and development, the nature of markets, and the appropriate role for government and policy.

Box 3.1

Four theoretical approaches to global finance and development

Neoliberalism: policymakers should focus on opening up and extending financial markets across and within developing countries to release the productive power of capitalist finance. The greater marginal returns to be had from investing in capital-poor countries will result in capital moving from rich countries to developing countries. This will increase their capital stock and (as long as governments get out of the way) increase allocative efficiency within the economy. A world of free capital flows and liberalised economies will enable poor countries to catch up and bring about a convergence in incomes. Government failure is the greatest barrier to development. The key to development, which is mainly about economic growth, is all about letting free markets get the prices right.

Liberal institutionalism: policymakers should embrace the productive power of free markets and democracy. But this is tempered by the recognition that markets don't emerge or work spontaneously, market failure means that governments must secure market-supporting institutions, such as legal systems, and help to steer the economy and financial system with careful fine-tuning at the margins. The key to development, which is about poverty reduction and welfare in addition to economic growth, is all about getting the institutions and policies right.

Critical reformism: policymakers should be sceptical of the welfare-enhancing claims of orthodox proponents of finance and instead systematically regulate the speculative elements of financial markets such that they are the servant rather than the master of industry. Contrary to liberal institutionalists, it is not so much that markets 'fail' but that they are fundamentally unstable. This is due to the role of uncertainty within the economy. The role of governments is to govern the market: managing and guiding investment to minimise the possibility of financial crises. This is because governments can take the long view and help to civilise capitalism. The key to development, which is about inequality and full employment in addition to poverty reduction and economic growth, is to directly harness financial capital and guide investment into socially valuable and productive enterprise.

Radicalism: policymakers should recognise the capitalist financial system for what it is: imperialist, patriarchal and exploitative. A globalised financial system extracts value from the periphery and recycles it into the core, actively perpetuating underdevelopment. This happens at a national level, from the rural hinterland into the urban centres, and globally from the periphery to the capitalist core. The view that finance is necessarily a force for good is a bourgeois ideology promoted by key international institutions. The goal of analysis is to develop a practical critique, and for developing countries to delink from the circuits of global capitalism and work towards new social, political and economic spaces outside the current finance-led system. The key to development, which is about emancipation from exploitative social relations, is not about civilising capitalism but rejecting it and adopting alternative and fairer forms of social organisation.

The problem of development and the solution of financing from the perspective of finance for development

As we saw in Chapter 1, from an FfD perspective the problem of development is a lack of money and credit to invest in capital accumulation and welfare improvements, and the lack of an efficient system to allocate financial resources to their most productive use. There's an old joke where a tourist, who has asked for directions from a local, gets the response, 'Well if I were you, I wouldn't start from here.' In characterising the problem of development as a lack of capital and investment as the solution, this book is definitely starting from 'here'. This is not a neutral starting point, nor one that all theoretical perspectives would accept. However, it is a starting point that helps to make sense of the evolution of development theory as well as allowing us to map out the competing interpretations of the relationship between global finance and development.

Gaps and traps: capital accumulation and classical development theory

The previous chapter introduced the UN Millennium Commission's poverty trap logic for investing in development to reach the MDGs (Sachs *et al.* 2004). A poverty trap is defined as 'any self-reinforcing mechanism which causes poverty to persist' (Azariadis and Stachurski 2005: 326). The view of a gap, and a trap, and the importance of a big push of investment for capital accumulation, are classic views of development and finance. That is to say, the amount of capital and labour available, and the productivity of these factors, are seen as the building blocks of economic growth and development. Figure 3.1 depicts the poverty trap. Where savings are below the amount lost to depreciation (wear and tear) and population growth (i.e. resources have to be spread around more people), then capital accumulation is negative and production and output will be low. The production curve, showing the amount of output produced (y axis) for a given input of capital (x axis), remains low initially because of a lack of infrastructure, institutions, education and so forth. But then it increases rapidly as investment opportunities mean a high return on capital, before flattening off again because of diminishing marginal returns (all of the easy gains from investment have been made). It's a simple but powerful diagram.

With more capital stock, a country is able to increase productivity and outputs by producing more goods and services, bring about structural change, create jobs, increase incomes and consumption, reduce poverty and deprivations and increase freedoms. On the other hand, low levels of capital accumulation create a poverty trap; without sufficient income, poor countries cannot save and therefore invest in their productive potential. So for capital goods or stock, think of machinery, factories, technology, roads and other infrastructure, as well as 'human capital' such as training and education. Capital refers to all goods or assets which are used to produce consumption goods and services. The emphasis in this book is on the role of *finance* capital in funding this process. Investment in capital accumulation boils down to the old adage that it takes money to make money.

Investment in capital goods has long been seen as the way to increase the total output of goods and services leading to long-run economic growth. Classic development theory identifies the role of savings and investment in breaking these traps. A lack of savings results in a lack of investment in a country's capital stock. As such, insufficient capital accumulation is both a cause and a consequence of poverty. To return to Sachs's poverty trap (Figure 3.1), investment in new capital, or upgrading existing stock, must outstrip that lost to wear and tear (capital depreciation) and population growth (which reduces capital per worker).

The emphasis on capital accumulation goes back to the classical political economists such as Adam Smith (1776), David Ricardo (1817), Thomas Malthus (1820), J.S. Mill (1848) and Karl Marx (1867, 1894). All of them identified the centrality of investment in capital accumulation to produce economic growth and development. Classic development theory built on this. The question, then, is what drives the processes of capital accumulation and increases in productivity? Growth theory has sought to uncover the combinations of labour, capital and productivity which result in economic growth (for the classic statements of this 'growth accounting' approach see Abramovitz 1956; Solow 1957). These accounts of growth help identify the drivers of development as well as the limits or constraints for development.

First, accumulating resources versus increasing factor productivity. Investment operates through two channels: (1) increasing (or augmenting) the *total stock* of physical capital (the number of machines, tools and infrastructure) and human capital (the number of skilled or unskilled workers), (2) increasing the *productivity* for each unit of capital or labour from technological advances – for example, updating machinery or through the efficiency gains from economies of scale such that there is an increase in the total output from each unit of input. In addition to more and better inputs, the knock-on demand for new capital goods from other sectors and industries helps to reinforce the impact of the initial investment throughout the economy. This is called the 'multiplier effect'.

An important cross-country analysis by Susan Collins and Barry Bosworth (2003) used a growth accounting framework to try and derive the relative contributions of physical capital, human capital and factor productivity towards developing countries' growth rates. Their results suggest that capital accumulation is indeed the main source of growth for developing countries. They found that productivity gains matter more during periods of rapid growth and for the HICs.

Second, labour versus capital. Traditionally the lack of physical capital, rather than labour, has been seen as the key binding constraint for developing countries. Why? Because capital is relatively scarce with respect to labour. This is best explained through the phenomenon of diminishing returns. Diminishing returns is depicted in Figure 3.1 as a flattening curve; this captures an intuitive notion. It suggests that for an individual or an economy at point A on the x-axis the benefit of an increase in capital is greater than for a similar increase in capital at point B. In essence, the productivity gains for an individual going from having no mobile phone or plough or laptop to having one, is greater than going from one to two, and adding a third matters even less. The concept of diminishing marginal returns is important for developing countries in general given that, by definition, they are capital-poor which suggests that the potential benefits of additional capital are initially large. Plus, as we will see, this is an important debate for aid (Chapter 5) and microfinance (Chapter 9).

Thus capital is modelled as the primary constraint on growth, and specifically the role of savings in capital accumulation. The reason why capital, not labour, is assumed to be the limit for development is based on the assumption that all developing countries were characterised by a 'dual economy'; that is to say, there is a modernising productive

Figure 3.1 *The poverty trap.*

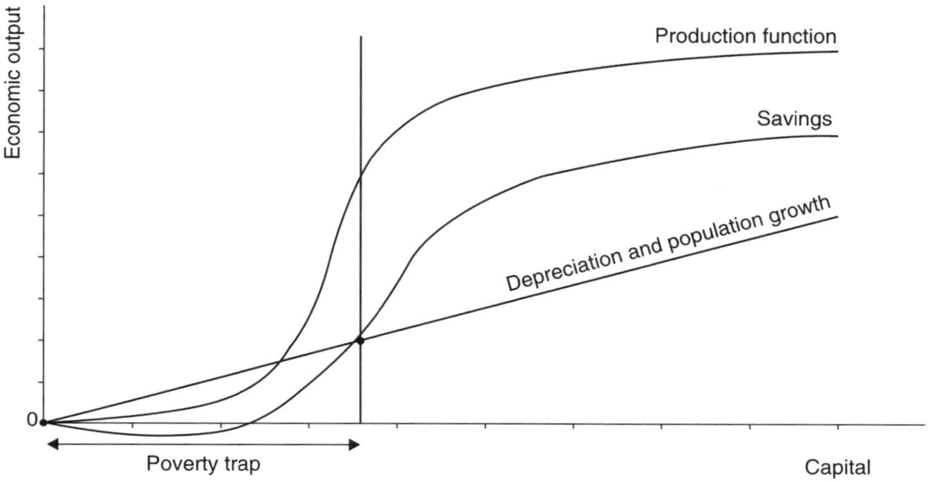

Source: Adapted from Sachs et al 2004.

industrialising sector and a traditional often rural, agricultural sector with a large pool of surplus labour (Lewis 1954; Fei and Ranis 1969). The trick, from this perspective, is to get growth in the modernising sector where the wealthy tend to save more. W. Arthur Lewis (1955: 155) provides the classic statement of the importance of capital accumulation and savings in development:

> The central problem in the theory of economic development is to understand the process by which a community which was previously saving and investing 4 or 5 per cent of its national income or less, converts itself into an economy where voluntary saving is running at about 12 to 15 per cent of national income or more. This is the central problem because the central fact of economic development is rapid capital accumulation (including knowledge and skills with capital).

Investment and savings

Where does the money to fund new investment come from? LICS and MICs have a number of potential sources for financing development. For example, generating income from production and trade, by the government raising finances through taxation, or domestic investment and capital accumulation and accessing domestic savings through the financial system (see Box 3.3 on savings and development).

One problem is that developing countries tend to lack access to some or all of these, hence a 'gap' and the turn to external sources of financing. The origins of a financing gap lie in the work of Roy Harrod and Evsey Domar, extended by Chenery and Strout

(1966) and their Two Gap Model (see Box 3.2). Chenery and Strout identify two potential gaps which act as a limit to growth: either an investment gap where the country saves less than necessary to fund investment, and/or a trade gap where a country exports less than necessary to fund its imports.

Box 3.2

Harrod-Domar and modern growth theory

Roy Harrod and Evsey Domar famously and independently developed the basis of modern growth theory, hence the Harrod-Domar model (Harrod 1939; Domar 1946). The Harrod-Domar model hammers home a couple of key lessons. First, Harrod and Domar began from the observation that labour is underemployed and is not the binding constraint on growth: lack of capital is. The model assumes a fixed level of productivity – that is to say, for each unit of input (capital) a certain level of output is produced. This is termed the capital-output ratio. As such, the Harrod-Domar model provides a basic way to calculate the amount of capital necessary to maintain steady growth. If the capital-output ratio is 4 and growth is 4 per cent, then the saving and investment rate needs to be 16 per cent per annum. If domestic savings is, however, 10 per cent, this means that there is an investment-savings gap to be filled..

Second, the Harrod-Domar model emphasises the instability of capitalist economies. Harrod (1939) took Keynes's (1936) insights about the potential for economies to get stuck in a position of structural unemployment and stagnation and turned them into a dynamic growth model. Harrod proposed three different growth rates: the actual growth rate, the warranted growth rate and the natural growth rate. The warranted growth rate is the one where planned saving equals planned investment, i.e. where an economy stays in equilibrium because consumption equals production. The problem illustrated by Harrod's model is that if actual growth goes above warranted growth, investment will be insufficient, incentivising greater investment and increasing growth and inflation. Likewise, if growth falls, there is a surplus of capital, and investment is discouraged. Hence the problem becomes worse. The crucial point to take from this is the fundamental instability of the economy (as opposed to the self-equilibrating orthodox view). The precarious nature of the warranted growth model has become known as the 'knife edge problem'

The model makes a number of rigid and unrealistic assumptions – for example, assuming fixed capital-labour and capital-output ratios, and not allowing for technological change and changing capital productivity. Nevertheless the model remains influential and important. For example, it is still used widely by the international financial institutions (IFIs). As the World Bank has stated, 'This so-called two gap model of the domestic saving and foreign exchange constraint to growth guided external aid and lending agencies in judging the extra resources that developing countries would need to finance imports and investment' (cited in Easterly 1999: 7). Similarly, in his discussion of poverty traps, Sachs (2005) presents a classical development account: emphasising the role of external interventions to help increase the capital available for those stuck in poverty traps.

So, if the key problem of low investment is low domestic savings rates, there are three solutions. First, to increase the level of domestic savings and the efficiency of the financial system. Second, to use the income from export earnings (in excess of imports). Third, to tap external savings by borrowing from foreign lenders.

Table 3.2 shows savings and investment rates as a percentage of GNI over the last 30 years across the key income groups. Two key points are demonstrated here. First, the

Table 3.2 *Average annual saving and investment rates as a percentage of GNI, 1980–2010*

Income group	Gross domestic savings rates	Gross domestic investment rates
LIC	9.2	18.7
LMIC	20.8	24.1
UMIC	28.2	26.8
HIC	21.3	21.4

Data: WDI 2013

lower levels of savings and investment in LICs relative to more developed countries. This points to 'financing gap' arguments about the implied required level of investment to foster sustainable growth and development. Because savings come from deferred consumption on the behalf of individuals, firms and governments, in conditions of poverty where the population is living on the edge of subsistence, savings rates tend to be low, thus domestic investment in capital formation is also low. Second, note the actual difference between investment and savings. The higher level of investment over and above domestic savings represents the role of external financing as capital inflows augment domestic savings. In short, external capital flows provide a way of bridging the financing gap and raising investment rates. We will see over the coming chapters the extent to which different types of capital do actually flow to all or just some developing countries, the drivers and stability of the different flows, the extent to which they are associated with improvements in development indicators, and the costs (economic, political and social) of relying on external capital.

From an orthodox perspective, increased international financial integration allows countries (and their governments, households and firms) to smooth fluctuations in their consumption and output. When a country is in recession – assuming all other countries in the world aren't also in recession – they can borrow internationally. And by investing in other countries, investors diversify their risk away from exclusively depending on domestic investments for growth. Thus, for countries with volatile growth and consumption, and largely uncorrelated with global growth, the potential welfare gains are large.

Box 3.3

Savings and development

Saving is generally done either from a precautionary motive (i.e. a form of self-insurance against future uncertainty) or to fund future investment (i.e. the idea of forgoing current consumption to save up and buy a new tractor or build a bridge). As we saw in the previous chapter, a financial system is designed to reallocate savings from sitting idly to more productive uses. This way, investment can by *financed* by other people's savings (in addition to the possibility of *funding* from one's own savings).

Savings and growth do tend, on average, to go hand in hand – though there are important cross-country differences, which are due to cultural, demographic and policy variation. Plus, the causal arrow flows in both directions; an increase in savings enables growth, but in addition growth also enables greater savings (Rodrik 2000) (up to a certain level, as it appears that in HICs saving rates tend to fall again). Compare, for example, Singapore's gross domestic savings rate which averaged at 46 per cent between 1980–2011, whereas Guatemala has managed just under 8 per cent (WDI 2012). Meanwhile, Guatemala's average annual growth rate over the same period has been less than half of Singapore's, at 2.75 and 6.9 per cent, respectively. As we will see, orthodox approaches tend to emphasise how the efficiency gains from financial deepening and higher savings rates mean more money for investment and more rapid capital accumulation. On the other hand, critical reformists question the automatic relationship from savings to investment and instead tend to emphasise the fact that investment must come first, which in turn leads to savings.

A big push

The policy consequences which flow from the classic theories of growth tended to emphasise the importance of government intervention to mobilise domestic savings and boost investment alongside an infusion of external capital as part of a 'big push' to help relieve the bottlenecks faced by developing countries. The work of Rosenstein-Rodan and Rostow characterises this well.

Rosenstein-Rodan's (1943) concept of a 'big push' is central to classic development theory. The problem he identified was that poor countries were stuck in a low-level equilibrium trap where actual growth was below warranted growth. Developing countries were, as per Lewis, characterised by a large surplus of agricultural labour. Rosenstein-Rodan's insight was that what was required was a big push to break the trap. Instead of frequent-but-small investments, Rosenstein-Rodan argued that one *big* injection would be much more effective because of the multiplier effect; that is to say, industries in one sector would be benefiting from the investment in other sectors through increasing aggregate demand, thus producing a positive cycle of growth (Rosenstein-Rodan 1943; Nurske 1953). Rosenstein-Rodan had identified a classic coordination problem and a tipping point or threshold above which capital investment needed to be in order to break the trap.

Meanwhile, Walt Rostow's (1960) five stages of growth remains the classic example of modernisation theory. His stages were: the traditional society, the preconditions for take-off, the take-off, the drive to maturity and the age of high mass-consumption. Crucially, the take-off stage requires an increase in savings and investment from 5 to 10-plus per cent of GDP, while the drive to maturity is characterised by investment of 10–20 per cent of GDP to bring about industrialisation and structural change in the economy. The move from traditional society to the drive to maturity required, in Rostow's view, two channels, i.e. increasing domestic savings and tapping international savings.

Channels: augmentation and financial development

There are two channels through which external development financing contributes to economic development. This draws upon the total stock and productivity distinction introduced above. First, there is the direct effect through augmenting the capital stock in the developing country. This can be increase the total capital stock available in the economy, or be targeted, for example through investments in development projects or infrastructure.

Second, depending on the nature of the external financing (e.g. private equity), a supplementary effect can be gained through the development of the financial system. Domestic financial development means that the financial system can act as a transmission belt for external capital to filter through more effectively and the developing economy can absorb more global capital. Plus, as the domestic financial system develops, it further increases the financial resources available by accessing domestic savings. This is an important distinction to bear in mind. For, while the capital augmentation route is most commonly associated with development financing in the popular imagination (e.g. giving money), because of the volume of private capital and the potential effects (positive and negative), the financial development route is important to consider.

The political economy of savings and investment

The question now becomes, given the framing of the development problem: how best to facilitate development? Of course, from a FfD perspective, how one fills the gap is the critical question – both the type of capital investment (aid or private capital?) and how to ensure this is maximised and effective (liberalisation or strong institutions?). As Arthur Lewis (1955: 376) pointed out many years ago:

> No country has made economic progress without positive stimulus from intelligent governments . . . On the other hand, there are so many examples of the mischief done to economic life by governments that it is easy to fill one's pages with warnings against government participation in economic life. Sensible people do not get involved in arguments about whether economic progress is due to government activity or to individual initiative; they know it is due to both, and they concern themselves only with asking what is the proper contribution of each.

The rest of this chapter is dedicated to unpacking how the four key theoretical approaches view the 'proper contribution' of the government and individual (or states and markets) and why.

Orthodox views of finance and development

Since the 1970s insights derived from neoclassical economics have defined the mainstream and determined the development policies of the key international financial institutions such as the World Bank, the IMF and most of the main donors. We refer to these ideas here as the development orthodoxy.

The neoclassical counter-revolution was a rejection of the Keynesian and classical growth theories introduced above and marked a return to the insights of the 'marginal revolution' (see Box 3.4). In contrast to the Domar-Harrod view of the world, which promoted a case for government intervention, neoclassical and then endogenous growth theory provided the intellectual basis for keeping government intervention to a minimum. Orthodox views are firmly grounded in the development of capitalism vision. More recent liberal institutionalist perspectives have tempered the extremism of neoliberalism, but still retain a strong role for free markets. We unpack neoliberal and liberal institutionalist approaches in this section.

Box 3.4

Marginalism

Neoliberal thought likes to hark back to the anti-mercantilist insights of the classical liberal political economists such as Adam Smith (1776), David Ricardo (1817) and John Stuart Mill (1848). However, it was the marginalist revolution at the end of the nineteenth century which really defined neoclassical economics, most especially the insights of Jevons (1871), Menger (1871) and Walras ([1954] 2003). Classical political economy was a largely macro perspective which concentrated upon the role of production and exchange in the wealth of nations. Crucially, the tradition mostly assumed that the value of a good or object was based upon the amount of labour that had gone into producing it.

Marginalism builds on a famous paradox articulated by Smith in Chapter 4 of Book I of *Wealth of Nations* (1776) where he notes: 'Nothing is more useful than water: but it will purchase scarce anything . . . A diamond, on the contrary, has scarce any value in use; but a very great quantity of other goods may frequently be had in exchange for it'. At one level, the difference is clear: Smith is describing the difference between the use value of a good and its exchange value. For the marginalists, *explaining* the difference required a shift in thinking about value away from 'total utility' to 'marginal utility' (marginal utility being the perceived *additional* utility of consuming an extra unit of a good). This meant that the value of goods (and the working of the economy more generally) is no longer about production and the use value of goods. These classical political economy notions were abolished and replaced by the concepts of scarcity and exchange (Walras ([1954] 2003).

So whereas the task of classical political economy was a macro one – to increase the wealth of nations and to focus on production – marginalism overturned this emphasis and instead focused on the consumer and switching the basis of economic analysis from macro concerns to micro foundations (Jevons 1871). A further significance of the marginal revolution was that it marked the key shift in thinking about the economy as a discipline, separate from politics and moral philosophy. The normative concerns of the classical political economists *qua* moral philosophers were replaced with the veneer of science. Orthodox economics became increasingly mathematical and theoretical, and increasingly divorced from the real world. Yet it continues to propose policy recommendations based on these abstract and idealised models.

Famously, in the 1950s, Arrow and Debreu (1954) 'proved' (mathematically) that, under conditions of perfect competition, there is a general equilibrium across all markets as prices ensure that supply equals demand. It is at this point that the utility of all consumers is maximised. The policy implication of this theoretical (strictly speaking, mathematical) insight is that markets should be expanded and deepened so that all goods and services can be traded. Previously untraded goods and services should be financialised so that they can be traded – for example, water, energy, carbon, mortgages. Asset prices on financial markets will serve as a signalling mechanism for the whole economy, thus improving aggregate welfare. The policy implications of this are huge – in short, more markets. So while some ideas of classical political economists were important – especially the emphasis on liberal values against mercantilist protectionism – it was the insights of marginalism that really created what we now think of as free-market economics.

Neoliberalism: capital liberalisation

The term neoliberalism comes from 1970s Latin America – *neoliberalismo* – and is used by pro-market economists to describe their approach (Steger and Roy 2010). Nowadays it tends to be much more of a prejorative label used by critics of the Washington Consensus. From the neoliberal perspective the problem of underdevelopment is not just a shortage of capital (as per classic development theory), but a problem of correctly allocating these resources.

The characteristics of neoliberal policies are well known. But to summarise, neoliberalism is the belief that development can be best achieved by releasing individuals' entrepreneurial drives in a system of free markets, limited government and private property rights (Harvey 2005). The government's role should be severely limited because governments do not possess the wisdom to make distributive decisions – the market is far better placed to do this by aggregating individuals' desires through the price system. Moreover, governments are also prone to capture by special interests with rent-seeking and corruption, and do not act in the general interest (Stigler 1971; Krueger 1974). As Ronald Reagan famously put it in his inaugural presidential address, 'In the present crisis, the government is not the solution to our problem. The government is the problem'. This means limiting the role of government to a minimal level, and to enforce property rights and provide essential public goods, such as national security. For neoliberals, if a government is more active than this, the likelihood is that its actions will be 'ineffective, unnecessary, or counterproductive' (Hoff and Stiglitz 2001: 415).

The term neoliberal is more useful here than neoclassical because it also implies a political project. More specifically, the attempt to bring the pristine textbook world of deregulated markets into reality. That is to say, because the abstract models of neoclassical economists have been used by policymakers then they begin to bring about the 'virtual' world that they describe (Carrier and Miller 1998). This means that it is crucial to interrogate the intellectual underpinnings of these policies. For 'virtualism' is a political project of imposing and institutionalising a market logic on developing societies. For example, Philip McMichael (1998) has documented the imposition of structural adjustment through the activities of the IMF and the World Bank (see Box 3.5). Such structural adjustment programmes have forced governments to cut spending, privatise public assets and limit their developmental interventions. But, as will be expanded upon in Chapter 6, this is a process that is intimately tied up with indebtedness and political power.

But it is also important to note that because neoliberalism is a political project – often about protecting privilege – neoliberal policies are often internally contradictory and can stand in contradiction with neoclassical economics. For example, the position of the wealthy world on trade tariffs: developing countries are subject to liberalising pressures while the agricultural sector in the EU, US and Japan remains subsidised and protected. This is also, of course, a consequence of the politics of the real world; a mixture of populism, coordinated lobbying, interests, power and wealth.

Box 3.5

Structural adjustment programmes

'Structural adjustment programmes (SAPs) are economic policies for developing countries that have been promoted by the World Bank and International Monetary Fund (IMF) since the early 1980s by the provision of loans conditional on the adoption of such policies. Structural adjustment loans are loans made by the World Bank. They are designed to encourage the structural adjustment of an economy by, for example, removing 'excess' government controls and promoting market competition as part of the neo-liberal agenda followed by the Bank. The Enhanced Structural Adjustment Facility is an IMF financing mechanism to support of macroeconomic policies and SAPs in low-income countries through loans or low interest subsidies. SAPs policies reflect the neo-liberal ideology that drives globalization. They aim to achieve long-term or accelerated economic growth in poorer countries by restructuring the economy and reducing government intervention. SAPs policies include currency devaluation, managed balance of payments, reduction of government services through public spending cuts/budget deficit cuts, reducing tax on high earners, reducing inflation, wage suppression, privatization, lower tariffs on imports and tighter monetary policy, increased free trade, cuts in social spending, and business deregulation. Governments are also encouraged or forced to reduce their role in the economy by privatizing state-owned industries, including the health sector, and opening up their economies to foreign competition.'

Source: WHO (2012)

In terms of the vision of development built into the neoliberal approach, it is a very narrow one. Neoliberal economic policies are focused on maximising economic growth and negative liberties (freedom from interference). Directly addressing poverty and inequality through redistribution should not be the focus of policy. The growth-not-redistribution mindset is best captured by Robert Lucas (2004) in the conclusion to an essay on growth and development:

> Of the tendencies that are harmful to sound economics, the most seductive, and in my opinion the most poisonous, is to focus on questions of distribution . . . [because] the vast increase in the well-being of hundreds of millions of people that has occurred in the 200-year course of the industrial revolution to date, virtually none of it can be attributed to the direct redistribution of resources from rich to poor. The potential for improving the lives of poor people by finding different ways of distributing current production is nothing compared to the apparently limitless potential of increasing production.

Traditionally, neoliberals have argued that inequality doesn't matter because rising inequality is a side-effect (and symptom) of development. Or, in a stronger argument going back to Lewis (1954), inequality actually accelerates successful development because capitalists tend to save more than poor rural farmers. So increasing the wealth of the wealthy will increase overall savings and therefore boost investment and development.

More recently Dollar and Kraay (2002) have maintained that economic growth is good for the poor insofar as it is sufficient to reduce poverty. Or, more precisely, that growth is distribution-neutral such that the incomes of the poorest quintile remained stable as an economy grows. This finding is then used to justify any and all policies which increase growth, which, according to the relationship between economic growth and the policy variables in their model, means increasing economic openness, liberalisation, but not government spending and growth (see Lübker *et al.* 2002; Amman *et al.* 2006 for critiques). In sum, it justifies the World Bank and IMF policy packages as set out in Box 3.5.

So, to summarise, for neoliberals, development is about optimal resource allocation, which is done by 'getting the prices right', i.e. freeing markets and reducing government distortions. This philosophy has been at the heart of the widespread liberalisation, deregulation, privatisation and expansion of markets in social and political life. What's the intellectual basis for this?

Neoclassical growth theory

The neoclassical response to Harrod's Keynesian model (see Box 3.2) came from Robert Solow (1956) and Trevor Swan (1956). Neoclassical growth theory is built around the primacy of 'economic fundamentals'; namely, preferences, resources and technology. The fundamentals are both independent from political and social forces and are more powerful than them. The neoclassical model rejects the knife-edge conclusion in favour of a self-equilibrating view of the economy. The crucial difference being that Solow's model assumes that prices are flexible and can help the economy return to equilibrium. Contrary to a fixed capital-output ratio in the Harrod-Domar model, Solow's model assumes that there are diminishing marginal returns to capital and, crucially, that labour and capital can be easily substituted for one another. Hence the price mechanism can incentivise a shift from capital (or labour) intensive production techniques depending on the factor endowments of a country.

Significant implications follow from this. First, governments should remove barriers or policies that distort markets and stop the smooth-functioning price mechanism. Second, diminishing marginal returns mean that financial capital will flow from capital-rich countries to capital-poor countries in order to take advantage of the greater returns available in developing countries. Third, this means that poor countries will grow faster than rich countries and ultimately there should be a convergence of income levels between poor and rich countries (*ceteris paribus*).

Chapters 6 and 7 look at private capital flows in more detail to see the extent to which capital does flow from rich to poor countries. But suffice to say it is much less than the theory predicts. As for the convergence in income per capita, it is clear that this hasn't come about either. In fact, over the period 1870 to 1990 the ratio of per capita incomes between the richest and the poorest countries increased by a factor of five (Pritchett 1997). There are two orthodox responses to the failure of the neoclassical predictions: endogenous growth theory and the new institutionalist economics.

Endogenous growth theory

Endogenous growth theory or 'new' growth theory argues that economic growth is best explained by dynamics inside the economic system itself, not dictated by exogenous factors. This is in contrast to the Harrod-Domar model where the savings rate was fixed, and the neoclassical model where the rate of technological change was outside the model (Romer 1986; Lucas 1988). In contrast the 'new' economic growth models placed innovation inside the model (Romer 1994). The implications of the difference are worth spelling out.

First, in contrast to Solow's diminishing returns, the production function displays increasing returns to scale. So, for every unit increase of capital or other factor of production there is *more* than a unit increase in output. This is because technological progress from research and development spill over and create positive externalities that are shared by other firms (Lucas 1990). Workers become more productive over time, resulting in increasing returns through specialisation and learning-by-doing (Arrow 1962). So, while diminishing marginal returns may well characterise capital accumulation, this is less relevant for productivity gains. Recent studies have suggested that technological and knowledge improvements are key in explaining economic growth (Hall and Jones 1999; Easterly and Levine 2003).

Significantly, endogenous growth theory can explain why income per capita can continue to grow in OECD countries and why there is no necessary reason to expect convergence. Rich countries can carry on getting richer through benefiting from research and development. Because richer countries can and do invest in research and development and continue to upgrade, a growing divide has appeared.

While endogenous growth theory is seemingly more attractive given that it doesn't assume exogenously-given technological progress, neoclassical theory remains vibrant. Growth accounting exercises like Collins and Bosworth (2003) suggest that technological progress is less important than capital accumulation in explaining variation in countries' growth rates. Similarly, Mankiw *et al.* (1992) find that 78 per cent of income variance is down to differences in savings rates and human capital.

Financial liberalisation

What are the implications of growth theories for finance and development? The policy implications of endogenous growth theory are twofold (Shaw 1992). First, if freer markets tend to be better at allocating capital than governments and international capital flows help fill idea and object gaps, then the policy implications are to ensure that capital flows freely across borders and that the economy is kept free from government interference. (The exception being government investment in human capital and innovation.) Put simply, this requires the removal of barriers to international capital flows (capital account liberalisation, see Chapter 3) and at the domestic level the extension and deepening of financial markets and the removal of non-market interference (financial development). Second, technological transfer is key to development. Romer (1993) famously complements the classic notion of development as the process of filling 'object gaps' (i.e. resources, goods, etc.) with the additional notion of 'idea gaps'. This is why proponents argue that FDI is particularly important for development as it fills both of these gaps (see Chapter 7).

From the mid-1980s many developing countries went through substantial processes of liberalising, deregulating, globalising and privatising their financial sectors (Bandiera et al. 2000). Orthodox policies tend to underline the importance of developing countries 'opening up' their economies to market forces. Liberalisation of interest rates, pro-competition measures such as lowering barriers to entry, a reduction in reserve requirements to increase funding available for lending, a reduction in the levels of directed loans banks are forced to undertake, privatising banks, deregulating securities and stock markets, and liberalising the capital and current account to free up international capital flows (Bandiera et al. 2000).

Box 3.6

Financial development

Financial development is thought to have beneficial effects through increasing the level of financial services and products available and used in an economy. This so-called financial deepening is held to lower the costs of borrowing (the real interest rate) (Rajan and Zingales 1998). Financial systems – through changing savings rates and allocation – increase capital accumulation, thus spurring economic growth (Romer 1986). But the improvements in economic output are not just down to increasing savings and the accumulation of capital, but also the efficiencies gained from an efficient and specialised financial system (total factor productivity growth) (King and Levine 1993a; Levine 1997; Beck et al. 2000; Benhabib and Spiegel 2000; Rao 2003). In addition, Demirgüç-Kunt and Levine (2008) have argued that access to the formal financial system by households and firms is an important determinant of opportunity. They show that access to finance can balance against the inequalities produced by parental wealth and social connections and have large implications for individuals' economic opportunities and welfare outcomes. However, the actual track record of financial liberalisation has been deeply disappointing.

The move to liberalise domestic financial markets was built on the seminal work of McKinnon (1973) and Shaw (1973), known as the 'financial repression hypothesis'. McKinnon's argument was that too much government intervention – specifically, too much control of where capital flowed, taxation and restrictions on interest rates – served to 'repress' the financial system. The setting of interest rates by central banks, along with excessively high reserve requirements and subsidised, directed credit resulted in low savings rates, credit rationing and poor investment. This was because credit was allocated according to bankers' discretion rather than guided by market signals and the productivity of investment projects. The solution, from this perspective, was to liberalise financial markets and let the price system allocate credit.

Much empirical work has been conducted on this issue. Liberal economists suggest a positive relationship between economic growth and financial development (King and Levine 1993b; Levine and Zevros 1993; Beck et al. 2000; Benhabib and Spiegel 2000; Levine et al. 2000). However, it is particularly tricky to ascertain the direction of causality – does financial development lead to economic development, or is it economic development that leads to financial development? The strong neoliberal view is that financial development leads to economic development (Jung 1986). For example, King and Levine (1993a) demonstrate that the relative size of the financial sector in

1960 is positively correlated with economic growth over the period 1960–89 for the 80 countries in their sample. Empirical tests of the hypothesis suggest that financial development contributes to economic growth in 85 per cent of countries, that its growth-promoting effects tend to be higher in lower income developing countries, and the impact of financial deepening is as great as expanding exports for developing countries (Odedokun 1996). However, other authors have suggested that causality runs in the other direction (Goldsmith 1969), or even that the direction of causality switches: when countries are beginning to develop the causal arrow is such that finance-leads-to-development, but above a certain threshold it reverses (Patrick 1966). Unfortunately, empirical work remains inconclusive, largely because it is built upon different datasets and different methods.

Failure of the Washington Consensus

The events of the 1980s and 1990s resulted in the Washington Consensus becoming a 'damaged brand' (Naím 2002). The 1980s and 1990s were lost decades in Latin America and sub-Saharan Africa despite (or because of!) the neoliberal structural reforms undertaken. Growth, unemployment, poverty and welfare indicators were all disastrous. The average per capita income growth in developing countries was 0 per cent between 1980–98 (Easterly 2001). This is in contrast to the success achieved in East Asia with distinctly non-orthodox policies (Hoff and Stiglitz 2001). Furthermore, financial liberalisation delivered financial crises rather than stability and growth: 'In sum, the real economic performance of countries that had recently adopted Washington Consensus policies, as opposed to the financial returns they were delivering to international investors or the reception their policies received on the conference circuit, was distinctly disappointing' (Krugman 1995: 41). Krugman argues that the Mexican crisis was proof that the free market and sound money basis of the Washington Consensus was greatly oversold and the developmental benefits of liberalisation actually represented 'a leap of faith, rather than a conclusion based on hard evidence' (1995: 29).

The 0 per cent average per capita income growth in developing countries in the 1990s was in contrast to 2.5 per cent between 1960–79 (Easterly 2001). Yet because financial depth had increased during the 1990s – as well as improvements in infrastructure, health and education – it would have been reasonable to expect income growth to have improved, not stagnated! Plus, all the standard policy variables had 'improved' from 1980, with more market-friendly policies – i.e. financial liberalisation, a move away from import substitution to outward orientation and away from government planning to 'getting the prices right'. From this, like Krugman, Easterly (2001: 154) concludes that the '1980–98 stagnation of LDCs was a major disappointment after all the policy reforms of the 80s and 90s'.

There are a number of interpretations of this. Most simply, the 'good' policies and the structural reforms of the Washington Consensus had failed. But more than this, other commentators have suggested that it's never possible to 'buy' good policy reforms through conditionality (Mosley et al. 1991; World Bank 1998a). As has been demonstrated since, at some length, successful policy reform can only come from within (Dijkstra 2005). Meanwhile, Easterly's (2001: 135) interpretation is that the stagnation of the LDCs had to do with 'worldwide factors like the increase in world interest rates, the increased debt burden of developing countries, the growth slowdown in the industrial world, and skill-biased technical change'.

Box 3.7

Anne Krueger

Anne Krueger (2004) gave a talk in New York in 2004 as the acting managing director of the IMF on policy reforms in emerging market economies. Her talk was called 'Meant Well, Tried Little, Failed Much'. The title comes from Robert Louis Stevenson's 'Here lies one who meant well, tried a little, failed much: surely that may be his epitaph, of which he need not be ashamed'. Krueger argued that failed reforms 'tend to be judged more harshly – by economists, of course, but above all by those who suffer as a result – the poor, the jobless and the hungry.'

Her interpretation was that the problem with liberalising reforms undertaken in the 1980s was that they weren't implemented far enough. She highlighted the notions of reform fatigue and failed implementation. The policy implication being that further, deeper liberalisation is the solution, not less. She pointed to the Asian crises of 1997–98, Argentina's crisis in 2001 and Turkey's of 2000–01 as examples of catastrophic policy failures. For Krueger the crises were the result of unconfronted structural weaknesses. Krueger noted that positive reforms were either intended, or more usually embarked upon, sometimes with success, but then not sufficiently followed through. For example, she noted how fiscal control in Argentina was undermined by off-budget spending, raising the government's deficit and borrowing requirements. This was partly a result of Argentina's decentralised system of government, meaning that provincial expenditures weren't controlled. At the micro level she highlighted the distortions in labour markets which weren't sufficiently reformed and remain too rigid with centralised collective bargaining and therefore higher unemployment and low to zero productivity growth. She went on to argue that the crisis in Argentina was made worse by limited reforms. Her explanation: 'The politician, understandably, has a different perspective. He shares the same long-term objective: sustained, rapid economic growth. But he wants to take the least painful policy measures consistent with achieving that objective'. Yet 'we do know what works, and what doesn't', the problem is this 'tension . . . between economists and politicians'.

Krueger's talk captured the neoliberal position well, identifying the power of economic liberalisation as well as the divide between (good) economics and markets and (bad) governments, politicians and regulation. And the sense that there is a set of structural reforms and adjustments that are necessary for successful development: the mantra of 'Stabilize, privatize, and liberalize' (Rodrik 2006a: 973).

The orthodox response has been either to argue that the reforms didn't go far enough (see Box 3.7 on Anne Krueger for such a view) or to turn to institutions. Levine *et al.* (2000) point to the importance of having well-designed and well functioning financial institutions and systems. For example, it is crucial to have well developed property rights, a functioning accounting system which operates to ensure transparency of performance, a developed system of banking supervision and a sufficiently strong legal environment to protect investors' rights, macroeconomic stability and the correct 'sequencing' of financial reforms (McKinnon 1991; La Porta *et al.* 1997;

Demirgüç-Kunt and Detragiache 1998). So, from this perspective, capital account liberalisation is not wrong per se, but just foolish to enact before domestic institutional reforms had taken place.

Liberal institutionalism: market failure

Neoliberalism, at least in the extreme and negative form it took during the 1980s, is now largely discredited, even inside the IMF and World Bank. A more moderate liberalism now defines the mainstream development agenda. This moderated liberalism is informed by the failures of structural adjustment and the intellectual responses provided by new Keynesian economics on market failure and new institutionalist economics on the importance of institutions and their combined lessons of sequencing and the importance of political and social preconditions for markets to flourish (Harriss et al. 1995).

Yet, just how far things have changed, and just how moderate the new mainstream is, remains contested, with many critical authors emphasising continuity in terms of a pro-market ideology. Liberal institutionalists build upon and recognise the value of neoclassical insights, but reject the neoliberal belief in the power of free markets as utopian. Liberal institutionalists emphasise the significance of market failure in addition to government failure and the necessity of political interventions to address market failures and build institutions. Institutions, history and distribution matter. Development is no longer seen only as the process of capital accumulation and deregulation but is now also a process of political organisation (Hoff and Stiglitz 2001; Rodrik 2006a). As such, this section shows that there is something distinctive about liberal institutionalism in contrast to neoliberalism. Yet, at the same time, liberal institutionalism retains the same neoclassical microfoundations which serve to limit its critique and differentiate it from a more critical reformist position. Liberal institutionalists do retain a belief in (long run) equilibrium, the superiority of markets and most especially the view of money as a neutral resource and politics as a technocratic process of good governance and a series of coordination problems.

Liberal institutionalism in many ways defines the current mainstream approach of the World Bank and the post-Washington Consensus. However, it is important to note that the school remains broad and many individuals are significantly to the political left of the mainstream and criticise the World Bank for not going far enough in embracing a more reformist agenda (e.g. Dani Rodrik).

Market, not government, failure

If it was a lack of capital that explained a lack of development, as neoclassical growth models predict, then the problem of development would resolve itself. The better returns to capital in developing countries would mean that capital flowed from north to south and help bring about convergence in incomes. This hasn't happened, which rather suggests that the problem is not self-correcting (Hoff and Stiglitz 2001).

Building on classical arguments of market failure, orthodox studies have sought to identify the characteristics that encourage a stronger and more successful relationship between finance and development. For example, the importance of investor protection, rule of law and protection of property rights, and high quality and transparent accounting standards (Levine *et al.* 2000). The World Bank (2001c) argues that setting up robust legal systems which protect investors and ensure that contracts are enforced and private property rights are protected are the *fundamentals* upon which an effective financial system can be built. Getting the legal and regulatory environment right should be a greater priority than establishing the financial institutions themselves and this should be the case 'whatever the level of income and regardless of the political and macroeconomic environment of the country' (World Bank 2001a: 79). If investors have confidence in the legal and regulatory systems they are more likely to invest in a country (Rao 2003). Notably, financial supervision was not in the original Washington Consensus. This, according to Williamson (2000), was a major oversight and a failure to recognise that finance does not self-regulate.

It was recognised and understood that markets fail, and without political and legal institutional development, financial liberalisation produces negative development outcomes. This was captured in documents from the World Bank (1998a) such as *Beyond the Washington Consensus: Institutions Matter*. It was also increasingly recognised that a significant safety net was needed as well. There was a focus on income inequality also, insofar as it is recognised that reducing poverty – by increasing human capital through education, land reform and enforcing property rights, and increasing access to finance through microfinance – will help reduce both inequality but also poverty (Kuczynski and Williamson 2003). For example, if income is unequally distributed within society, poor individuals cannot utilise their entrepreneurial skills which limits productivity and acts as a drag on growth. Wealth distribution, history and institutions matter; matter in the sense that they affect the equilibrium assumptions of neoclassical models. For example, because poor individuals cannot save and bequeath money to their children, inequality is reproduced (Banerjee and Newman 1993).

New institutionalist economics

The leading light of new institutionalist economics, Nobel Prize winner Douglass North, has stated, 'Privatisation is not a panacea for solving poor economic performance' (North 1995: 25). Why? Because markets (as well as the economy and society more generally) require institutions to work. North builds on Ronald Coase's (1937, 1960) work on transactions costs. Transaction costs exist because individuals have to search for information and enforce contracts. Institutions emerge as the solution to these problems of uncertainty and coordination. Institutions, following North (1990), are the 'rules of the game' which involve formal rules, informal constraints and enforcement mechanisms. As North argues, the neoliberal assumption that institutions don't matter doesn't make any sense.

Mancur Olson (1996) uses classic growth accounting techniques to verify North's claim. He shows that variations in economic institutions and policies are the main determinants in international differences in income levels rather than a country's stock of productive knowledge or access to capital markets. As Olson points out, on the one hand the lack of convergence in income levels invalidates neoclassical growth theory, and on the other the fact that the fastest growing countries are always a sub-set of LICs invalidates endogenous growth theory. Olson concludes that it must be institutions – which provide the coordination and social rationality necessary to help correct market failures – which explain why some countries develop and some don't. This is not a new insight. Many years earlier, Adelman and Morris (1965) demonstrated that two-thirds of the variation in economic development across countries was down to non-economic differences. And of course this was precisely the premise of classic development theory, that developing countries faced specific challenges which had to be addressed (Rostow 1952, 1960). The neoclassical counter-revolution in development economics had erased institutions, and it was only now rediscovering them (Abramovitz 1986; North 1990).

Since then, studies based on the initial insights of new institutionalism have set the intellectual agenda (see especially the work of Hall and Jones 1999; Acemoglu *et al.* 2001; Rodrik *et al.* 2004). Essentially these authors have developed a new theory of growth which incorporates institutions. Instead of focusing on the proximate causes of growth in the form of inputs, new institutionalists look for the element of growth which cannot be explained by neoclassical theory; for example Rodrik *et al.*'s (2004) seminal argument that 'Institutions Rule'. As the title suggests they demonstrate that institutions are the most important determinant of variation in countries' income levels. They explicitly contrast this with two other positions: the importance of market openness and geography (which is close to the position of Sachs *et al.* 2004 in Chapter 1). Their results suggest that integration has no direct effect on incomes, geography has a weak effect (and mainly via institutions), whereas property rights and rule of law have a significant positive effect on income levels (see also Acemoglu *et al.* 2001; Easterly and Levine 2003; Hausmann *et al.* 2005).

The aim of Rodrik, Subramanian and Trebbi is to show that capital accumulation is only the proximate cause of growth and we need to go deeper to ask the prior question, 'What causes accumulation and innovation in some countries and not others?' (Rodrik *et al.* 2004: 133). Their results suggest that good institutions are six times more important for physical accumulation than human capital and over three times more important than productivity improvements. Their interpretation is that institutions play the important role of preventing expropriation of property which incentivises investment. So capital accumulation and financial investment do still matter, but it is institutions which determine whether the investment will be more or less productive or even happen in the first place. In sum, the shortage of capital is a symptom, not a cause, of underdevelopment (Hoff and Stiglitz 2001; Rodrik *et al.* 2004).

Post-Washington Consensus?

These different trajectories – the failure of the Washington Consensus, the insights about market failure and institutions – have combined to produce the so-called post-Washington Consensus. The post-Washington Consensus (see Table 3.3) recognises the importance of markets and protecting private property rights, but also identifies the wider importance of institutions for economic growth and is less ideological on the positive role of external finance and liberalisation of the capital account. The important difference between neoliberalism and liberal institutionalism is the extent of institutional support needed for a well-functioning economy. Neoliberalism tends to focus on the existence of important 'market-creating' institutions such as private property rights, contract enforcement, an independent judiciary and so forth. Liberal institutionalism goes beyond just market-creating institutions and also underlines the importance of market-regulating, market-stabilising and market-legitimising institutions (Rodrik 2005; Bhattacharyya 2011). The overall point here is that markets aren't self-sustaining. Market-regulating institutions are needed to deal with the problems of market failure, such as the under-provision of public goods. Market-stabilising institutions maintain monetary and fiscal stability as well as prudential regulation and supervision. And market-legitimising institutions of social protection and insurance are needed to cater for those negatively affected by the operation of markets. As such, this marks a return to the position that governments are part of the solution and not just the problem. The policy implications of this position have been to update and adapt the list of best practices and structural reforms needed for successful growth (rather than do away with them). For example, the Monterrey Consensus, in contrast to the 10 points of the Washington Consensus, contains '63 action points, including not just aid and economic issues, but also governance, corruption, and human rights' (Clift 2003: 9). This focus on political and legal reforms has been dubbed 'second-generation reforms' (Naím 2000a).

Table 3.3 *The augmented Washington Consensus*

Original Washington Consensus	'Augmented' Washington Consensus, the previous 10 items, plus:
1 Fiscal discipline	1 Corporate governance
2 Reorientation of public expenditures	2 Anti-corruption
3 Tax reform	3 Flexible labor markets
4 Financial liberalisation	4 World Trade Organisation (WTO) agreements
5 Unified and competitive exchange rates	5 Financial codes and standards
6 Trade liberalisation	6 'Prudent' capital-account opening
7 Openness to FDI	7 Non-intermediate exchange rate regimes
8 Privatisation	8 Independent central banks/inflation targeting
9 Deregulation	9 Social safety nets
10 Secure property rights	10 Targeted poverty reduction

Source: Rodrik 2006a: 978

This view is also embedded in the World Bank. Justin Lin, when he was the World Bank's chief economist, identified the need for state intervention to help the market mechanism along in the process of economic growth (Lin 2012). Lin argues that the process of structural change associated with the transformation from largely agrarian to an industrialised economy demands a proactive role from governments to solve coordination problems (Lin and Monga 2010). It is the government that needs to provide infrastructure and information to encourage and enable firms in promising industries to thrive, because without government support firms have no incentive to be pioneers into new industries. A government is necessary to build ports or a national highway which, as public goods, will be undersupplied by the market. Likewise the extra skills and education needed to develop and profit from industrialisation require investment in education. But, critically, the support must only be temporary:

> the market should be the basic mechanism for resource allocation, but government must play an active role in coordinating investments for industrial upgrading and diversification and in compensating for externalities generated by first movers in the dynamic growth process.
>
> (Lin 2012: 5).

Lin's recipe is for governments to follow the market, not try to buck it. They should identify industries in which they have a 'latent comparative advantage' and then identify a 'compass' country not too far up the development ladder and seek to emulate it, moving into labour-intensive before more capital-intensive industries.

A further consequence of liberal institutionalism is that it has opened the door for a renaissance in international development aid. This is because it was recognised that markets alone won't solve poverty traps. But second, it was now possible to show that with the right institutions and polices, aid can work. In an important report called *Assessing Aid* the World Bank (1998a) captured and publicised this new received wisdom on aid: that it can work, but only with the right polices and institutions (open trade, secure private property rights, the absence of corruption, respect for the rule of law, social safety nets, and sound macroeconomic and financial policies). Under these conditions aid boosts growth, catalyses and encourages private investment and reduces poverty. This has led to a shift away from the conditionality associated with neoliberalism to aid selectivity. Notably, the report's lead author was David Dollar, one of the Bank's senior economists (see Chapter 5 for a discussion of Burnside and Dollar's seminal article).

To conclude, there has been a move away from the one-size-fits-all policies of the Washington Consensus. Rodrik (2006a) argues that the Washington Consensus period was the historical anomaly. The initial euphoria and assumed positive relationship between financial globalisation and development has been replaced by a concern with getting the domestic institutions right, getting the sequence of policy and institutional reforms right, limiting the levels of indebtedness and dependency on external capital flows, and excessive exposure to foreign exchange risk. For example, Hausmann and Rodrik (2003) make a strong case for institutional diversity and alternative development models and routes. This view is well captured by the recent World Bank's

Commission on Growth and Development (2008) which is based on the analysis of 13 cases of sustained high growth. While the report suggests there is no magic bullet and each country's growth path is and should be country-specific, it does identify the importance of economic openness and macroeconomic stability, activist governments in terms of investment in infrastructure, health and education, activist but limited industrial policies, maintaining high rates of saving and investment, having committed, credible and capable governments, but using markets to allocate resources.

Despite the very important changes within the mainstream of development thinking, there remain some important continuities. New institutionalist economics doesn't reject but 'builds on, modifies, and extends neo-classical theory' (North 1995: 17). While there may not be a policy consensus, there is something of a consensus over a set of universally good economic principles, even if their institutional embodiment varies with the national context (Rodrik *et al.* 2004; Rodrik 2007). So, it is absolutely not the case that anything goes, but there are 'higher-order principles of neoclassical economics' (Rodrik 2007), and to be successful countries must remain committed to the '"big principles" that growth requires – sound money, property rights, openness, free markets' (Lin 2012: 43). Critics argue that this means that the list of reforms (in Table 3.3) is more of an *augmented* than *post*-Washington Consensus (Chang and Grabel 2004).

So, while the changes in policy and tone are recognisable and largely progressive, there are very important continuities from neoliberalism, primarily via the theoretical commitment to its neoclassical microfoundations (Mankiw 1989; Gordon 1990). The emphasis on 'good governance', good policies and the liberal institutions of protecting property rights, as a fix for market failure, is a way of sneaking government failure back in. Admittedly it is a more positive role for government where instead of just 'getting out of the way', government is required to actively work with the market. However, there is nothing, as Rodrik (2007) would say, heterodox about this. This is in contrast to critical reformism's prescriptions where the government must play a market-*replacing* not just market-creating, regulating, stabilising or legitimising role.

Heterodox views of finance and development

There are two key issues that divide orthodox from heterodox approaches that concern us here. First, orthodox approaches have a neutral view of money and finance, whereas heterodox approaches see finance as actively shaping the economy. Second, orthodox approaches see economies as self-equilibrating systems, whereas heterodox approaches do not. Heterodox approaches show that free markets consistently result in 'non-equilibrium' outcomes which are detrimental for development, such as structural unemployment, booms and crashes, gluts and overproduction of goods, structural inequalities and ongoing relations of dependency of poor countries on rich countries.

Instead of long-run convergence predicted by orthodox approaches, heterodox scholars highlight the so-called 'Matthew effect' (Merton 1968): 'For to all those who have, more will be given, and they will have an abundance; but from those who have nothing, even what they have will be taken away' (Matthew 25:29). What we are confronted with is an international economic system which works to a logic of accumulated advantage where global inequalities undermine any advantages of liberalisation and worsen poverty and underdevelopment (Wade 2004).

The orthodox belief that economies are equilibrating (either naturally in the case of neoliberals, or with careful support for liberal institutionalists) can be traced back to Jean-Baptiste Say, the nineteenth-century economist (see Box 3.8). In Say's orthodox world, markets clear and money acts as the neutral token in this process, merely allowing the exchange of products for other products and not affecting anything else. Critical reformists and radicals argue that the economy doesn't work like this because money isn't like that (see Box 2.1, p. 21).

In the following we identify two key heterodox approaches. First, 'critical reformists' are those who follow John Maynard Keynes and work 'within-but-against' the system of financialised capitalism. Second, there are 'radicals' who, following Karl Marx, believe that the only solution is to replace the capitalist system. In the remainder of this chapter we deal with them in turn.

Box 3.8

Say's Law, equilibrium and money

Say famously argued that supply creates its own demand. His logic was as follows. When goods or services are bought, this transfers money to the seller and producer of the good who, in turn, buys other goods which they desire, and so on. Thus, goods and services, by being supplied in the first place, create demand for other goods: supply creates its own demand. The orthodox notion that markets 'clear' – i.e. that prices help supply and demand to move into equilibrium – can be traced back to Say's Law. The perspective acknowledges that there may be local gluts if the wrong goods are produced, but in a free market system the laws of supply and demand ensure the right goods are produced in the long run. As such, there can be no recession or general glut of products.

For Say, and orthodox approaches, money, and finance more generally, plays a neutral role in this process. Money is just a token facilitating 'real' economic activity: 'Money performs no more than the role of a conduit in this double exchange. When the exchanges have been completed, it will be found that one has paid for products with products' (Say [1803] 1832: 139). This is in line with the commodity view of money introduced in Box 2.1. This turns the focus back onto savings as the supply of finance and ignores underconsumption and lack of effective aggregate demand (Turner 1989).

Critical reformism: civilising capitalism

> 'Finance must be the servant, and the intelligent servant, of the community and productive industry; not their stupid master' (National Executive Committee 1944).

Critical reformists are consistently critical of financial liberalisation. But they do not reject the positive potential of financial capitalism – they just maintain that it cannot be trusted nor left unmanaged. This is true for both domestic financial liberalisation and external capital flows. As John Maynard Keynes famously put it: 'Ideas, knowledge, science, hospitality, travel – these are the things which should of their nature be international. But let goods be homespun whenever it is reasonably and conveniently possible, and, above all, let finance be primarily national' (Keynes 1933: 758).

Finance needs to be seriously reined in to ensure that it plays the servant role to development, not be the master of it. As such, critical reformists see a much larger role for government than liberal institutionalists do. This is because they have a different view of how the economic system functions, one indebted to Keynes (see Robinson 1956, 1974; Davidson 1972, 1992, 2003, 2007, 2009; Minsky 1986; Gordon 1990; Thirwall 2007; Skidelsky 2009; Keen 2011). Other insights of critical reformists come from the lessons of 'developmental state' successes (Amsden 1989; Wade 1990; Woo-Cummings 1999).

Critical reformists see the economy in need of (1) direction and management and (2) protection from international capitalism. Critical reformists advocate strict management of international capital inflows (protection) and are against domestic financial liberalisation, preferring a state-owned or guided developmental bank model (direction). Economic development, they argue, needs to be directed and managed to overcome collective action problems. The premise is that it is only the state that can potentially take an economy-wide and long-term perspective. For example, Meredith Woo-Cummings (1999: 1) opens her excellent collection on the developmental state with a quotation from Alexander Hamilton: 'Capital is wayward and timid in lending itself to new undertakings . . . the State ought to excite the confidence of capitalists'. Thus, the state needs an industrial policy that can efficiently and effectively coordinate investment (Chang 1994, 1999).

The notion of protecting an economy has a long heritage within the infant industry arguments of Friedrich List (1841) and Alexander Hamilton (1791). Ha-Joon Chang (2002, 2007) has famously and persuasively shown just how important protectionist policies of tariffs, FDI regulation, as well as the industrial policies of subsidies and state-owned enterprises, were in the development of almost all industrialised countries during the nineteenth and twentieth centuries. Tariff protection allowed developing countries, such as the German states in the nineteenth century, to grow and catch up. Protection allowed fledgling industries to grow until they could compete, otherwise the benefits of free trade accrued to the most powerful states which were already industrialised, i.e. England at the time. Moreover, not only did competitor states use trade and industrial policy to develop, but England responded with non-liberal, non-free trade policies of its own, for example banning the emigration of skilled workers, banning the export of tools and utensils and financing industrial espionage (Landes 1969; Chang 2003; Yin and Mongo 2010). Chang's historical analysis shows that the industrialised states did not succeed because of laissez-faire policies, quite the opposite. Moreover, even now, the US's current advantage is on the back of large-scale ongoing state support for key strategic industries such as aerospace, pharmaceuticals and so forth (Chang 2003). As such, understanding successful development is understanding how governments protect and direct their economies.

Box 3.9

Justin Lin and Ha-Joon Chang

Simon Maxwell, in introducing the debate between Justin Lin and Ha-Joon Chang, notes that an easy take on the debate would be to interpret their positions as 'neo-liberal theory which eschews intervention . . . [versus] more structuralist policies which favour government support and extended infant-industry protection. The debate is more subtle than that, however. Both protagonists favour government intervention, but in different ways and for different purpose' (Lin and Chang 2009: 483). This is indeed correct. And the debate captures the differences between liberal institutionalist and critical reformist positions very nicely. Lin foregrounds the importance of market-facilitating intervention, but warns about the limits of this. And Chang highlights the importance of uncertainty, animal spirits and the 'hitchbound' nature of development. The following are their key points which go to the heart of the theoretical divide between the two (with page numbers in brackets):

Lin:

- 'industrial upgrading and technological advance are best promoted by what I call a facilitating state' (484)

- 'the state needs to take the lead in development, because the facilitating-state approach requires government to do much more than a pure laissez-faire approach would allow. Developing economies are ridden with market failures, which cannot be ignored simply because we fear government failure' (484)

- 'the positive externalities of firm entry and experimentation and needs for co-ordination can justify government intervention, and do so in a way that is perfectly compatible with neoclassical economic theory' (485)

- 'by distorting market signals and shifting resources from competitive to noncompetitive sectors, high levels of protection and subsidies slow the country's accumulation of physical and human capital. They also encourage firms to divert their energies from productive entrepreneurship into rent-seeking, which corrupts institutions and further slows capital accumulation' (487)

- 'rather than serving as midwife to healthy new industries, it [the state] is likely to find itself becoming a long-run nursemaid to sickly infant industries that never mature' (487)

Chang:

- 'I believe that comparative advantage, while important, is no more than the base line, and that a country needs to defy its comparative advantage in order to upgrade its industry' (489)

- 'factor accumulation does not happen as an abstract process. There is no such thing as general 'capital' or 'labour' that a country can accumulate and that it can deploy wherever necessary. Capital is accumulated in concrete forms, such as machine tools for the car parts industry, blast furnaces, or textile machines' (491)

- 'Japan had to protect its car industry with high tariffs for nearly four decades, provide a lot of direct and indirect subsidies, and virtually ban foreign direct investment in the industry before it could become competitive in the world market' (491)

- 'a good neoclassical economist may be tempted to argue that a country should do a cost-benefit analysis before deciding to enter a new industry, weighing the costs of technological upgrading against the expected future returns, using comparative advantage as the measuring rod. However, this is a logical but ultimately misleading way of looking at the process. The problem is that it is very difficult to predict how long the acquisition of the necessary technological capabilities is going to take and how much 'return' it will bring in the end. So it is not as if Nokia entered the electronics industry in 1960 because it could clearly calculate that it would need to invest such and such amount in developing the electronics industry (through cross-subsidies) for exactly 17 years but then would reap huge future returns of such and such amount. Nokia probably did not think that it would take 17 years to make a profit in electronics. It probably did not know how large the eventual return was going to be. That is the nature of entrepreneurial decision-making in a world with bounded rationality and fundamental uncertainty' (491)

- 'the rational-choice, individualistic foundation of neoclassical economics limits its ability to analyse the uncertain and collective nature of the technological learning process . . . In the real world, firms with uncertain prospects need to be created, protected, subsidised, and nurtured, possibly for decades, if industrial upgrading is to be achieved' (501)

All page numbers come from Lin and Chang (2009).

Uncertainty, money and animal spirits: Keynes's insights

To understand why the state is the necessary economic coordinator for development, it's useful to turn to Keynes's theory of investment, demand and development. Keynes rejects the orthodox view that markets fail to clear because of supply-side problems. For example, a supply-side explanation for why unemployment persists would be that wages are above the equilibrium level because of, say, minimum wages, union strength or burdensome labour market regulation. The orthodox solution is therefore to remove government interference in the market. The alternative to this is a demand-side explanation. For Keynes (1936) it is not the 'stickiness' of prices (including wages) that stops an economy reaching equilibrium. Instead, it is the very nature of money that creates a structural failure of demand and development (Davidson 1992). This is worth unpacking.

In orthodox theory – from Say onwards – money is a neutral token in lieu of bartering. However, for Keynes (1936: 293) the 'importance of money essentially flows from its being a link between the present and the future'. Money's key role is as a liquid store of wealth. So, Keynes argued that, in addition to the standard transactions motive for holding money, there is also a precautionary motive. If economic agents lack confidence about future prospects they prefer to hold savings in an easily accessible or liquid form such as cash rather than long-term investments. This is because the future is fundamentally uncertain, in the sense that 'there is no scientific basis on which to form any calculable probability whatever. We simply do not know' (Keynes 1937: 214). Thus 'our desire to hold money as a store of wealth is a barometer of the degree of our distrust of our calculations and conventions concerning the future' (Keynes 1937: 216). As a result, actors defer investments as a way of avoiding the costs of 'unknowledge' (Shackle 1983). And in turn, the desire of individual consumers to hold back from spending now creates what Keynes called the 'paradox of thrift'. Essentially, what might be rational for individuals is collectively disastrous. That is to say, it may well be rational for individuals to hold off spending or investing, but if everyone does this at the same time then the aggregate consequences are calamitous: nobody buys or invests anything and the economy stagnates. So to summarise, the failure of effective demand and investment is driven by uncertainty, money and liquidity preference.

This is the basis for Keynes's most famous insight: when individuals and businesses lack confidence in the future (lacking in 'animal spirits') they tend to save and underspend (Keynes 1936: 162). Thus saving and hoarding reduce aggregate demand and consumer confidence, increasing uncertainty and furthering demand for liquidity. Meaning there is a tendency for an economy to move away from equilibrium, not towards it. Hence we return to the Harrod-Domar view of the world (see Box 3.2).

The notion of 'animal spirits' captures 'our innate urge to activity which makes the wheels go round' (Keynes 1936: 163). So, when thinking about investment, in an uncertain economic environment:

> our decisions to do something positive, the full consequences of which will be drawn out over many days to come, can only be taken as the result of animal spirits – a spontaneous urge to action rather than inaction, and not as the outcome of a weighted average of quantitative benefits multiplied by quantitative probabilities . . . Thus if the animal spirits are dimmed and the spontaneous optimism falters, leaving us to depend on nothing but a mathematical expectation, enterprise will fade and die.
>
> (Keynes 1936: 161)

This also means we need to reverse the orthodox view of savings leading to investment. Keynes was clear that savings are not the source of finance for investment, a position called the 'loanable funds fallacy' (Bibow 2009). Investment is accompanied by increased saving, but not preceded by it. Keynes's famous comment was that investment is the parent, not the twin, of savings (Keynes 1939). Or as Joan Robinson (1952: 86) put it, 'where enterprise leads finance follows'. This means that investment depends upon the expectations of entrepreneurs – the famous 'animal spirits' – rather than a society's propensity to save (Davidson 1992). To use Schumpeter's phrasing, the 'classic' views of Mill and Say generate an 'essentially hitchless' model of economic development where all saving is automatically converted into investment goods (Schumpeter 1954: 545). In contrast, the Keynesian view is a 'hitchbound' model of economic development. It is the tendency for developing economies to become 'hitched' and stagnate that justifies an activist role for government.

There is a legitimate role for interventionist public policy in supporting investment for business and infrastructure as well as limiting the animal spirits in financial markets (Dow and Dow 2011). This is because if animal spirits become overheated, especially in speculative financial activity, they can also be the undoing of the economy. Building on Keynes, Hyman Minsky famously postulated that the seeds of crisis were sown in the good times. As speculative bubbles are built the financial structure comes to regulate a giant 'Ponzi scheme' (Minsky 1982). The role for financial regulators, then, is to take an activist stance and try to control liquidity, excessive innovation and speculative trading (Nesvetailova 2007). In contrast, orthodox approaches see the government's role as much more limited: to public investment, setting of tax rates and monetary policy, ensuring price and wage flexibility domestically and exchange rate flexibility at the international level, but not economic planning and social control (Turner 1990). In short, a coordinator and a facilitator of competitive markets (Davidson 1992).

In contrast, the developmental state model addresses this uncertainty and the potential 'hitches' between savings, profits, investment and growth, and links them. The developmental state does so by directly intervening in the economy – protecting, directing and investing – allowing the 'animal spirits' to sustain capital accumulation.

The developmental state

A state's means of raising and deploying financial resources tell us more than could any other single factor about its existing (and its immediately potential) capacities to create or strengthen state organizations, to employ personnel, to co-opt political support, to subsidize economic enterprises and to fund social programs.

(Skocpol 1995: 17)

The developmental state literature addresses the question: Is it possible – through state intervention – for developing countries to accelerate their investment and development? Proponents of this approach also call it 'governed market theory' (Wade 1990: 27) or the 'late industrialising model' (Amsden 2001) or 'alliance capitalism' (Gerlach 1992). The developmental state acts as an entrepreneur or venture capitalist, coordinating firms and industry investment plans as part of a national development strategy, managing conflict and steering markets. The classic statements in the developmental state literature deal with the cases of Japan (Johnson 1982), Taiwan (Wade 1990) and South Korea (Amsden 1989; see also Evans 1995).

In her book *The Rise of 'the Rest'*, Amsden (2001) addresses the late industrialisation of China, India, Indonesia, South Korea, Malaysia, Taiwan and Thailand in Asia; Argentina, Brazil, Chile and Mexico in Latin America; and Turkey in the Middle East. From this Amsden summarises the four functions of a successful developmental state as: 'developmental banking, local-content management, "selective seclusion" (which is the process of opening some markets to foreign transactions while keeping others closed), and national firm formation' (Amsden 2001: 125). To carry out these functions, the state intervenes with subsidies which are deliberately intended to distort market prices and stimulate economic activity in the industries the state has identified for development. The government directs credit and investment to key sectors and firms via state-owned banks, subsidies and tax breaks. Developmental banking and finance institutions are central to the success of the developmental state in order to raise funds and channel investment. Instead of the invisible hand of the market, commentators talk about the 'visible hand' of the state (Lauridsen 1995). Famously, this means that state-led developmental banking entails a commitment to 'getting relative prices wrong' (Amsden 1989: 8, see also Wade 1990: 29). (See Box 3.9 for a debate between liberal institutionalist Lin and critical reformist Chang to see how the differences flesh out.)

The close bank-firm relationships of alliance capitalism mean that governments have liberalised their trade and capital accounts very cautiously. For example, in order to maximise the benefits for the domestic economy, transnational corporations (TNCs) have been accepted but with stringent rules on who could act as suppliers to them and committing them to using domestic content in the manufacturing process. As Wade (2000) has noted, there is a very good reason for keeping a partially and strategically-closed capital account: the government needs to limit external finance in order to maintain its monopoly power over domestic industry. As Wade (1990: 367) expands:

> the government must maintain a cleavage between the domestic economy and the international economy with respect to financial flows. Without control of these flows, with firms free to borrow as they wish on international capital markets and with foreign banks free to make domestic loans according to their own criteria, the government's own control over the money supply and cost of capital to domestic borrowers is weakened, as is its ability to guide sectoral allocation.

Hence, the close ties and the high levels of corporate debt which drive the success of alliance capitalism are fundamentally dependent on keeping capital inflows regulated.

How was this regulation and control successfully achieved? Beyond identifying a generally 'authoritarian' character to the politics of the developmental state (Leftwich 1993), two more specific points are worth making. First, as both Amsden (1989) and Wade (1990) note, governments used performance criteria, which were tough and very much enforced, to regulate the subsidies provided to firms (Wade and Veneroso 1998). Thus market discipline was replaced with performance target discipline; and the enforcement of strict discipline is crucial for the model to work (Chang 1994). Second, the developmental state must be able to resist the pressures from external private forces, such as TNCs, pushing for their own short-term gains, and resist domestic particularist interests in order to act in the national interest. Peter Evans (1995) has described the optimal state organisation required as 'embedded autonomy'. The notion captures need for the government to be autonomous from particularist demands, with a strong and independent bureaucracy and to be capable of making policy. But, simultaneously, the government needs to be sufficiently embedded within the relevant business and industrial sectors to work with them responsively. The importance of this embeddedness is captured by Linda Weiss (1998: 39) when she notes: 'Policies for this or that industry, sector or technology are not simply imposed by bureaucrats or politicians, but are the result of regular and extensive consultation and coordination with the private sector'. It is only when embeddedness and autonomy are joined together that a state can be called developmental (Evans 1995). Where this doesn't happen, and the state lacks an effective bureaucracy, such as Mobutu's Zaire, the state model is 'predatory' rather than developmental.

So in sum, the selectivity and management by the state meant that FDI flows and TNCs are carefully handpicked, and financial flows are carefully regulated and limited. In terms of whether this is a suitable model for others to adopt, in his study of Taiwan, Wade (1990) argues that consistent and effective government interference in the market is both sensible and possible to emulate. However, there are a number of cautionary issues worth noting.

First, it is not clear how well this can work outside of the few states it has been successful in. The regional specificity of the 'developmental state' is not coincidental. The authoritarian features of East Asian governments, and specifically their 'autonomy' from special interests which would threaten capture and serve vested and particular interests, have been key to promoting the rapid national development behind the catching-up of late industrialisation. However, consider the case of Botswana which seems to fit many of the features of a developmental state (Taylor 2005) as well as the examples of Turkey and those countries in Latin America in Amsden (2001). Second, because of political pressures, rent-seeking and corruption the model does run the risk of leading to over-accumulation and excessively high levels of corporate debt (Amsden 2001).

Third, there are key elements of the model which are under increasing pressure from international liberalisation and the institutionalisation of global rules. For example, the industrial policies and protectionism of the developmental state struggle in the face the trade rules of the WTO such as the GATS provisions which demand developing country financial markets to open up and liberalise (Wade 2003). Indeed, the East Asian financial crisis has its roots in the liberalisation of East Asian economies. Contrary to politicised and shallow attempts by orthodox commentators to discredit the model and explain its failure in terms of corruption and 'crony capitalism', critical reformists emphasise that it was the opening of the capital account in the 1990s and the surge in private capital flows which preceded the crisis that were part of the developmental model's undoing (Wade and Veneroso 1998). This is because, as noted above, capital account liberalisation undermines the government's ability to control interest rates or direct credit (Wade 2000). There is seemingly a fundamental incompatibility between open financial markets and government autonomy (cf. Weiss 1998). We return to these issues in the following chapter in the shape of the 'impossible trilemma' (Chapter 4).

Radicalism: capital as power

> Money abuses all the gods of man – and turns them into commodities. Money is the universal self-established value of all things. It has, therefore, robbed the whole world – both the world of men and nature – of its proper value. Money is the alienated essence of man's work and existence, and this alien essence dominates him, and he worships it.
>
> (Marx 1844: 49)

Radicals, as the name suggests, offer a radical critique of finance, capitalism and to a certain extent development itself. A number of different theoretical schools of thought fit under the 'radical' umbrella. There is more diversity and disagreement within this group of approaches than any of the preceding three. It includes Marxism, theories of imperialism, world systems theory, dependency theory and theories of underdevelopment more generally as well as post-development approaches. Indeed, although the radical approach to global finance and development tends to be heavily indebted to Marxist thought, it is not exclusively Marxist. Whereas Marxist and dependency scholars have focused on developing a critique of capitalism, the post-development theorists' target is development itself (see Box 3.10).

There are strong disagreements among critical scholars from these different theoretical traditions. But there is a shared view of development as a form of imperialism. Like imperialism, development involves coercion, oppression and often violence and conquest by external actors, usually Western. Even though, like civilisation, development *qua* progress seems like an intuitively good thing, difficult to be against, radicals would suggest that there is nothing inherently progressive about development and absolutely not in the way in which it is practised. Radicals emphasise that, whether we are talking about development, progress or civilisation, these processes tend to involve conquest, violence, coercion and oppression (Diamond 1974). Consider the following blistering paragraph from Gavin Kitching (1989: 195) which challenges any cosy liberal optimism:

> . . . development is an awful process. It varies only, and importantly, in its awfulness. And that is perhaps why my most indulgent judgements are reserved for those, whether they be Marxist-Leninists, Korean generals, or IMF officials, who, whatever else they may do, recognise this and are prepared to accept its moral implications. My most critical reflections are reserved for those, whether western-liberal radicals or African bureaucratic elites, who do not, and therefore avoid or evade such implications and with them their own responsibilities.

In sum, radical approaches share a common rejection of the Western, liberal development project and with it a rejection of the positive potential of external interventions. Instead, development is part of existing international hierarchies. Unequal power relations and cultural and economic imperialism mean that exploitation is the outcome of development and economic and financial integration, not increasing freedoms.

Box 3.10

Post-development

Non-Marxist radical accounts tend to agree on the generally exploitative and imperialist nature of global finance and the financialisation of life under capitalism, but disagree on the causal processes, emphasising the cultural and discursive valorisation of financial capitalism as opposed to the Marxist emphasis on the material workings of capitalism, specifically the role of commodities, production and the labour theory of value. Post-development scholars include the likes of James Ferguson (1994), James Scott (1998, 2009) and probably most celebratedly Arturo Escobar's *Encountering Development* (1995) (see also Sachs 1992; Rahnema and Bawtree 1997).

Escobar's argument is that development, as outlined in Harry Truman's 1949 speech, is a dream or an illusion. 'Instead of the kingdom of abundance promised by theorists and politicians back in the 1950s, the discourse and strategy of development produced its opposite: massive underdevelopment and impoverishment, untold exploitation and repression. The debt crisis, the Sahelian famine, increasing poverty, malnutrition, and violence are only the most pathetic signs of the failure of forty years of development' (Escobar 1995: 4). For post-development authors, this failure is not a coincidence, nor does it represent the failure *of* development, but instead debt crises, famine, poverty, malnutrition and violence are the logical outcomes of the 'development project'. Escobar's approach is to highlight the ways in which poverty and underdevelopment came to be understood as a 'problem', a problem to be fixed; 'if the problem was one of insufficient income, the solution was clearly economic growth' (Escobar 1995: 24). As such, the starting premise – that there are countries and peoples who are somehow lacking what the West has – means that it is the discourses and practices of development which create the very problems they purport to try and solve. The result was a 'top-down, ethnocentric, and technocractic approach' (Escobar 1995: 44). Radical scholars tend to highlight how 'development' – *qua* cultural Westernisation and homogenisation – results in (and requires) the loss of indigenous culture – psychological and environmentally rewarding ways of life (Sidaway 2008). It is an imperialistic and a totalitarian mode of thought which does not respect local, indigenous alternatives. Hence, from a radical perspective, development is seen as a form of and a continuation of colonial patterns of intervention, transformation, control and rule. As such, in the words of Wolfgang Sachs (absolutely no relation to Jeffrey), 'it is not the failure of development that is to be feared, but its success' (Sachs 1992: 3).

Marx and Marxism

Karl Marx's work – arguably the archetypal radical – speaks strongly to the issues of development and the role of capital. Marx, like the other classical political economists, held that the accumulation of capital was at the heart of economic development. However, Marx and Marxists emphasise the exploitative nature of capital accumulation. The wake of the global financial crisis has seen a return to prominence for Marx and Marxism. His view of development is actually relatively straightforward, but his work on money is more difficult and contested. This is partly because it was developed in many places and changes over time (Marx 1970, 1975, 1986), but also partly because Marx's view of money is quite different from other approaches. For Marx money is a social relation (Fine and Lapavitsas 2000). To understand what this means and the implications it is necessary to look at Marx's theories of value and exploitation.

First, what was Marx's view of money and finance? Marx's view of finance was that the credit system emerged as a way of mobilising idle capital generated from industry and channelling it back into the circuits of capital accumulation (Marx 1885, 1894). Money was also a commodity itself – something that allowed for exchange. Thus far, then, nothing extraordinary here. But the key to understanding Marx – as always – is that his theory of money is grounded in the labour theory of value. Hence, the value of a good or commodity is based on the labour time which has gone into it. Crucially, for Marx, the price that the commodity is sold at is routinely higher than the value of the effort gone into it; the difference between what workers are paid and what a good is sold at represents 'surplus value'. Surplus value is the profit that goes to the capitalist, a result of his control over the means of production (i.e. capital, tools, savings, raw materials). As such, surplus value is what makes capital accumulation possible as well as, crucially, a measure of and evidence for the exploitation of labour that is at the heart of capitalism. So surplus value is key to understanding the process of capital accumulation and development, plus the exploitation that makes this possible. And it is at this point that money comes in.

For Marx, money is a commodity, albeit a special commodity. It is the 'universal equivalent' which allows commodities to circulate, and the separation of production and exchange because it acts as a store of value. But more interestingly and importantly, money is also a 'value-form'. More specifically, money is objectified and alienated abstract labour. This is because money is the thing which enables the abstraction and objectification of labour value (and therefore exploitation). This objectified labour value can then be stored, exchanged and ultimately hidden. It facilitates the extraction of value and exploitation over time and space and as such is absolutely not a neutral resource, nor is the financial system which moves money around. Both are part of an elegant (but violent) social system that alienates, exploits and empowers.

Second, what is Marx's view of development? Essentially it's an evolutionary, linear one, not unlike Rostow's (which isn't a coincidence given Rostow's subtitle). His view was that capitalism would spread universally to create a single capitalist system before the final revolutionary transition from capitalism to a fully socialist society. Marx's theory of history was that each mode of production grows until it reaches its limits and then there is a shift to a new mode of production: from feudalism, to capitalism, to communism. For Marx it was always changes in the economy (the base), especially changing technology, that determined changes in society and politics (the superstructure). For example, the steam-engine, mule and spinning jenny were the productive forces that made the abolishment of slavery possible, and serfdom disappeared with the agricultural revolution. Marx viewed capitalism as historically progressive because of the huge leaps in productivity it allowed, but socially abhorrent because of the inequalities it produced and its reliance on exploitation (Sutcliffe 2008). He also felt that capitalism was doomed, given its inherent contradictions. But crucially, ever since Marx, those who have followed in his footsteps have actually sought to understand why capitalism is, in fact, so resilient. Many authors point to how capitalism has 'fixes' (see Box 3.11 on David Harvey). One such fix is the geographical expansion of capitalism – i.e. imperialism, colonialism and globalisation.

Box 3.11

David Harvey

David Harvey is one of the best living exponents of Marx's vision. His books, including *The Limits to Capital* (1982), *The New Imperialism* (2003) and *The Enigma of Capital* (2010), apply Marx's work to questions of global capitalist finance. Harvey focuses on capital, arguing that it is the lifeblood of the economy. His most recent, and popular, book, *The Enigma of Capital* examines the global financial crisis. Harvey is clear that the crisis is a normal, inevitable and indeed necessary feature of capitalism. It is structural. But as such, the solution is that the system itself has to be reformed.

In this and his earlier work he highlights how the globalisation of capitalism is intimately tied up with it working through its own crisis tendencies. Harvey calls this the 'capital surplus absorption problem'. To explain, Harvey returns to Marx's arguments in the *Grundisse* to show that the crisis of capitalism emerges from the contradiction of capitalists earning surplus value at the expense of labour or their suppliers when these are simultaneously the capitalist's customers.

There are two fixes to this. First, there is what Harvey calls the 'spatial fix', which is the expansion of capitalism to new markets and exploitation of new sources of labour and raw materials. This is similar to the view outlined by Rosa Luxemburg ([1913] 1951) and other theorists of imperialism.

And second there is the 'credit fix', which is essentially a temporal fix. How, given the relative availability of labour, low and stagnant wages, do capitalists still manage to find a market for their products? If workers are getting paid less and less, how can they still buy products? The question being raised here – which comes from Marx – is that if surplus value is required now, but none is available, where does it come from? The answer is by creating 'fictitious capital' – i.e. credit and increased indebtedness. And this is what was behind the extension of the sub-prime mortgages noted in Chapter 1. New credit and buying power is created to solve the immediate problem of overaccumulation and delay the crisis. Similarly, developing countries can access financial markets. However, turning to the capitalist 'credit system simultaneously makes territories vulnerable to flows of speculative and fictitious capitals that can both stimulate and undermine capitalist development and even, as in recent years, be used to impose savage devaluations upon vulnerable territories' (2003: 118).

This interaction of spatial and temporal fixes is best expressed by Harvey himself (2003: 109) from his *The New Imperialism*: 'Overaccumulation within a given territorial system means a condition of surpluses of labour (rising unemployment) and surpluses of capital (registered as a glut of commodities on the market that cannot be disposed of without a loss, as idle productive capacity, and/or as surpluses of money capital lacking outlets for productive and profitable investment). Such surpluses may be absorbed by (a) temporal displacement through investment in long-term capital projects or social expenditures (such as education and research) that defer the re-entry of current excess capital values into circulation well into the future, (b) spatial displacements through opening up new markets, new production capacities and new resource, social and labour possibilities elsewhere, or (c) some combination of (a) and (b).'

Imperialism and capitalism's contradictions

Clearly capitalism failed to develop the world and capitalism's 'gravediggers' failed to materialise. As such Marx's followers sought to understand how the inner logics of capitalism failed to produce the predicted crisis. One answer to this conundrum was the phenomenon of imperialism (Brewer 1990). For Marxists, imperialism is not simply a political, colonial system of government, but is instead tied up with the logics of capitalist expansion.

In order to succeed, capitalism depends upon increasing profits. This is achieved through the extraction and appropriation of surplus value from labour. The contradiction at the heart of capitalism is that each individual capitalist is trying to reduce the wages of their workers, but at the same time these same workers are also consumers. The problem for capitalism and capitalists is, as wages and purchasing power falls, how can consumption be maintained and overproduction avoided?

Rosa Luxemburg ([1913] 1951) was one of the first to identify the role of the global 'periphery' in solving this puzzle. Theories of imperialism showed how the build up of contradictions within the capitalist countries in the nineteenth century led to European expansionism and war, as the capitalist countries scrambled for colonies in order to access cheaper labour and secure new markets and raw materials (Hilferding [1910] 1981; Luxembourg [1913] 1951; Lenin 1916). But in addition, it also helped stave off revolution in the capitalist core as cheap production and products essentially 'bought off' consumers. Thus surplus value, and the exploitation it depends upon, were extracted from the periphery. Notably it was Lenin and Hilferding's analyses that highlighted the role of finance in the processes of imperialism.

Lenin (1916) famously referred to imperialism as the highest stage of capitalism. His analysis drew heavily on Hilferding's work on imperialism and finance capital. Like Marx, Hilferding ([1910] 1981) argued that the financial system emerges to sustain capital accumulation. But it was his analysis of 'finance capital' that was crucial. Hilferding identified a process whereby there was monopolisation of the so-called 'holy trinity' of industrial, commercial and banking capital. To explain, as industrial production became increasingly large-scale at the turn of the twentieth century, industry needed access to larger and larger amounts of credit. So much so that there was an amalgamation of industrial and banking capital, creating finance capital to meet this need. Crucially the state was beholden to this monopolised finance capital. Finance capital came to dominate the organisation of the economy in its own interests. Hilferding explains how the globalisation of capital from the core to the periphery served to 'stunt' the development of the periphery because most of the profits were returned to the core, limiting any accumulation in the periphery: 'As economic tributaries of foreign capital, they [the periphery] also became second-class states, dependent on the protection of the great powers' (Hilferding [1910] 1981: 330). So according to this view, access to finance serves to immiserate rather than develop develop countries.

Underdevelopment and dependency theory

More recently, a set of theorists, from Latin America especially, developed the notions of dependency and underdevelopment. Picking up from the theories of imperialism, they divided the world into advanced capitalist economies (core) and underdeveloped economies (periphery) (Baran [1957] 1973). At the heart of dependency theory is the contention that countries on the periphery of the world economy cannot achieve autonomous development. Dependency theorists argue that the capitalist system has gone from being the engine of economic development to the biggest hurdle for development. This is because it is in the interests of capital to keep developing countries as the 'indispensible hinterland' to extract economic surplus and resources from. As such, from this perspective, and this is important to underline, underdevelopment is a direct function of development in the industrialised core.

In part this is a legacy of imperialism and colonialism. As Samir Amin (1974) has noted, colonialism's legacy was to tailor production for exporting to the core European countries, resulting in many developing countries depending upon a monoculture for export (e.g. bananas, cocoa, sugar). This specialisation inhibits development, as it leaves the country dependent on an undiversified export portfolio which is dominated by low value-added raw materials, and tends to experience low and unstable prices.

André Gunder Frank (1969) took Baran's ideas and developed and popularised them as a direct attack on Rostow's modernisation theory. Frank's insights came from the observation that during the periods when the international system was less integrated,

Latin American economies enjoyed autonomous industrialisation and growth. But during those periods when the international economy became more integrated, the Latin American economies suffered more sluggish growth and de-industrialised. Frank reinforces Baran's point that underdevelopment is not a premodern 'stage' to travel from, but an active condition which is reproduced by the system: prosperity in the core is a function of underdevelopment in the periphery. Frank's term for this was 'the development of underdevelopment'. Others have called it dependency. The most widely cited definition of dependency was one that was put forward by Dos Santos (1970: 231):

> By dependence we mean a situation in which the economy of certain countries is conditioned by the development and expansion of another economy to which the former is subjected. The relation of interdependence between two or more economies, and between these and world trade, assumes the form of dependence when some countries (the dominant ones) can expand and can be self-sustaining, while other countries (the dependent ones) can do this only as a reflection of that expansion, which can have either a positive or a negative effect on their immediate development.

While early dependency theorists tended to focus on the role of trade relations (e.g. Prebisch 1950; Singer 1950), many increasingly examined the role of financial ties between the core and the periphery and the processes of financial exploitation (Cardoso 1972; Sweezy 1972). By financial exploitation they mean the ways in which multinational firms invest in countries in the periphery, but then repatriate the profits back to the core from the host economy. Under these circumstances the developing country fails to accumulate capital and develop. In an empirical test of this thesis, Christopher Chase-Dunn (1975) ran a regression analysis on longitudinal data to adjudicate whether penetration of foreign capital resulted in financial exploitation and the development of underdevelopment. His results suggest that there is a statistically significant negative correlation between the dependence of a country on foreign investment and borrowing and its subsequent economic performance. In short, finance and development are negatively correlated. It's worth pointing out that, in his later work, Prebisch (1981) also came to see investment and financial networks as key mechanisms of underdevelopment, not just trade. In Chapter 6 we look at the role of the private Western banks (overseen by the World Bank and the IMF) in reproducing international inequalities. They became the new main channels of capital movement as they granted loans to the capital-starved periphery; a process which eventually precipitated the Third World debt crisis.

It's important to note that there are a number of disagreements and debates between different dependency, underdevelopment and world systems theorists. First, there are important debates over whether the unit of analysis should be nation states or classes (Brenner 1977; Wallerstein 1979). Second, whether the contradictions are in the international system, or exist within developing countries themselves (Warren 1980). For example, Frank argues that the metropolitan core exploits and extracts from the periphery, but then *within* developing countries there is an urban metropole which extracts and underdevelops the rural hinterland. So it is not just a process of 'external' exploitation. Third, and related, there are those who place a much greater emphasis on the role of political agency, identifying the crucial role of a liaison elites or 'comprador' class which cements the core-periphery relations, instead of dependency being a structural and economic law (Galtung 1971; Frank 1972). The comprador class or 'lumpenbourgoisie' help sustain dependency because they come to see their interests as being more closely tied with Western capitalism than the fate of their own populations. And fourth, there are disagreements over the extent to which integration within the capitalist system makes development impossible. Some argue that the only solution is to 'delink' and pursue autonomous or south-south development (Amin 1990), while others maintain that successful national industrial and development strategies are possible (i.e. closer to the critical reformist developmental state position). But, to conclude, all radicals agree that the relationship between global finance and development is an imperialistic and largely negative one.

Conclusion

This chapter has introduced the range of theoretical positions in the debate about global development and finance. These have been grouped into four broad approaches: (1) neoliberal, (2) liberal institutionalist, (3) critical reformist and (4) radical. These distinctions will provide the framework for the rest of the chapters in the book to organise and understand the debates around aid, the bond markets, bank lending, portfolio equity, foreign direct investment, remittances and microfinance.

One way of reading the differences in the approaches is using a traditional right-left political spectrum. And this is indeed probably accurate, but it is also superficial. A deeper way of thinking about it, and one which is grounded in a financial 'ontology', is to look at their theory of money and finance and whether or not they view the economy as a self-equilibrating system. It turns out that this provides a way of understanding the foundations of the different views. For neoliberal and liberal institutionalism, based in neoclassical economics, the problem of development is all about the free movement of capital and ideas and the appropriate institutional framework to enforce rules and economic markets. The issues of exploitation, or power, or unequal ownership, tend to get ignored. And, by introducing and explaining the framework we have illustrated how the mainstream agenda has moved from a crude neoliberalism to a more moderate and nuanced liberal institutionalism, but also that these two approaches share a common set of neoclassical assumptions and treat money in a similar way.

A couple of things – 'real world' and 'academic' – have combined to shift the mainstream away from neoliberalism to what has been termed the post-Washington Consensus. Or, probably more accurately, an augmented Washington Consensus (Chang and Grabel 2004) insofar as it is less of a rejection of the core tenets of free markets than the recognition that 'free markets' are, by themselves, not enough. First, a set of empirical puzzles increasingly suggested that the neoliberal view was wrong: the failure of neoclassical convergence (Pritchatt 1996); the failure of capital to flow from rich to poor countries (Lucas 1990); and the increasingly widespread and uncontentious recognition that the neoliberal, structural adjustment agenda had failed (Naim 1990).

Explanations for these failures varied, but intellectual trends were already pointing the way towards issues of good governance and market failure (World Bank 1999). The influence of endogenous forms of growth theory had begun to highlight a limited role for government action. But this role was significantly increased with the growth of new institutionalist economics. Authors such as North (1990) and Rodrik *et al.* (2004) had been making important strides in showing the importance of the right institutions for successful development, identifying these as the deeper causes of development which underpin the proximate explanations for growth and capital accumulation (such as savings and investment).

Meanwhile, heterodox approaches have continued to develop more critical analyses. Critical reformists work from a position of general scepticism towards capitalist finance but, notably, not a rejection of it. Keynes, after all, is probably best understood as a 'capitalist revolutionary' (Backhouse and Bateman 2011). The critical reformist view is that finance needs to be seriously reined in and returned to the role of servant of the economic system, not the master. Although there are superficial differences between liberal institutionalism and critical reformism, these are just that: superficial. Critical reformists do not see market failure as an unfortunate deviation from the norm. It is the norm. This is illustrated very nicely by Patrick Spread's (2012) critique. He argues that 'market failure' has become a 'catch-all explanation for the many misalignments of neoclassical theory with observed behaviour', but these misalignments are inevitable because neoclassical textbook markets don't exist: 'It is like describing a camel as an ugly consequence of "unicorn failure"' (Spread 2012: 49).

Finally, the radical perspective provides much food for thought. Consistently marginalised within mainstream accounts of finance and development, it has enjoyed something of a renaissance in recent years given the global financial crisis. But it has always been a consistent and valuable perspective within development studies more generally and the Global South especially. Although there are important differences between many different theoretical schools, there is a consistent emphasis upon global finance as an imperialistic system, one that extracts wealth from developing countries and helps to maintain international inequalities.

Almost all theories and authors can be placed into one of these four categories – though you will find this is more usually done by other authors than by themselves. Some might be closer to the edges of their 'box' and even travel back and forth a bit. Like any framework it is merely a way of simplifying the complexity of reality and a shorthand for presenting this. Dividing the world up is always a dangerous task and can hide as much as it can reveal; as the statistician G.E.P. Box (1979) memorably put it, 'Essentially, all models are wrong, but some are useful'. The fourfold division presented here, I think, is a useful place to begin. Schumpeter is also useful, again, to note at this point:

> the setting up of such types is an expository device. Though certainly based upon provable facts, neither must it be taken too seriously or else what is intended to be a help for the reader turns into a source of misconceptions . . . types are useful only so long as this is remembered.
>
> (Schumpeter 1954: 52)

As such, the framework should be used as a heuristic device which helps emphasise the key points of disagreement within debates about global finance and development. It should be seen as a guide, not a cage. On further reading you will discover that some authors or theories don't fit neatly inside a single school and bleed across the boundaries; furthermore, there will be some disagreements *within* the approaches that are at least as large as those between the different schools. We will flag these up where relevant.

Discussion questions

1 Do, in your view, savings naturally lead to investment and development or is it investment that leads to development and savings?

2 Are governments capable of (and to be trusted with) steering the economy and financial markets better than the price mechanism?

3 What is the difference between a liberal institutionalist's view on the government playing a proactive role and the critical reformist's view?

4 Which of the four theoretical approaches do you feel is the most convincing, and why?

5 Is finance the cause or solution to all of life's problems?

Further reading

- For an accessible and superb introduction to the best of new institutionalist political economy, read Acemoglu and Robinson's (2012) *Why Nations Fail*. It's a tour de force of how economics, institutions and politics interact to produce development outcomes, or failures. It builds upon years of top-quality research, but is written for the general reader.

- Ha-Joon Chang's (2007) *Bad Samaritans* is also an accessible introduction, presenting a distinctively critical reformist perspective. Chang emphasises the way in which the rich countries, having successfully developed, have effectively 'pulled up the ladder' after themselves by denying developing countries the tools and benefits they had.

- For a really comprehensive and excellent text offering an introduction to development economics, which tends to view the world from a thoughtful liberal institutionalist perspective, take a look at *Economics of Development* by Dwight H. Perkins, Steven Radelet and David L. Lindauer (2012).

- For a different, but equally comprehensive and excellent text, but one that comes at development economics from a true Keynesian perspective, take a look at *Economics of Development* by Tony Thirwall (2011).

Useful websites

- *Finance & Development* is the IMF's quarterly magazine which is free to access and covers a range of issues on development and the international economics system. For example, the stability, growth and regulation of the international financial system, governments' monetary policy and theory, the nature of economic development, poverty reduction and other related issues. It also carries a useful 'Back to Basics' regular feature which introduces and explains important economic issues – for example, inflation, fiscal policy, econometrics, LIBOR, etc. Well worth taking a look at and/or subscribing to, although it does tend to privilege orthodox economic perspectives. Available at: http://www.imf.org/external/pubs/ft/fandd/fdinfo.htm

- *David Harvey and a Marxist view*. David Harvey gave a really interesting talk on the 'Crises of Capitalism' at the RSA in London. The talk is accompanied by a fabulous animated video that can be viewed online at http://youtu.be/qOP2V_np2c0. Harvey also has a website offering a free set of lectures providing a close reading of Marx's *Capital*. It's a wonderful and brilliant resource. Highly recommended. Available at: http://davidharvey.org/reading-capital/

4 The international monetary and financial system

Learning outcomes

At the end of this chapter you should:

- Appreciate the centrality of the international monetary system to the functioning (or failure) of the world economy in terms of trade and investment flows.
- Know the historical emergence of the international monetary system, the role of ideas and the domestic and international geopolitical pressures which helped to shape it in different eras.
- Understand the economic logic behind the key forms of the international monetary system over these eras – i.e. the gold standard, the gold-exchange standard of the Bretton Woods system, and the current system of floating exchange rates.
- Understand the trade-offs a government faces in deciding whether to prioritise monetary policy autonomy, exchange rate stability and access to international capital.
- Show awareness of the consequences of the increasingly liberal international financial system in terms of the volume of capital and the frequency of financial crises and global imbalances.
- Know the purpose and operation of capital controls and capital account liberalisation and the debate about their merits and disadvantages.

Key concepts

International monetary system, gold standard, gold exchange standard, trilemma, monetary policy autonomy, fixed and floating exchange rates, capital mobility, capital controls, capital account liberalisation

Introduction

The international monetary and financial system provides the necessary framework for the transfer and flows of money and credit. If money and credit are the lifeblood of the economic system, then the international monetary and financial systems are the veins, arteries and heart. For example, when companies, individuals or governments want to trade goods and services – to import or export – they need to be able to convert money from their national currency (their rupiah, dong or dollar) to the currency of the counter-party (their shilling, dirham or euro). And, when producers and investors wish to fund production overseas, they also need a system of creating and moving money and credit across national boundaries.

The two systems – monetary and financial – are analytically distinct, but in reality are deeply entwined (O'Brien and Williams 2010). The international monetary system is the way in which national currencies are exchanged for one another. The credit system, as it suggests, is the creation and distribution of credit across borders. The later chapters deal with different types of credit – loans, grants, equity, debt instruments – but this chapter focuses on the international monetary and financial system.

The international monetary and financial system has undergone a series of significant transformations over the past three centuries. These represent different ways of organising the economic and political interface between domestic and international financial systems. These changes have been influenced by changes in economic thinking but also by domestic politics and geopolitical power. The different regimes and the shifts from one to another are deeply political. First, in the sense that decisions to move to, support or undermine the financial system are driven by governments, and so reflect their political power, interests and ideas, but also the interests and influence of financial and business elites. Second, in the sense that the precise configuration of how the domestic and international systems interact – and there are a number of different ways of organising this – has profound political implications for who benefits more and whose interests are being served best.

Putting the current international monetary system in its place

Before evaluating contemporary cross-border capital flows and their relationship to development, it is necessary to consider the historical evolution of the international monetary and financial system. International political economists and economic historians have provided a series of excellent studies on the form, function, shifts in, and politics of the international monetary and financial system and its governance (Cohen 1977, 1998, 2008; Kindleberger 1978; Eichengreen *et al.* 1992, 2008, 2011; Helleiner 1994, 1996; Germain 1997, 2010; Langley 2002; Porter 2005). This chapter draws on this body of work to provide the political economy background to the international monetary system and begins to draw the links with developmental outcomes.

The current system is characterised by the absence of a multilateral exchange rate system (a 'non-system') and is highly globalised and dominated by private finance. Yet this is neither unprecedented (earlier eras were just as, if not more, globalised) nor has it always been this way (fixed exchange rates used to be the norm – and many governments continue to manage their exchange rate – and states, not markets and corporations, used to have the primary responsibility for the creation and distribution of credit). As such it is useful to trace the emergence and structure of the current system by putting it into its historical and political context.

The importance of highlighting the historical evolution of the international monetary system is several-fold. First, it is essential for students of global finance and development to understand where the current system has come from, in terms of identifying the economic and political pressures that combined to produce the contemporary system. The current financial system didn't emerge from nowhere, it emerged from within pre-existing historical legacies. Certain key elements of the financial system's design are the consequence of decisions taken in the past which reflected the balance of power at the time and have become locked in; what political scientists call 'path dependency'. As such, one cannot understand the current system without understanding where it came from. Second, explaining the historical evolution of the international monetary system helps illustrate the different trade-offs that exist in adopting different systems, such as fixed or floating exchange rates, capital controls or a liberalised capital account, as well as the implications for governments' policy autonomy, economic growth, poverty and development more generally. Third, and related, it shows that current debates – about the use of capital controls, currency areas and the role of the dollar – are not new debates, nor do we lack historical and empirical references to help evaluate them. Drawing historical lessons from previous experiments – if cautiously made – is helpful. Fourth, and finally, it reinforces and underlines the fact that the international monetary and financial system is a political construct and not an economic given. The international monetary system has taken different forms over the years, and has always reflected the ideas, interests and power of governments, businesses and international organisations. So the system can (and will) change and we should not underestimate the suddenness of changes when they come about, but nor should we underestimate the difficulty of bringing them about.

As such, in this chapter we cover the different forms that the international monetary and financial system has taken. We also unpack the underlying logic of individual systems; for example, the trade-offs between domestic and international priorities. This also allows us to highlight the benefits and disadvantages of different individual systems from a developmental perspective. The key questions for the international monetary system are: How does it work? What are the adjustment mechanisms? Are they automatic, built into the system, negotiated? Who bears the primary responsibility for adjustment? What are the domestic trade-offs and space for setting policy autonomously? What are the implications of the system for development? How stable is the system and to what extent does it promote economic efficiency, growth, the allocation of capital to where it is needed and/or financial crises?

Box 4.1

The governance of global finance

The number of actors involved in monetary and financial governance has diversified, while the relationships between them have become more complex (Baker *et al.* 2005). These changes are typically and somewhat inevitably tied to discussions of globalisation. It is commonly argued that there is a 'growing disjuncture' between the patchwork of political sovereignty and the overarching reach of global capital. This is in contrast to some of the more alarmist suggestions that with the 'retreat of the state' an absolute vacuum of authority has developed in the international economy (Strange 1996).

So, in addition to the state, the literature identifies at least four further categories of structures of authority which come together to govern global finance: international multilateral institutions, transgovernmental regulatory networks, regional regulatory cooperation and private authority (Baker *et al.* 2005). The first category highlights the role of the IMF, the World Bank and the OECD. There is also the 'gaggle of Gs' (G7, G8, G10, G20, G22, G24, G30) (Baker 2000). These multilateral financial institutions are responsible for preserving financial stability through surveillance, formulating codes of best practice and responding to crises. Second, there are the transgovernmental regulatory networks, which are composed of national regulatory agencies working in a given specialist area of financial governance. The aim of the networks is to develop and disseminate standards of best practice among national regulators. The most important example is the Basel Committee for Banking Supervision (BCBS) which oversees the regulation of international banking, but within the realm of securities there is the International Organisation of Securities Commissions (IOSCO). Third, there has been the development of regional regulatory authority, mostly notably within the EU. Fourth, and finally, there has been the shift towards looking at structures of private authority.

It is the work that has begun to interrogate the significance of private actors in the structures of governance that is the most exciting (Cutler *et al.* 1999; Higgott *et al.* 2000; Hall and Biersteker 2002). Within finance there are a number of self-regulating industry associations, such as the World Federation of Stock Exchanges (FIBV) (created as the International Federation of Stock Exchanges), the International Securities Market Association (ISMA), and the International Council of Securities Associations (ICSA). Similarly, international standards for auditing and accounting have been developed primarily by private sector organisations. The International Accounting Standards Board (ISAB) (formerly the International Accounting Standards Committee (IASC)), is charged with continuing to develop global rules for financial reporting.

Globalisation, capital mobility and the structural power of capital

The power that finance is said to possess is most often expressed in terms of the structural power of capital (Przeworski 1985; Gill and Law 1989). The power that capital is reputed to have over governments is by virtue of its mobility, in opposition to the rooted nature of territorially bound and confined governmental authorities. Because states have to compete among themselves in order to attract capital, we have witnessed over the past 40 years or so a 'new hegemony of financial markets' (Cerny 1994, 1997).

These insights have become popular and persuasive when taken up with discussions of globalisation (see Box 4.2). It is suggested that technological advances, economic liberalisation and integration associated with globalisation have further empowered capital at the expense of state control and labour. In a more popular rendition of the capital mobility hypothesis, Thomas Friedman talks of international investors as the 'electronic herd'. He argues that they keep government policy within strict bounds, a so-called 'golden straitjacket' (Friedman 2000). Specifically, the straitjacket imposes a tight monetary policy to ensure low inflation, a balanced budget (both of which mean minimal government spending) and market-friendly economic policy more generally. The rest of this chapter documents the emergence of globalised capital markets and examines these claims.

Box 4.2

Capital mobility and state capacity

Capital mobility – alongside Coca-Cola and the internet – is often heralded as *the* archetypal illustration of globalisation. Consider a passage from Anthony Giddens' (1999) BBC Reith Lectures, in which for him the biggest difference between the globalised and the pre-globalised world: 'is in the level of finance and capital flows. Geared as it is to electronic money – money that exists only as digits in computers – the current world economy has no parallels in earlier times. In the new global electronic economy, fund managers, banks, corporations, as well as millions of individual investors, can transfer vast amounts of capital from one side of the world to another at the click of a mouse. As they do so, they can destabilise what might have seemed rock-solid economies as happened in East Asia . . . I would have no hesitation, therefore, in saying that globalisation, as we are experiencing it, is in many respects not only new, but revolutionary'.

These sorts of accounts are often impressionistic and characterised by a 'casual empiricism' (Busch 2000). They are often something of a 'caricature', and an 'initial over-reaction to the revival of global finance' (Cohen 2002: 439). There are a number of reasons why the globalisation – or capital mobility hypothesis – is not all-powerful. Capital is far from perfectly mobile. Various metrics of capital markets integration – for example, interest-rate parity, capital costs, national savings-investment ratios – all suggest limited globalisation (Berger and Dore 1996; Boyer and Drache 1996; Hirst and Thompson 1999; Watson 1999). Nevertheless, while sometimes overstated, exaggerated and stylised, accounts of global capital do capture something of the dilemmas of accountability and control now faced by governments, especially in developing countries. The best analysis focuses on the limited efficacy of (particular) macroeconomic policies that remain feasible for state authorities (Sassen 1996: Ch. 2; Germain 1997; Weiss 1998; Pauly 2000). For example, Layna Mosley's (2000, 2003, 2005) excellent work demonstrates that individual governments – in developed and developing countries – retain room for manoeuvre in the face of global capital markets. As Mosely demonstrates there is significant variation in tax rates and public spending and the type of welfare regimes present in many countries. These variations are a result of domestic political processes – alliances, party politics, the organisation of the poor and trade unions, and institutions – which affect the priorities and trade-offs taken.

The policy issues that governments face

Before going any further it's worth flagging up a few key issues that will recur in this chapter and others. These concepts are monetary policy, balance of payments, exchange rate regimes and capital controls.

Monetary policy is one of the two types of macroeconomic policy – the other being fiscal policy. Monetary policy allows the government, via its central bank or other relevant regulatory body, to control the money supply and therefore influence the level of aggregate demand in the economy. Through the buying and selling of assets, or by changing the levels of reserves banks have to hold, the central bank influences the interest rate. A reduction in interest rates makes saving less attractive and borrowing more attractive, which stimulates spending. And vice versa. Plus, changes in the rate of interest can affect the exchange rate. For example, an increase in the interest rate is more attractive for short-term foreign investors – as it increases the returns on their investment. This can increase the demand for the currency which will, in turn, reduce demand for exports as they become relatively more expensive (and increase imports) as the currency appreciates. So a country's monetary policy is one important lever for governments to be able to influence growth and unemployment – trying to avoid overheating and hyperinflation on the one hand, and stimulating enough growth to create jobs on the other. For example, the central bank in South Africa introduced inflation-targeting in Februrary 2000 and currently attempts to keep inflation between 3 and 6 per cent. But other priorities also guide monetary policy elsewhere, such as targeting the exchange rate or employment. Orthodox scholars place a great deal of importance on a government's control of the money supply. Heterodox approaches are less impressed by the effectiveness of monetary policy in stimulating economic demand.

The balance of payments is the sum of all money coming into and out of a country. It is divided into the current account and the capital account. The current account is what we tend to think of as goods and services, essentially trade. The capital account is comprised of capital flows – investments and profits. Technically speaking, countries' accounts should balance. However, in a situation where a country is importing more than it is exporting, it produces a deficit, or when exporting more than it imports, a surplus. Such deficits can be sustained for as long as a country's savings allow, or by borrowing from overseas such that inflows of capital can pay for the deficit. For example, in 2007–08 many commentators were worrying about India's balance of payments; the global financial crisis had forced both its current and capital accounts into deficit as capital flowed out of developing countries back to the US at the same time as exports were falling. Whereas in the past India has drawn on development aid, this time it drew upon its own substantial savings of foreign currency reserves. By the end of 2012, India's balance of payments had stabilised, though it remained weak. The implication is that this can lead to crisis: very large or very prolonged trade deficits can begin to weaken investors' confidence in a country's ability to repay in the future and can lead to a balance of payments and currency crisis as investors rapidly pull out their money, forcing a rapid depreciation of the currency. This is precisely what happened in Mexico in 1994–95. A rapid growth in imports (because of liberalisation) was sustained for a while by large capital inflows (encouraged by high interest rate rises). But the glut of investment flows led to bad bank lending, falling private savings and domestic investment. Eventually international investors lost confidence in the Mexican peso and sold their assets, forcing the Mexican government to abandon its fixed exchange rate system.

There are a range of different exchange rate regimes that a government can use to manage the value of its currency, from a pure floating to a pure fixed system, and forms of pegged or 'dirty' (managed) forms in between. Floating exchange rates are where the value of the currency (its exchange rate) is determined by market forces, i.e. the demand and supply of a currency is based on trade, competitiveness and attractiveness for investors. There is, as we will see, a separate debate about the extent to which investors are rational and market-determined exchange rates are stable and accurately reflect the underlying health of an economy. A fixed, managed or pegged exchange rate is where the value of a currency is determined by the government. The government maintains the value of the currency within a certain band by intervening in the currency markets, buying up currency to keep the rate high, or selling the currency when there is increasing demand in order to increase supply and prevent the value of a country's currency going higher. For example, Barbados has had a fixed exchange rate since the 1970s in order to maintain currency stability. Barbados has done this by pegging the value of its currency at roughly 2:1 against the US dollar. Although floating exchange rates have become more and more popular, it is rare that a country has a completely free exchange rate system. For example, Indonesia used to have a managed float (1978–99), but has more recently allowed its exchange rate to be determined by the market. Nevertheless, during times of currency fluctuation the central bank will intervene in order to stabilise the exchange rate.

Capital controls are government restrictions on the free movement of capital either into or out of an economy. For example, this could be done through a tax on inflows (or outflows), or by stipulating a minimum amount of time that investments have to be kept in the country, or by placing restrictions on the foreign ownership of firms or property. All of these effectively slow down or limit the influence of international capital flows on the domestic economy. Many countries, especially in East Asia and Latin America, have used capital controls to minimise the influence of short-run destabilising capital flows that rush in and/or rush out of an economy, excessively heating it up and/or enabling capital flight. For example, in September 1998, Malaysia used emergency capital controls to limit the damage from the East Asian financial crisis. Chile is the most famous proponent of capital controls. Chile's *encaje,* used in the 1990s, was designed to limit potentially volatile capital inflows and minimise macroeconomic disruption and currency appreciation caused by excessive capital inflows overheating the economy.

These concepts and issues provide the essential foundation for understanding how government policies are used to manage the interaction between the domestic and international economy. The classic way of understanding how governments prioritise and think about the trade-offs between the various goals of managing the exchange rate, setting interest rates, accessing international finance, or not, is provided by something called the unholy trinity or impossible trilemma.

The impossible trilemma or unholy trinity

The trilemma comes from the Mundell-Fleming model which is based on an orthodox model of the economy which is extended to include the international system (Fleming 1962; Mundell 1963). As such, the model makes a number of simplifying assumptions. Nevertheless, the trilemma is a useful heuristic for understanding the trade-offs individual governments face, and useful for explaining the different designs of the international monetary system (Pauly 1997: 23–29).

The Mundell-Fleming model highlights the relationship between three factors or government priorities:

1   Capital mobility: the free flow (or not) of capital across borders.
2   Exchange rate stability: the decision to have a fixed or floating exchange rate, i.e. whether the government controls it or it is set by the market.
3   Monetary policy: whether a government enjoys the ability to set its own interest rate or not.

As already suggested, these are all important issues. The ability of a government to choose its monetary policy and set its interest rate such that it can accelerate lending and borrowing, economic growth, employment and poverty, or to slow down an overheating economy, is clearly crucial. Both from an economic perspective, but also in terms of democratic functioning and accountability. Likewise, whether or not a government is willing to delegate the setting of its exchange rate to the markets or prefers to maintain control over it is key; but also whether a government *can* set and defend its own currency is, as we shall see, a live issue. For example, many countries which try to maintain a fixed exchange rate find themselves subject to speculative attacks and are forced to abandon a peg, as happened in East Asia in 1997–98 or to the UK in 1992. Capital mobility can just as much determine whether a government can access international money to increase its investment as whether an economy is exposed to the rapid inflows and outflows of capital that routinely overheat and crash economies.

The lesson of the trilemma is that governments can pursue two, but never three, of the following policy goals: (1) exchange rate stability, (2) national monetary policy autonomy and (3) capital mobility. For example, if a government decides it wants to prioritise free movement of capital, in order to access international financial markets and investment, and a fixed exchange rate, to try and maintain stability, then it *must* sacrifice its monetary policy autonomy. Why? Because if a government tries to lower its interest rate (to boost lending and economic activity) it will trigger outflows of capital. This is because the lower interest rate is less attractive to investors who will look to move their money elsewhere to get a better return, and because of the government's commitment to capital mobility there is nothing to stop them doing so. In turn, the capital outflows will result in increased selling by investors of the country's currency which translates into downward pressure on the exchange rate, thus, undermining the goal of exchange rate stability. So, in order to follow an independent expansionary monetary policy, a government needs to either stop capital leaving (through imposing capital controls) or allow the exchange rate to fluctuate, i.e. sacrifice one of the original priorities. Figure 4.1 summarises the priorities and trade-offs that emerge from the Mundell-Fleming model.

As was noted above, the impossible trinity allows us to both understand individual government choices, but also to understand the design and priorities of different international monetary systems. These are summarised in Table 4.1 and the following section will elaborate on each system in turn. For example, the priorities of the gold standard era were exchange rate stability and capital mobility – meaning that governments' domestic policy autonomy was minimal. The priorities of the Bretton Woods gold-exchange standard were fixed (but adjustable) exchange rates and policy

Figure 4.1 *The impossible trinity.*

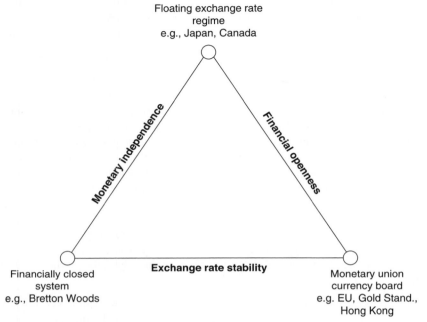

Floating exchange rate
regime
e.g., Japan, Canada

Monetary independence

Financial openness

Exchange rate stability

Financially closed
system
e.g., Bretton Woods

Monetary union
currency board
e.g. EU, Gold Stand.,
Hong Kong

Source: Adapted from Aizenman and Ito (forthcoming).

Table 4.1 *International monetary systems and the trilemma*

	Dates	Description	Priorities	Sacrifice
Gold standard era	1870–1914 and 1925–1931	Gold standard Fixed exchange rates	Capital mobility Exchange rate stability	Policy autonomy
Interwar period	1914–25 1931–45	Fragmented system Floating rates Capital controls Currency blocs	Policy autonomy Competitive currency setting	Capital mobility
Bretton Woods system, gold-exchange or dollar-exchange standard	1945–71	Multilateral system Gold exchange standard Adjustable peg Capital controls	Exchange rate stability Policy autonomy	Capital mobility
Post-Bretton Woods system	1971–	Fiat system based on US$ Floating exchange rates Open capital accounts	Capital mobility Policy autonomy	Exchange rate stability

autonomy – meaning that capital mobility was restricted. The priorities of the post-Bretton Woods system have been free capital mobility and either monetary policy autonomy (generally OECD countries) or fixed rates (within a regional arrangement or using capital controls).

To understand the current system of floating exchange rates it's necessary to understand the Bretton Woods system that operated from the 1940s to the 1970s. And to understand the Bretton Woods system, it is necessary to understand the gold exchange standard that operated before that. We unpack each in turn.

Gold standard

The extent of contemporary globalisation is by no means unprecedented. At the end of the nineteenth century, by some measures, the global financial system was more integrated than it is today (Helleiner 2011). The international monetary system that coordinated cross-border flows of capital and exchange rates was called the gold standard. The gold standard was a fixed-exchange rate system. The system was in place up until 1914, and again briefly in the 1920s.

The theory behind the system was famously set out by David Hume in his description of the price-specie flow mechanism. The mechanism was a self-regulating system whereby external imbalances between countries (i.e. trade deficits or surpluses) led to domestic changes in prices and wages until the internal imbalances were corrected. For example, if a country was running a trade deficit then it would be spending more money on imports than it was earning on exports – as such the supply of money in the country's economy would shrink. This would reduce domestic wages and prices as the economy adjusted to the smaller money supply. This, in turn, would also make the country's exports cheaper relative to other countries, making it more competitive, and eventually helping to restore its trade balance. That's the theory, at least.

To see the problem we need to return to the theory of money (Chapter 1). Hume's view is in line with the orthodox view of money (as a token) rather than seeing it as a system of account controlled by governments. In reality, because governments controlled the money supply through the issuance of bank notes and deposits – i.e. the domestic money supply wasn't just composed of gold coins – the mechanism was not automatic, as assumed by Hume. Instead, governments had to play by the rules of the game. Playing by the rules meant that when facing a trade deficit governments should take steps to deflate their economy – for example, by raising the interest rate in order to increase the cost of borrowing, thereby reducing the amount of money in circulation. This interest rate rise would also attract short-term capital flows looking for a better return – thus helping to bridge the payments shortfall. So, the operation of the system fundamentally depended upon governments choosing to follow the rules of the game rather than it being an automatic mechanism. And in fact, the historical evidence shows that these rules of the game were consistently broken by governments (Eichengreen 1985), and for good reason: deflating an economy comes with a great deal of pain for domestic producers and labour.

Prior to World War I, economic and monetary policy was largely insulated from popular and democratic pressures. In developing countries, monetary policy was commonly set by the colonial power. Meanwhile, within the core economies a number of factors worked against democratic (or populist) monetary policymaking: the public franchise remained narrow, many central banks were not public bodies, and there was an intellectual commitment to liberal internationalism and a belief in the automatic mechanisms underlying the gold standard. However, as the franchise was widened in many countries after 1914, and the power of organised labour increased, and new economic ideas which proposed greater interventionism in the domestic economy to prevent unemployment such as Keynesianism became popular, governments became less and less wedded to their commitment to maintaining the fixed currency peg when faced with a trade deficit and/or capital outflows (Eichengreen 1992, 1996). These pressures ultimately undermined the gold standard.

The gold standard was disrupted in 1914. Facing war, many countries abandoned the gold standard, allowing their exchange rates to 'float' and the value of their currencies to move freely against one another to reflect the demand for a currency. Governments' rationale was that floating exchange rates increased their policy autonomy, because they could devalue their exchange rate and gain a competitive advantage. Furthermore, the use of capital controls limited the outflows of capital, meaning that governments could pursue expansionary monetary policy to stimulate the economy. Moreover, the other advantage of floating was that the exchange rate adjusted to domestic prices and wages rather than the other way around, making adjustments less socially and politically costly. However, what was rational for individual governments was not collectively rational.

Although many countries did return to the gold standard in the 1920s, and trade and capital flows resumed, this was short lived (Pauly 1997). In 1931, facing the Great Depression and a financial crisis, many governments chose to abandon the gold standard. What emerged was a system of three competing currency blocs – namely the sterling bloc which was based around Britain and the countries in the Empire or close trading allies (such as Egypt, Iraq and Portugal), the dollar bloc led by the US and including Canada and the Latin American countries, and the gold bloc which was led by Germany and consisted of countries that adhered to the gold standard (France, Belgium, the Netherlands, Switzerland, Italy and Poland) (Helleiner 2003). These currency blocs extended from the imperial power to their colonial territories. The values of the different currencies fluctuated against one another. And while there was some lending within each of these blocs, there was only very limited lending between them. The period witnessed increasingly tighter regulation of cross-border flows on capital which limited the ability of private actors to convert currencies. This tripolar system wasn't particularly stable (Eichengreen 1996). By abandoning the gold standard, and allowing their currencies to depreciate, countries made their exports cheaper and reduced imports – thus helping to boost their domestic economy. International investors understood this and had much less confidence in governments' commitments to maintain the value of their currency. And because investors understood that currency instability would be the rule, capital flows became increasingly short-term and speculative. Crucially, the lessons drawn from this period of competitive devaluations and beggar thy neighbour policies formed the intellectual foundations for the multilateralism behind the Bretton Woods system.

The Bretton Woods system

The Bretton Woods system ran from 1945–71, and enjoyed its heyday between 1958 and 1971 (Helleiner 2011). It was named after the conference held in the Mount Washington Hotel in Bretton Woods, New Hampshire, US in 1944 where the papers were signed. The Bretton Woods conference – officially called the United Nations Monetary and Financial Conference – was also significant because it was the place where the IMF and the International Bank for Reconstruction and Development (the World Bank) were born.

The Bretton Woods conference was famously attended by John Maynard Keynes as the British delegate and Harry Dexter White for the US. Much of the shape of the international system that emerged was down to the intellect, conflict and compromise between these two men, the lessons that were taken from the interwar period, as well as the underlying power of the US and Britain in the international system of the time. Ultimately, the US was in a much stronger position than the British post-World War II, since Britain was in debt to the US having had to borrow to finance their war effort (Block 1977).

The planning and design of the system was underway well before the end of the war, for example the Atlantic Charter of 1941, which was the outcome of a secret meeting between Churchill and Roosevelt on a ship in the North Atlantic. It was widely agreed at the time that the Great Depression and World War II were a result of the beggar thy neighbour policies and the economic protectionism and unilateralism of the 1930s (Nurske 1944; Cohen 2002). The US – as the newly-emerging leader – was committed to setting up a multilateral liberal order. However, while it was decided that the new international monetary system had to prevent the unilateralism, protectionism and instability of the 1930s, it also had to provide sufficient policy autonomy for governments to pursue domestic goals such as full employment and promoting social welfare as they saw fit. This compromise – between an open international and multilateral liberalism on the one hand, and a socially democratic domestic order on the other – was famously dubbed 'embedded liberalism' by John Ruggie (1982).

The design and operation of the Bretton Woods system

How did the Bretton Woods system work? The system was characterised by three things: (1) a gold exchange standard, (2) a fixed (but adjustable) peg exchange rate regime, and (3) current account currency convertibility alongside capital controls (see Figure 4.2).

The first of these three things – the gold exchange standard – placed the US dollar at the centre of the system. The dollar provided the anchor to gold. The US dollar was convertible to gold at a fixed rate of $35 per ounce. This convertibility and the fixed exchange rate between dollars and gold was the way of providing everyone with faith in the stability of the system. Hence rather than it being a pure gold standard, it was a gold exchange or gold-dollar standard.

Figure 4.2 *The Bretton Woods System.*

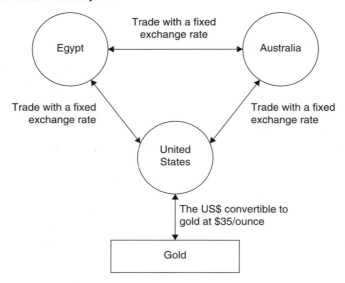

The second aspect of the Bretton Woods system also helped secure this stability and faith in the system: all other currencies were pegged to the US dollar. This was done in order to avoid the threat of competitive devaluations and speculative financial flows that were associated with the floating exchange rates of the 1930s. Though, importantly, the designers of the system also sought to include some flexibility into the exchange rate mechanism. There was an option within the system for countries to adjust their currency's par value in a situation of 'fundamental disequilibrium'. So, if a country's productivity underwent a sudden change, say, because it was destroyed in conflict or an earthquake, then the country's currency value could be lowered, thereby maintaining its competitiveness and balance of payments, and support the expansion of economic activity again.

Third, and finally, although countries' currencies were freely convertible (after 1958) for trade and commercial purposes, governments had the right to control speculative capital, which was deemed to be non-equilibrating and non-productive. This was made possible through the use of capital controls (see Box 4.3). It meant that currencies remained convertible for all trade payments (i.e. the current account), but that governments could limit flows of private capital – slowing their ability to undermine economic stability and domestic policy autonomy. Keynes and White wanted to protect governments from capital flight, i.e. to enable them to pursue independent interest rate policy. This commitment was captured by Keynes's famous dictum to 'let finance be primarily national' (Keynes 1933: 758; and see Chapter 2).

Box 4.3

Capital controls and capital account liberalisation

Capital controls are any measure taken by a government to limit the free flow of capital into and out of a country, for example, by taxing it, requiring a minimum length of time that an investment stays in the country, or stipulating which sectors can have foreign investment. The main purpose of capital controls is to slow down and limit the volume of capital inflows and outflows to prevent international financial flows from disrupting an economy and governments' policies by pouring in or suddenly departing. By contrast, capital account liberalisation is the process of removing these barriers to capital. A list of example controls is reproduced below, taken from Gallagher *et al.* (2012).

Inflows	Outflows
· Unremunerated reserve requirements (a proportion of new inflows are kept as reserve requirements in the central bank)	· Mandatory approval for domestic agents to invest abroad or hold bank accounts in foreign currency
· Taxes on new debt inflows, or on foreign exchange derivatives	· Mandatory requirement for domestic agents to report on foreign investments and transactions done with their foreign account
· Limits or taxes on net liability position in foreign currency of financial intermediaries	· Prohibition or limits on sectors in which foreigners can invest
· Restrictions on currency mismatches	
· End use limitations: borrowing abroad only allowed for investment and foreign trade	· Limits or approval on how much non-residents can invest (e.g. on portfolio investments)
· Limits on domestic agents that can borrow abroad (e.g. only firms with net revenues in foreign currency)	· Restrictions on amounts of principal or capital income that foreign investors can send abroad
· Mandatory approvals for all or some capital transactions	· Limits on how much non-residents can borrow in the domestic market
· Minimum stay requirements	· Taxes on capital outflows

The Bretton Woods system was in many ways a tremendous success. It achieved its primary aim of facilitating the emergence of a liberal, open, multilateral international economic system. Plus, the stability it provided also presided over a historic 'golden age' of economic growth and development – at least for the major powers. Exchange rates remained stable, growth was rapid, inflation remained moderate and world trade expanded rapidly (Bordo and Eichengreen 1993).

Nevertheless, the Bretton Woods regime collapsed in the 1970s. Two points need to be made about the system's success. First, the actual operation and success of the system was underwritten by the US rather than the IMF and the World Bank, and second, the system had a fatal flaw. Ironically, this flaw arose from the system's necessary, as it turned out, dependence on the US. This flaw is called the Triffin dilemma.

Meanwhile, the Bretton Woods institutions – the IMF and World Bank – obviously continue to exist and remain important, though their specific roles have changed and adapted to the new international monetary regime – see Box 4.4 on the IMF.

Box 4.4

The International Monetary Fund (IMF)

The IMF was designed to manage and regulate the multilateral Bretton Woods currency system – to maintain stability, ensure that countries current accounts were open and manage the fixed par rates between countries. As part of this the IMF was to serve countries' needs for short-term borrowing to cover temporary balance of payments deficits. The idea was that, unlike under the gold standard, governments should have more breathing space to turn around their deficits rather than being forced to immediately deflate their economy. However, with the end of the Bretton Woods multilateral system of pegged exchange rates the IMF's future looked uncertain, despite the fact that it was crucial in legalising the switch to floating exchange rates through the 1978 amendment of its Articles of Agreement. As part of this, the IMF managed to 'repurpose' itself to carry out a surveillance role – the IMF increasingly focused on countries' balance of payments difficulties, but sought to resolve them with orthodox policy advice assuming that they were the consequence of underlying structural problems with countries' economies. While IMF lending has always come with conditions, introduced in the 1950s, during the neoliberal era conditions became increasingly harsh, intrusive and coercive. The imposition of 'structural adjustment' was strengthened and enabled by the emergence of the Third World debt crisis where many developing countries in Latin America and Africa found themselves with structural balance of payments problems and having to borrow from international organisations to cover the shortfalls (see Chapter 6). Conditionality was often brutal, damaging and resented, as well as being bad for growth and poverty (Vreeland 2003, 2007).

The IMF has 188 members (nation states) and is governed by a managing director (traditionally a European, while the World Bank president is usually an American, reflecting the balance of power in 1944). There is also an executive board of 24 executive directors, of which eight are appointed to represent the large countries (the US, Japan, Germany, France, the United Kingdom, China, the Russian Federation and Saudi Arabia). The other 16 represent constituencies of the smaller economies (anywhere between 4 and 24 countries in size). Finally, the highest level of governance is the board of governors. The board meets once a year. Each member has a governor which is usually a minister of finance or head of the central bank. The board of governors has delegated most of its powers to the executive board, but still approves quota increases, special drawing right (SDR) allocations, the admittance of new and compulsory withdrawal of members, as well as amendments to the Articles of Agreement.

Each member state is allocated a quota based on the size of its economy. The quota determines how much the member has to pay in subscriptions, how much the member can borrow if needed (members can borrow up to 200 per cent of their quota per year and 600 pe rcent in total at any one point). The quota also, crucially, determines the voting power of the member. Voting power varies from 16.76 per cent of the total vote for the US to 0.03 per cent of the vote for Tuvalu. For more information on the IMF, its role, its history, its governance and the various types of lending programmes it operates see <http://www.imf.org/external/about.htm>. For a more critical evaluation see James Vreeland's (2007) *The International Monetary Fund: Politics of Conditional Lending*.

The collapse of the Bretton Woods system

The collapse of the Bretton Woods system is actually the end of two different aspects of the system: (1) the end of dollar-gold convertability, and (2) the end of the adjustable peg exchange rate. In August 1971 the US suspended the convertibility of the US dollar into gold. And by 1973 the value of the US dollar, and other currencies, was essentially determined by the markets rather than in Washington, DC.

The end of dollar-gold convertibility and the Triffin dilemma

The 'Triffin dilemma' was named after a Belgian economist Robert Triffin (1960) who first identified the problem. It describes the contradiction at the heart of the Bretton Woods system which arose as a consequence of relying on the US, rather than an international body, to manage global demand and provide the world's currency. Essentially, using the US dollar as the world's reserve currency meant that there was always a fundamental conflict of interest between the US's domestic economic objectives and the needs of the international economy. Put simply, the international economy needed a constant injection of liquidity (i.e. money and credit) into the system to maintain investment and growth. But, increasing the amount of dollars in the system – i.e. the US running large current account deficits – eventually put increasing pressure on the US dollar, as the excess supply of dollars served to undermine its value. This wasn't how it was meant to be.

The logic behind the IBRD (the World Bank) was to provide loans to inject money, or liquidity, into the international economic system – more specifically to help rebuild Europe. Unfortunately, the IBRD was not up to the task and the US – motivated by Cold War politics and a need to stimulate export markets for US exporters – stepped into the breach and funnelled money through what was know as the Marshall Plan (Wood 1992). The Marshall Plan transferred some $13 billion into Western Europe. Incidentally, the organisation set up to manage these funds – the Organisation for European Economic Co-operation – eventually became the OECD, the Paris-based body which remains behind the bilateral aid regime (see Chapter 5).

Thus the amount of money available in the world economy (liquidity) was directly tied to the US authorities: for the global economy to grow it was necessary for the US to run a balance of payments deficit, pumping more US dollars into the world economy. If too few dollars were injected into the international economy it would have acted as a brake on economic activity, but if too many dollars were in circulation this would undermine confidence in the value of the US dollar and the convertibility of dollars into gold. The reason being that if the US provided more and more dollars in order to finance international trade and investment then the amount of dollars in circulation would exceed the amount of gold held by the US. The circumstances would be set for a classic bank run if holders of US dollars increasingly believed they wouldn't be able to convert their dollars into gold.

In the event the dilemma was 'resolved' by choosing the path of liquidity (and collapse of the system). Notably, this was the path more closely aligned with US interests. Printing more dollars allowed the US to pay for the costs of the war in Vietnam as well as the Great Society programmes, the domestic welfare spending programmes designed to end poverty and racial injustice.

Despite helping to meet the its short-term interests, by printing more dollars the US now faced severe long-term vulnerabilities. It was facing the very real possibility of having to convert its gold into the dollars that were held by overseas governments, firms and investors. In addition, the fixed exchange rates were undermining the competitiveness of the US's exporting firms. In the event the US government managed to put off a run on the dollar through intergovernmental negotiation with Germany and Japan and setting up the London Gold Pool which was designed to preserve the value of gold at $35 per ounce. Crucially, the background of the Cold War helped galvanise support for the US's position. At least initially. However, growing private currency speculation and French intransigence eventually undermined this. The French left the Gold Pool in 1968, arguing that the US was abusing its position as issuer of the world's currency to fund its domestic and foreign adventures – the 'exorbitant privilege', as France's Charles de Gaulle called it. All of which meant that speculative pressures built until 15 August 1971 when President Nixon unilaterally closed the gold window – ending the link between the US dollar and gold. The dollar floated and immediately devalued which, alongside a 10 per cent tax on imports, served to restore US competitiveness.

The move to floating exchange rates

Following the 'Nixon Shock', a series of attempts were made to establish a new set of par values, but successive waves of private currency speculation undermined this. By February 1973 the Bretton Woods system was over and exchange rates began to float against one another. This was eventually formalised in 1978 with an amendment to the IMF's Articles of Agreement which allowed countries to set exchange rates independently, thereby allowing the use of floating exchange rates. The following decade was one of massive monetary instability and record high levels of inflation.

What is clear is that the fixed peg system had been undermined by increasingly large private capital flows. Speculative attacks and increasing capital mobility made it harder and harder for governments to defend the value of their currency in a fixed or pegged system (Eichengreen 1996). Plus, in the event, the fixed-but-flexible qualities of the Bretton Woods system turned out to be more fixed than flexible. Governments proved unwilling to undertake much adjustment of their currency's peg in circumstances of 'fundamental disequilibrium' because it required too much domestic and international negotiation and expenditure of political capital. Even with a multilateral system, international cooperation clearly had its limits. There had also been a significant shift in the intellectual policy climate whereby the benefits of floating exchange rates were re-evaluated. Orthodox proponents, such as Milton Friedman (1953), argued that floating exchange rates allowed the market to set exchange rates more efficiently and accurately than governments. Friedman argued that if the 1930s was a time of exchange rate instability this was because the underlying economic fundamentals were unstable, not because of the exchange rate system per se.

The globalisation of finance

The third thing to change in 1973 was a gradual opening of countries' capital accounts. Even industrialised countries tended to have capital controls to limit the inflows and outflows of financial capital because unfettered capital flows make defending a fixed exchange rate very difficult. As noted at the outset of the chapter, the current international monetary and financial system is characterised by its global and liberalised nature. Plus, as also noted, it was increasing capital mobility, especially the volume of private capital, that did for the Bretton Woods system. What was behind the increasingly liberalised, privatised and globalised financial system? In short, a combination of financial and technological advances combined with political decisions and ideological shifts towards an increasingly neoliberal worldview.

First, telecommunications and financial engineering – especially the proliferation of various derivatives products – have lowered the costs of financial transactions. This has made much higher volumes of trading possible and increased the ability of private investors to circumvent various regulatory limitations. At the same time, there have been growing volumes of international trade and production which have increased the demand for international capital flows to finance these activities. But there have also been other changes of a more political nature which have shaped the ways these innovations and demands have been allowed to develop.

In London there was the growth of the 'euro-market' during the 1960s. The euro-currency markets refers to any market in which trades are conducted in currencies other than of the country the market is located in. So for example, in New York, trades using pounds sterling count as euro-currency, but equally so do trades in US dollars in London, which is where the practice started. The history of it was that the USSR wanted a market where it could trade in dollars outside of the US, which in combination with London wanting to re-establish its position as a key global financial centre meant that there was a confluence of interests in making this happen. The advantage for traders and investors was that these markets were largely unregulated. In 1973, with the oil price hike, the volume of trades in these unregulated, offshore markets increased exponentially.

Further deregulation followed the breakdown of the Bretton Woods system. The US removed its capital controls in 1974 and the UK did so in 1979. Other countries followed suit (Abdelal 2007). It was largely an ideological thing, a neoliberal belief in the power of the free market to act as a useful discipline on governments. But it was also a dynamic of political competition between countries seeking to attract mobile capital (Cerny 1994). This was especially the case for the UK and US who were trying to bolster London and New York as the leading financial centres. But this has also been the case in countries such as Cayman Islands whose offshore status has helped attract business and money, as well as offering a way for companies and individuals to avoid paying tax and more easily laundering money (Palan *et al.* 2009; Shaxson 2011) (see Box 4.5). As radicals and critical reformists have both argued, it is also clear that the financial industry, as the main beneficiary of the new system, has lobbied and pushed to maintain its elite advantage. The industry has worked hard to create increasingly sophisticated ways of making money out of financial trades and helping to deregulate the markets (Gill and Law 1989).

Box 4.5

What is offshore and why does it matter?

For Nick Shaxon, tax havens are the 'great untold story of globalisation'. The following passage describes his brilliant book (2011) *Treasure Islands*:

'Tax havens are not exotic, murky sideshows at the fringes of the world economy: they lie at its centre. Half of world trade flows, at least on paper, through tax havens. Every multinational corporation uses them routinely. The biggest users of tax havens by far are not terrorists, spivs, celebrities or Mafiosi – but banks . . . Tax havens are the ultimate source of strength for our global elites . . . [in these] nodes of secret, unaccountable political and economic power, financial and criminal interests have come together to capture local political systems and turn the havens into their own private law-making factories, protected against outside interference by the world's most powerful countries . . . Tax havens aren't just about tax. They are about escape – escape from criminal laws, escape from creditors, escape from tax, escape from prudent financial regulation – above all, escape from democratic scrutiny and accountability. Tax havens get rich by taking fees for providing these escape routes. This is their core line of business. It is what they do. These escape routes transform the merely powerful into the untouchable. "Don't tax or regulate us or we will flee offshore!" the financiers cry, and elected politicians around the world crawl on their bellies and capitulate. And so tax havens lead a global race to the bottom to offer deeper secrecy, ever laxer financial regulations, and ever more sophisticated tax loopholes. They have become the silent battering rams of financial deregulation, forcing countries to remove financial regulations, to cut taxes and restraints on the wealthy, and to shift all the risks, costs and taxes onto the backs of the rest of us. In the process democracy unravels and the offshore system pushes ever further onshore. The world's two most important tax havens today are United States and Britain. Without understanding offshore, we will never understand the history of the modern world. Poverty in Africa? Offshore is at the heart of the matter. Industrial-scale corruption and the wholesale subversion of governments by criminalised interests, across the developing world? Offshore is central to the story, every time. The systematic looting of the former Soviet Union and the merging of the nuclear-armed country's intelligence apparatus with organized crime, is a story that unfolds substantially in London and its offshore satellites. Saddam Hussein used tax havens to buttress his power, as does North Korea's Kim Jong-Il today. Prime Minister Silvio Berlusconi's strange hold over Italian politics is very much an offshore tale. The Elf Affair, Europe's biggest ever corruption scandal, had secrecy jurisdictions at its core. Arms smuggling to terrorist organisations? The growth of mafia empires? Offshore. You can only fit about $1 million into a briefcase: without offshore, the illegal drugs trade would be a fraction of its size. Private equity and hedge funds? Goldman Sachs? Citigroup? These are all creatures of offshore. The scandals of Enron, Parmalat, Long Term Capital Management, Lehman Brothers, AIG – and many more? Tax havens lay behind them all. The rise of multinationals, the explosion of debt in advanced economies since the 1970s is substantially an offshore tale. Complex monopolies, frauds, insider trading rings – these corruptions of free markets always have tax havens at their heart. As Treasure Islands explains in vivid, thrilling, horrifying detail, every big financial crisis since the 1970s – including the great global crisis that erupted in 2007 – has been a creature of the tax havens.

Source: http://treasureislands.org/the-book/

The case for capital account liberalisation

By 1998 capital account liberalisation had become such a powerful norm that the IMF executive board and interim committee were in the process of rewriting the IMF's Articles of Agreement (Abdelal 2007). The notion that capital should be able to move unhindered was at its apogee. Stanley Fischer (1997) as the IMF's first deputy managing director, set out the case for capital account liberalisation arguing for a capital account amendment to the IMF's Articles of Agreement. In doing so he set out the classic neoliberal view of the advantages of capital account liberalisation and the free movement of capital:

> free capital movements facilitate a more efficient global allocation of savings, and help channel resources into their most productive uses, thus increasing economic growth and welfare. From the individual country's perspective, the benefits take the form of increases in both the potential pool of investable funds, and the access of domestic residents to foreign capital markets. From the viewpoint of the international economy, open capital accounts support the multilateral trading system by broadening the channels through which developed and developing countries alike can finance trade and investment and attain higher levels of income. International capital flows have expanded the opportunities for portfolio diversification, and thereby provided investors with a potential to achieve higher risk-adjusted rates of returns. And just as current account liberalization promotes growth by increasing access to sophisticated technology, and export competition has improved domestic technology, so capital account liberalization can increase the efficiency of the domestic financial system.

As we can see, capital account liberalisation is designed to increase the mobility of capital and thereby increase the financial integration of countries into the world economy and their access to international capital flows. The orthodox view – as expressed by Lawrence Summers (2000) – is that the gains from the movement of capital from industrialised to developing countries will result in 'enormous social benefits'. This is because the marginal productivity of investment in developing countries is so much higher, i.e. international investors will benefit from higher returns as well as developing countries benefitting from increased access to international financial flows.

Prasad *et al.* (2003) argue that increased financial flows to developing countries could, 'in principle', increase economic growth rates. They identify a number of direct and indirect channels through which this might occur. The direct channels of greater financial integration are: (1) an augmentation of domestic savings, (2) through lowering the costs of capital because of a better risk allocation, (3) a transfer of technology (e.g. FDI), and (4) the development of the domestic financial sector. Plus there are indirect effects too: (1) through the promotion of production specialisation because of improved risk management, (2) inducement for better policies, both macroeconomic and institutions because of the competitive pressures released by free market forces, (3) and finally because of these improvements an enhancement of capital flows by signalling better policies.

But other studies, from a liberal institutionalist perspective, have found that capital account liberalisation only becomes positively related to growth as income levels increase, and that the relationship either doesn't exist or is negative in LICs (Edwards 2000; Eichengreen 2000). The emphasis from a liberal institutionalist perspective has been to get the sequencing correct (Eichengreen *et al.* 1998). That is to say, a steady and orderly liberalisation of the capital account would be something like this: liberalise long-term flows before short-term flows and FDI before portfolio flows, get the domestic accounting and information regimes and prudential regulation right first, remove explicit and implicit government guarantees which make investors take unwarranted risks in the knowledge that they will be bailed out, and the adoption of appropriate monetary and exchange rate policies and targets (given that capital mobility is now one of the government's objectives). The current received wisdom is that capital account liberalisation should proceed *after* getting these things in place, rather than it being a catalyst that will help bring about these domestic reforms.

Even the bastions of liberal free market thinking – such as the IMF and Jagdish Bhagwati – have tempered the neoliberal optimism about the benefits of capital account liberalisation. Bhagwati (1998) has argued that the case for capital account liberalisation is purely ideological and not based on empirical evidence of the economic benefits of liberalisation. Instead, the case for financial liberalisation and unfettered capital flows is a myth, based on little more than assertions perpetuated by the Wall Street-Treasury complex. The revolving doors between Wall Street, the Treasury Department, the State Department, the IMF and the World Bank have created a powerful lobby which is guided by a mix of self-interest and ideology: specifically, hijacking the arguments about the benefits of free trade and applying them to free capital mobility. So, instead of *assuming* the benefits of capital account liberalisation, the reality is that 'the weight of evidence and the force of logic point in the opposite direction, toward restraints on capital flows. It is time to shift the burden of proof from those who oppose to those who favor liberated capital' (Bhagwati 1998: 12).

Where are we now?

The emergence of the system of floating exchange rates and increased capital mobility provides an opportunity to evaluate them. First, it is clear that, unlike the 1930s, the current regime of floating exchange rates since the 1970s has not resulted in financial autarchy and the collapse of cross-border trade and investment. However, second, the assumed smooth and efficient adjustments to currencies and the underlying economies has not come about. Ongoing systematic imbalances plus frequent and damaging crises have characterised the post-Bretton Woods system. So much so, that one seasoned observer, the late Susan Strange (1986), titled one of her books *Casino Capitalism*, describing the volatile and speculative nature of many financial transactions. We deal with financial crises in a moment, below. But first, how have governments responded to the new international and financial system? What strategies and priorities have they followed?

One thing that the trilemma makes clear is that – under a strong version of the thesis – although governments can't pursue all three objectives, they do have a choice, either collectively and multilaterally – as in the case during the gold standard and dollar-exchange standard eras – or individually, as is currently the case. We see a variety of responses to the trilemma, which we review below. The success or otherwise of the different choices is also noted.

Government strategies

Under the current system what are the choices made by governments? Some countries have embraced floating exchange rates. For example the UK, Australia and the US have prioritised the free movement of capital and the ability to set independent monetary policy. These two priorities rule out having fixed exchange rates. It is worth noting that this is the most common strategy within the core of the OECD countries. But it has also been the preferred strategy of the IMF – as we have seen – and has been adopted by a small and shrinking number of developing countries, for example Zambia.

Meanwhile, some countries have adopted a peg system, for example, Jordan, Belize, Barbados, China and other East Asian countries all have pegged currencies. The way this works is that the government links the value of its currency to another currency or a basket of currencies at an appropriate rate of exchange. It is often the case that these are its major trading partners. Developing and small economies tend to find a fixed or pegged exchange rate attractive for a number of reasons (Svensson 1994). It can provide stability by reducing exchange rate volatility and exchange rate risk, which can help facilitate international trade and investment without the fear of frequent changes in value. It can also signal credibility to international investors that the government and its central bank will follow a 'more disciplined country', as it were, effectively importing policy prudence. However, there are costs. A pegged currency certainly doesn't rule out speculative attacks and may in fact encourage them. Plus the central bank needs to hold sufficient stocks of foreign exchange to buy and sell to keep the demand and price of the currency at the desired value. This comes with an opportunity cost since these reserves could be invested more productively. Also, a fixed peg means that a trade deficit isn't easily resolved by simply allowing the value of the currency to depreciate.

In general, most countries have chosen to access international capital flows and then tried to strike a balance between monetary autonomy and exchange rate stability (O'Brien and Williams 2010). For example, as Indonesia and the likes of Trinidad and Tobago have done, they mostly let the market set the exchange rate, but intervene to minimise unwelcome volatility. Yet, some countries have taken the opposite position and chosen to limit capital mobility through the use of capital controls – for example, Malaysia used capital controls during the Asian financial crisis in order to limit capital mobility, stabilise its exchange rate and exercise some domestic monetary autonomy; Chile, Colombia and Thailand have also used them. And since the instability following the global financial crisis a number of key states, including Brazil, Taiwan and South Korea, have been experimenting with capital controls. Capital controls have, for many years, been frowned upon by mainstream economists, the IMF and the World Bank. However, the global financial crisis has prompted some rethinking and created the space for a more positive view of capital controls. The IMF, very significantly, in 2010–11, performed a reasonably public U-turn on the use of capital controls. It published a 'Staff Position Note' in 2010 on the effectiveness of capital controls, followed by a set of guidelines in 2011, which were approved by the IMF board, on how developing country governments could use capital controls (Ostry *et al.* 2010; IMF 2011).

Finally, some countries have looked to create a currency union. For example, within the Association of Southeast Asian Nations (ASEAN), the Chang Mai initiative in 2000, the Asian Bond Markets Initiative in 2003 and the Asian Currency Unit (ACU) in 2006, are following, at a distance, the precedent set within the Euro zone. There is an economic argument for sharing a common currency; namely, that it reduces the costs of conducting trade. But there is also an external aspect to the logic: sharing a common currency provides a greater level of protection from the vagaries of international capital markets. Similarly, absent being able to create a monetary union with the US, some countries – for example, Ecuador and El Salvador – have chosen to go down the route of adopting the US dollar as their official currency; a process called dollarisation. Another example was Argentina's use of a currency board, which effectively tied its currency to the value of the US dollar over the period 1991–2001. The main disadvantage of monetary union or dollarisation is a loss of monetary sovereignty. The government's monetary policy is essentially set in Washington, DC. Interest rate decisions are made for the US economy, not for the interests and needs of Ecuador, El Salvador and so forth. Nevertheless, politicians in some countries have decided that this is an acceptable trade-off for the stability it provides and for the confidence it gives international investors.

Financial crises

Charles Kindleberger (1978: 1) opens his *Manias, Panics and Crashes* with the line, 'There is hardly a more conventional subject in economic literature than financial crises'. And it is notable that banking and currency crises have become much more frequent since the 1970s, under more liberalised conditions (Rodrik and Velasco 1999). Crisis frequency since 1973 has been double that of the Bretton Woods and classical gold standard periods and is rivalled only by the crisis-ridden 1920s and 1930s (Bordo *et al.* 2001). Contrary to the technical approach and concerns about sequencing from liberal institutionalists, there is a long tradition in the literature that identifies the endemic nature of crisis to the financial system.

A series of financial crises such as in Mexico (1994–95), East Asia (1997–98), Russia (1998), Brazil (1999), Turkey (1994 and 2001), and Argentina (2002) saw capital inflows suddenly reversed with catastrophic human and economic effects. In 1997 the change in capital flows to the five East Asian countries at the heart of the crisis (South Korea, Indonesia, Malaysia, Thailand and the Philippines) went from inflows of $93 billion in 1996 to outflows of $12 billion, which Griffith-Jones and Kimmis (2003) calculate to be a turnaround equal to 11 per cent of the pre-crisis GDP of these countries.

The economic and human costs of crises are huge. A World Bank paper has estimated that the aggregate cost of the banking crises from 1980 to 2000 was over US$1 trillion. Furthermore, on average, developing country governments in the sample spent 14.3 per cent of national GDP to clean up their financial systems. And the costs of individual crises were much higher, as much as 40–55 per cent of national GDP in the cases of Argentina and Chile (Honohan and Klingebiel 2000). And what begins as a currency, banking, then more generalised economic crisis, transmits through a number of channels to disproportionately affect the poorest in society – for example, through lost jobs, the increased cost of imports, the drying up of credit and welfare cuts. These changes have been shown to increase the incidence of poverty as well as damage other indicators of human welfare such as infant and preschool mortality, school attendance and literacy rates (Griffith-Jones and Kimmis 2003). How do orthodox and heterodox approaches understand the causes of these financial crises? The East Asian crisis provides a useful example to outline the difference between neoliberal and liberal institutionalist explanations on the one hand and critical reformist and radical understandings on the other.

Explaining the East Asian financial crisis

The East Asian financial crisis which engulfed East Asia from 1997 onward was a classic and particularly virulent example of financial crisis (Henderson 1998; Jomo 1998) (see Box 4.6). During a few short weeks the currencies of the East Asian newly-industrialised countries (NICs) collapsed, with the national stock markets going the same way. There is little arguing with the suggestion that the financial markets played a key role in the events which engulfed East Asia from 1997 onward. However, where there *is* disagreement over the extent to which we should look at states or markets in locating the source, and thus blame, for the crisis. It is possible to identify two dominant narratives that appear within both popular and academic accounts of the East Asian crisis – one orthodox and one heterodox. The orthodox accounts lay the blame for the crisis firmly at the door of the East Asian states. Such accounts are most commonly articulated through the lens of 'crony capitalism'. On the other hand, more heterodox accounts lay the blame for the crisis at the door of the financial markets, or more accurately the community of investors that collectively make the decisions in these markets.

Box 4.6

The East Asian financial crisis

The story of Mayuree is a story of the East Asian financial crisis (Bullard 1998). As a young school leaver she decided to move from her village in central Thailand and travel north up to Bangkok.

The city was booming and Mayuree found a job easily with a finance company near the Silom Road – Bangkok's Wall Street. Mayuree describes her company's business in straightforward terms: it was borrowing money from abroad and then lending it to Thai property developers at a mark-up of around 100 per cent. Mayuree worked in the collections department, where she described her work thus (Bullard 1998).

'No one really cared if clients were behind in their repayments. I just had to keep the records of everyone who hadn't made their loan repayments, send them reminder letters and occasionally ring them to find out what was going on. We didn't take anything very seriously. We talked about our boyfriends and our families and shopping. The boss was very relaxed – he seemed more interested in his latest car than what we did in the office.'

However, this all changed in May 1997 when Mayuree's boss suddenly started taking a lot more interest in who was, and crucially who was not, repaying their loans. The company's non-performing loans stood at 40 per cent. The floating of the Thai baht on 2 June 1997, and its subsequent devaluation, made matters worse since the company had to pay back its foreign loans in dollars, yen and pounds; the result was inevitable. The company was suspended and then closed by the Finance Restructuring Authority in December 1997. Mayuree was, like so many in the region, unemployed (Bullard 1998).

'What am I going to do? Some of my friends have gone back to their parents and others are working in bars. But I don't want to do that. I studied hard and was good at my job and now it's all gone. I now sell some clothes with a friend at weekend markets, but it's hardly enough to make ends meet. My boyfriend worked in the same company and he also lost his job . . . Some of my friends blame foreigners. They believe that the crisis was created so that foreign companies could come and buy up the country. Others say it's because of corruption, or because people gambled on the stock exchange, or because rich people wasted their money on overseas trips. I don't know what to think. It's all very confusing. All I know is that I want a new job so that I can get on with my life.'

In a sense, although Mayuree's is not a happy story, she was one of the luckier ones. In South Korea, which had been the eleventh largest economy before the events of 1997 onward, the crisis hit home hard. In 1998, there were around 10,000 people losing their jobs and about 90 businesses failing each week. In the first three months of the year, around 2,300 people committed suicide; so many were jumping from the Han River bridge in Seoul that it had to be greased to prevent people scaling it (Frei 1998). The BBC correspondent Matt Frei suggests that contrary to the feelings of anger that we would expect to see in the West, the Confucianism of Korea served to produce a deep and personal sense of shame:

> 'Mr Yang keeps up appearances meticulously. But he dresses for work that he no longer has. In April he went bankrupt and his fertiliser factory employing 27 people closed down. For two months he was too ashamed to tell his wife. Every morning he commuted to a Seoul park with hundreds of other men also pretending to go to work.'

As Nicola Bullard writes, 'There are still several unanswered questions at the back of my mind. How could this export-led model of capitalist development, endlessly peddled and lauded by the IMF and the World Bank, collapse like a pack of cards? Miracle to meltdown in a matter of months.'

Orthodox explanations of crisis: 'crony capitalism'

The dominant orthodox understandings of the crisis were both from the West and distinctly Westernised too, in the classic occidental manner (see Said 1978 and Higgott

1998). There is a distinctive and consistent liberal belief in the superiority of the rationality of the market mechanism over the machinations and corruption of politics. But also an implicit racism and slur upon the appropriateness of 'Asian values' for operating in the contemporary system of liberal-democratic capitalism. The familial and corporate links that undergirded the economies were castigated as representing a 'crony capitalism'. The shift was surprising, for only recently was the East Asian developmental model still being praised for its dynamism and growth (World Bank 1993).

However, one of the earliest critiques of the East Asian growth model was from the economist Paul Krugman (1994). He argued that because growth was based upon factor accumulation rather than productivity growth it would be subject to diminishing returns. The argument was that the financial markets had successfully identified the structural contradictions of the economies and had moved against them accordingly, speculating against their eventual demise.

This explanation is a reflection of the assumptions of market efficiency and rationality. As long as all the relevant information is available, the market is able to aggregate it, through the price mechanism, to provide accurate assessments of economic value. However, even with his prescience, Krugman was expecting a gradual slowdown in the region rather than the crisis that actually occurred. And there are other problems with the dominant account in terms of timing.

Heterodox explanations of crisis: 'irrational speculation'

The tendency among orthodox commentators is to conceptualise the emergence of financial risk as an exogenous shock. And then, at most, contagion effects and herding behaviour within financial markets are seen as mere transmission mechanisms (Wellink 2002). However, there are some problems with this position. For, in the case of East Asia, most (though not all) of the fundamentals were alright; fiscal balances were being maintained, monetary expansion and inflation were both limited (Jomo 1998). What did become a problem, as was suggested by Mayuree (in Box 4.6), was the excessive amount of short-term investment in the region, causing larger current account deficits; and this was already a region that was blessed with high domestic savings rates. Inevitably this led to increased risk-taking in seemingly high return investments such as property.

Hence it appears that it was financial liberalisation that operated as the source of the crisis alongside the quasi-peg the countries were operating. The quasi-pegging, against the US dollar, encouraged unhedged borrowing from abroad but it also meant that the currencies became an increasing target for speculation as market sentiment moved against the NICs. This leads on to the second problem with the orthodox arguments: the fundamentals varied across the different economies in the region, yet the crisis hit them all. So, while short-term borrowings were an issue in Thailand, they were much less so in Indonesia, Malaysia and the Philippines (Montes 1998). Chapters 6 and 7 will take up the arguments again in more detail looking at the debates between the orthodox and heterodox perspectives on the causes of and solutions to financial crises.

Global imbalances

The issue of balance of payment imbalances has risen back up the international public policy agenda over the past decade or so, mainly because of the imbalances between China and the US – though these are a specific bilateral instance (with political spice) of a broader set of imbalances between East Asia and many OECD countries, and between Germany (as a surplus economy) and much of Europe.

In theory, according to the neoclassical orthodoxy, capital should flow from rich countries to poor countries. This is because of the much higher returns to capital available in developing countries. However, the opposite is true. Developing countries have been financing rich country borrowing. This has become known as the 'Lucas paradox' (Lucas 1990). Lucas's own argument was that developing countries suffered from a lack of human capital which meant that financial capital couldn't be as productive as the neoclassical assumptions predict. Other – orthodox and popular – interpretations emphasise the 'interference' of governments and politics in the foreign exchange markets. For example, China stands accused of deliberately undervaluing its currency in order to keep its exports more competitive. The Chinese government – it is alleged – is buying up foreign exchange to keep the value of dollars high vis-à-vis their own. The growing reserve holdings, mainly of dollar-denominated assets, is pointed to as evidence for this contemporary 'mercantilism'.

What's the problem? Well, some have argued very little. For example, Dooley *et al.* (2003) have argued that because the imbalances serve political interests on both sides they actually represent a relatively stable equilibrium. The US benefits from cheap imports (effectively exporting inflation) as well as lots of cheap financing through issuing government bonds which continue to get bought up. China and East Asia more generally, benefit from exports, industrialisation and jobs. However, more critical observers see the growing imbalances as inherently unstable, and predict that at some point the system will break and the world economy will be plunged into economic chaos similar to the 1930s. A further developmental cost – in addition to the instability of privatised international finance that produces speculation and crises – is the deflationary tendencies the system produces and the opportunity cost of governments holding reserves (Hudson 2010).

The developmental costs of global imbalances

Developing countries – especially, but by no means exclusively, the so-called emerging markets – have enjoyed current account surpluses which have enabled them to build up large foreign exchange reserves. Dani Rodrik (2006b) documents that, prior to the global crisis, reserves had risen to almost 30 per cent of developing countries' GDP from just 6–8 per cent during the 1970s and 1980s. A huge increase. Tellingly, industrialised countries' reserves have remained constant at 5 per cent of GDP since the 1950s. Reserve levels in developing countries have risen to the equivalent of 8 months of imports, whereas the traditional rule of thumb for central banks was always 3 months of imports. Plus, it is not just China and it is not just emerging markets that are increasing their reserves: Africa's reserves also stand at 8 months of imports (Asia's are at 10 months).

The popular explanation for reserve accumulation is that it is a byproduct of governments – especially in Asia – limiting currency appreciation as part of mercantilist, export-led growth strategies (Dooley *et al.* 2003). However, an alternative argument is that developing and emerging markets have had to learn how to insure themselves against the sudden stops, capital flight and volatility associated with financial crises, rather than relying on the IMF for protection (Feldstein 1999). The startling growth in reserves since the mid- to late-1990s – i.e. after the East Asian and Mexican crises – would seem to suggest that precautionary motives are behind the increases. In testing the competing explanations, Aizenman and Lee (2005) find that trade openness and exposure to financial crises are both statistically and economically important in explaining reserves, whereas lagged export growth and deviations from predicted PPP are statistically but not economically significant in explaining the patterns of reserve holdings. Hence they conclude that precautionary demand is a better explanation for reserve holdings than mercantilist interpretations.

Feldstein (1999), while recognising the costs of 'self-help', argues that the benefits of not having to rely on the IMF and independence from austerity outweigh the costs of reserve holdings. However, more recent analyses suggest that the costs are significant – both in terms of direct costs to individual countries holding excess reserves and collectively in terms of lost global aggregate demand. Critical reformists emphasise how the costs are a function of the uncertainty of a liberalised financial system which creates incentives to self-insure, taking capital away from productive investment (Bibow 2009). Thus the direct costs are opportunity costs of forgoing domestic public investment, such as on education, public health, social security and physical infrastructure. Instead of such public investment, central banks hold their reserves in the form of low-yielding short-term US treasuries and other securities. It has been estimated that this reserve accumulation has caused Asian (excluding Japan and China) investment rates to fall by some 9 per cent of GDP, and 3 per cent of GDP in Latin America (Eichengreen and Park 2006). Rodrik (2006b) has calculated the opportunity cost of excessive reserves as equal to 1 per cent of countries' GDP – a huge number, especially in relation to the amounts expended by governments on poverty alleviation programmes. Yet, despite the clear developmental costs of precautionary reserves, there is still a sense in which they are rational given the costs of financial crises which are saved. The current system creates these individual and collective costs; costs which, for example, capital controls could alleviate (Cruz and Walters 2009). Regulating capital flows, through capital controls or a financial transactions tax such as the Tobin Tax (Eichengreen *et al.* 1995), would be part of any solution.

Conclusion

In this chapter we have reviewed the importance of the international monetary and financial system for the world economy, highlighting its role in facilitating the cross-border flows of money and credit – allowing firms, individuals and governments to trade with one another. Also, a stable international monetary system is essential to encourage economies to trade, invest and produce within. The chapter has reviewed the different ways of designing and organising the world's international monetary and financial systems, including the gold standard, the gold-exchange standard and the post-Bretton Woods system of floating exchange rates that has been in existence since the early 1970s.

The Mundell-Fleming framework helps highlight a fundamental trilemma that all governments face; essentially governments can achieve only two of the three following policy goals: a stable exchange rate, access to international capital markets and domestic monetary policy autonomy. One of these goals has to be sacrificed. Under the current 'non-system' this means that some countries (often the wealthier OECD ones) tend to allow their exchange rates to float and others (often smaller or developing countries) seek to fix or peg their exchange rates to try and maintain a degree of stability for their currency, as well as build credibility with international investors. Sometimes this works, but on other occasions, as we have seen, it exposes currencies to attack by international speculators. The consequences of these currency attacks can be severe in terms of economic growth, poverty and human development.

Despite this, the overwhelming trend has been for developing countries to be exposed and open to international capital flows. There are economic and political elements to this. The liberalised and privatised financial system proved overwhelming for the previous system of fixed exchange rates and for individual countries that now seek to stabilise the value of their currency. At the same time, developing countries are often compelled – either through a lack of domestic saving and investment or by the international financial institutions – to follow a path of financial liberalisation, opening up the capital account and encouraging capital mobility. There is an increasingly large evidence base to suggest that liberalisation isn't always the best policy. Even the IMF has now officially, in its institutional statements in 2012, accepted that capital controls are sometimes desirable and offer countries an important extra policy tool which should be used. The following chapters will now move on from the monetary system to look at the various types of capital flows that constitute the global financial system.

Discussion questions

1   Why is it important to understand the historical evolution of the international monetary system?

2   Given the trilemma faced by developing countries' governments, which priorities, in your view, offer the optimal strategy?

3   In your view, are orthodox or heterodox views the most convincing account of financial crises? Are crises an inevitable part of development (as both neoliberal and radical scholars might argue) or can they be minimised through domestic reforms (liberal institutionalist) or international regulation (critical reformist)?

4   Are economic or political factors the primary drivers of financial liberalistion?

5   Do financial crises or global imbalances represent the greatest threat to development?

Further reading

- For a background on the evolution of the international monetary system take a look at Barry Eichengreen's *Globalizing Capital* (second edition, 2008). It is a masterful history of the international monetary system from 1850 to the present day. At only 230 pages it's an impressive feat. Well worth a read.

- For a fascinating set of case studies detailing the political economy of exchange rate decisions – i.e. the influence of political interests from politicians, trade unions and businesses, the role of elections, international pressures, and economic fads and fashions – take a look at the report of Worrell *et al.* (2000) 'The Political Economy of Exchange Rate Policy in the Caribbean', produced for the Inter-American Development Bank. It examines the experiences of Jamaica, Trinidad-Tobago, Antigua, Guyana, Barbados and Grenada.

- For an insightful account of the role the IMF has played in developing countries, especially exploring the contradictions between the IMF's self-perception as a technical 'best practice' organisation and the actual political and social implications of structural adjustment and political conditionality, take a look at Ben Thirkell-White's (2005) *The IMF and the Politics of Financial Globalisation: From the Asian Crisis to a New International Financial Architecture*. Thirkell-White uses the East Asian financial crisis to reveal these dynamics, plus the role of the US in the cases of South Korea, Thailand, Malaysia and Indonesia. Excellent stuff.

Useful websites

- Just published in 2012, the transcripts from the 1944 Bretton Woods Conference give a unique inside perspective of what participants at this major international gathering said behind closed doors. Kurt Schuler found the transcripts while browsing in a section of uncatalogued material in the library of the US Treasury Department. A table of contents, preface, introduction and sample transcript can be viewed here: http://www.centerforfinancialstability.org/bw/BWSample.pdf.

- The Pardee Center launched an important report titled *Regulating Global Capital Flows for Long-Run Development* in 2012. The report argues that the global financial crisis has underlined that capital controls (or capital account management) are both necessary and desirable. In addition the report provides some useful empirical detail on countries that have employed various types of capital controls. The report can be downloaded from here: http://www.bu.edu/pardee/2012/03/07/task-force-march-2012/ and a video presentation from the report launch can be watched here: http://www.bu.edu/pardee/2012/04/19/pardee-capital-controls-video/.

- The IMF's 'Back to Basics' paper on 'Capital Account Liberalisation' by M. Ayhan Kose and Eswar Prasad is available online: http://www.imf.org/external/pubs/ft/fandd/2004/09/pdf/basics.pdf. It explains what the capital account is, motives for controlling capital and liberalising the capital account, and the empirical evidence as well as what governments might want to do.

5 Development aid

Learning outcomes

At the end of this chapter you should:

- Be able to describe what aid is, its internationally agreed definitions and the accounting standards used to define what is and what isn't official development assistance.
- Know the history and geography of the international aid regime.
- Understand who gives development aid (donors), to whom (recipient or partner countries), and how (via a discussion of the different modalities or types of aid).
- Know the aid effectiveness debate, i.e. whether and how aid has an impact on development through economic growth and welfare measures.
- Understand the critical and radical critiques of aid as a form of imperialism and as perpetuating the dependency of low income countries.
- Be aware of the contemporary policy issues surrounding the international aid regime such as aid effectiveness, donor coordination and transparency.

Key concepts

Aid, official development assistance (ODA), modalities, determinants of aid, aid effectiveness, fungibility, transparency, absorptive capacity, OECD Development Assistance Committee (DAC), Paris Declaration

Introduction

International aid is the transfer of resources from rich to poor countries. Peter Bauer (1981: 87) famously opined that 'The Third World is the creation of foreign aid: without foreign aid there is no Third World'. Bauer's views on aid were famously negative and conservative – viewing it as a source of patronage, corruption, stifling market forces and entrepreneurship, all issues picked up later in the chapter. As such, for Bauer, foreign aid serves to maintain poverty and underdevelopment. But, more generally, Bauer's comment also suggests the centrality of aid to the development project; aid is almost constitutive of development and is indeed what most people think of if you stop them in the street and ask them about 'development' (see Box 5.1 on public attitudes to aid). Even the etymology of the term 'aid' conjures up images of unconditional help from a beneficent Good Samaritan; but the reality is often less clear cut (Gronemeyer 1992).

The web of actors involved in international aid is huge and complex. Nearly every single country in the world is either an aid donor or recipient, and some, such as India, China, Brazil and South Africa, are both (Riddell 2007). The traditional actors in the aid regime are the bilateral donors who operate under the auspices of the OECD. There are currently 24 members of the OECD's DAC. The DAC is the most high profile and established group of donors and acts as the traditional forum for bilateral coordination and cooperation with developing countries with respect to aid and aid effectiveness. However, recently the centrality of the DAC within the aid universe has been challenged by the emergence of new donors such as Saudi Arabia, China, India, Russia, Brazil and South Africa, who are becoming increasingly important. In addition to the traditional and emerging bilaterals, the third group of donors are the international organisations and IFIs such as the multilateral development banks and various organs of the United Nations, for example the World Bank's International Development Association (IDA), the African (AfDB), Inter-American (IADB), and Asian Development Banks (ADB), the IMF, United Nations Development Programme (UNDP) and so on. There is also a growing and increasingly influential collection of private donors. On the one hand these include the international NGOs and charities commonly associated with development assistance. NGO aid is upwards of $24 billion per year, which is in excess of 30 per cent of total ODA (Riddell 2007: 259). On the other hand there are the huge new private and philanthropic funds such as the Bill and Melinda Gates Foundation.

Like all the flows of capital that we will consider in this book, there are two key debates within the development and academic communities. The first is about quantities of aid – how much is promised, delivered, and where and to whom does it go, and why? – and the second is about the quality or effectiveness of this aid, i.e. to what extent it promotes development, or not – essentially how much 'bang for the buck' donors and recipients get for each dollar of aid. There are important policy implications which flow from both of these debates and which serve to shape summit discussions and institutional design, for example the Paris Declaration on Aid Effectiveness (2005).

Ignoring, for the moment, the geopolitical motivations which have shaped the aid regime, from a financing for development starting point, aid has a dual purpose. First, in terms of capital accumulation, aid is held to augment savings and finance investment to help growth; to increase productivity through investments in a population's health and education; and to transfer expertise and technology through technical assistance (Radelet 2006: 8). Because, unlike private flows of investment which are driven by profit opportunities, aid can be delivered to the poorest and marginalised and target poverty traps (Sachs *et al.* 2004). But aid is not just about increasing economic growth and productivity. Aid supporters, such as Jeffrey Sachs, feel that aid's successes do not get reported enough. Sachs has publicly and tirelessly argued for large increases in development aid to end poverty (Sachs 2002, 2005). He highlights the role that aid has played in reducing African child mortality from 229 per 1,000 births in 1970 to 146 per 1,000 births in 2007 and increasing adult literacy from around 27 to 62 per cent over the same period (Sachs and McArthur 2009). Plus, he argues that aid has been instrumental in the 'Green Revolution', eradicating river blindness and introducing oral rehydration therapy. It is certainly a fact that in the 40 years since aid has become widespread, health and education indicators have 'risen faster than during any other

40-year period in human history' (Radelet *et al.* 2005). The question arises, then, as to whether this is because of aid or merely coincidental?

This chapter is structured as follows. First it is necessary to define and disaggregate aid in order to provide some precision about what types of aid we are referring to when seeking to understand why aid is given, who receives it and how effective it is, as well as the resulting policy implications. The first section describes how development aid is defined and accounted, introducing the different categories of aid and how important they are. It also covers who gives aid and who receives aid. The second section reviews the debates about the determinants of aid – mostly why states give aid, to whom. Section three examines the debates around the impact of aid: does aid support development or promote dependency and maintain economic backwardness, and why? The final section considers some key policy debates in development cooperation that emerge from these discussions and have served to shape contemporary institutions and aid practices.

Box 5.1

'Bottomless begging bowls'? Public attitudes towards development financing

Aid is indeed what most people think of if you were to stop them in the street and ask them about development financing. If you were to ask those same people some further questions about aid, a striking picture about public understandings, or misunderstandings, would likely unfold. Foreign aid is an issue that tends to divide opinion. Polls conducted across Europe and North America have regularly shown support for aid to be between 65 and 90 per cent, which is amazingly high (Hudson and vanHeerde-Hudson 2012). However, it is equally the case that comments such as Jesse Helms's famous analogy of foreign aid being akin to throwing 'money down a rathole', and Tom DeLay's soundbite of 'putting Ghana over Grandma', also capture common sentiment (Kristof 2006). Consider the following portrait of aid from one UK respondent as reported in a study of public perceptions of development (Henson *et al.* 2010: 11):

> My views about world poverty are very non 'PC'. If you live in a failed state (mostly in Africa) and then have 10 kids and expect white people to pay for them I call that irresponsible, not my blooming burden – people in the so called 'UK' don't owe a living to aids infested African baby machines with bottomless begging bowls. When does it stop? Malthus got it right (just not the time scale). It is coming to fruition soon. Geldoff said 30 years ago that there was famine in Ethiopia. Funny the population of said country is twice what it was then. By contrast the population of Ireland is still only half what it was in 1845 when a real famine was enforced by an alien neighbour (England). The world is full of bulldropiness and most of it comes out of the gobs of these lying African lovers who lie and lie to get kind hearted people in Europe to give them money when all they really need is contraception. We do no favours to Africa with 'aid'. We infantilise, immobilise, paralyse, restrict any hopes of developments in that blighted part of Africa South of the Arab lands. Let them stand on their own feet, work to feed their children (or else don't have them). I've got one grown up child – she has none. We don't go begging to feed 10 unnecessary babies. It's these people who will eventually destroy the whole human race with their sheer postulating [*sic*] numbers.

The conclusion from the Henson study is that public opinions tend to be strongly held, regardless of how well informed people are or think they are. A well-known survey in the US asked respondents what proportion of the government's budget they thought went to foreign aid (PIPA 2001; van Heerde and Hudson 2010). On average, Americans thought just under 25 per cent of the US federal budget was allocated to foreign aid. The reality is approximately 1 per cent. When asked what the appropriate percentage of the federal budget would be for foreign aid (if any), the average response was less than 14 per cent. Once told the actual amount that the US spends on foreign aid, 37 per cent of respondents thought this was too little, 44 per cent thought it was about right, and only 13 per cent thought it too much. Furthermore, other research into public perceptions of foreign aid has shown that when asked about development aid, people tend to think of humanitarian assistance or disaster relief (McDonnell *et al.* 2003). For most people, when they think of aid, they are thinking of the aid given in emergencies such as the famine referred to above and the 1985 Live Aid concert, or an earthquake in Haiti, or a tsunami in the Indian Ocean, or after a civil war or internal displacement. Similarly, most people think of development NGOs and charities when they think of foreign aid. Yet humanitarian aid is but a small part of the overall volume and purpose of aid; in 2008 humanitarian aid only accounted for 7 per cent of all donor countries' aid commitments (OECD 2009b). So, it is clear that the public holds strong views about development aid when asked about it, and in general people show concern about global poverty, but it's unclear how often they think about development aid in their daily lives, the extent to which views are well-informed, and what, if any, action they take on global poverty.

Trends and geography of aid

Aid is the generic term for describing the transfer of resources from rich to poor countries; it can come in a variety of forms, such as financial resources (both grants and loans), technical know-how, military hardware, physical goods or food aid. But not all aid is development aid.

From a developmental perspective, the key category of aid is ODA (Official Development Assistance). ODA is defined as those grants and loans provided at concessional terms, by public agencies, to developing countries, for the expressed aim of promoting economic development and welfare. It includes emergency assistance and technical assistance, but excludes aid for military purposes or loans which are not concessional (see Box 5.2 for an explanation of what counts as aid).

This definition of ODA was first agreed by the DAC in 1969 and refined in 1972 (Riddell 2007). But defining development aid is by no means politically neutral; it has been a historically contested enterprise. Technically defining something, fixing it as a 'truth', is a very elegant form of power given it appears to depoliticise it (Hopwood and Miller 1994). Although ODA is now relatively fixed, it is important to note that what counts as aid has changed considerably over the years in line with trends in political fashion (Thérien 2002). A genealogy of aid reveals that what is now taken for granted is the result of a series of political struggles rather than a technical refinement of the concept. Notably, prior to the 1960s, private capital counted as official aid, as did military expenditures. Neither of which are currently allowed to count as ODA under the definition provided by the OECD's DAC. Furthermore, the 'generosity' of concessionality reflects interest rates that prevailed in the 1970s (see Box 5.2). Having flagged up these cautions the rest of this section will document how much development aid is provided, by whom, how and to whom.

Box 5.2

What's aid? Defining and measuring official development assistance

The primary category of aid is ODA. The precise definition of ODA provided by the OECD DAC in its *Statistical Reporting Directives* (OECD DCD/DAC 2007a: 12) is:

> Official development assistance is defined as those flows to countries and territories on the DAC List of ODA Recipients and to multilateral development institutions which are:
>
> i. provided by official agencies, including state and local governments, or by their executive agencies; and
> ii. each transaction of which:
>
>   a) is administered with the promotion of the economic development and welfare of developing countries as its main objective; and
>
>   b) is concessional in character and conveys a grant element of at least 25 per cent (calculated at a rate of discount of 10 per cent).

First, to be eligible as ODA the funding must be delivered to a developing country or territory or multilateral development institution recognised by the DAC. That is to say, the country has to fall into one of the three World Bank developing country income groups (the OECD also reports using the UN's LDC category).

Second, aid must be provided by official agencies. That is to say, it should be provided by government bodies – whether at the central, federal, state or local level – using public funds (raised from taxation or borrowing). If funding is provided by official agencies, but does not meet the other two criteria it is counted as 'other official flows' (OOF).

Third, and intuitively, ODA must aim to promote economic development and welfare (ii, a). However, in practice this isn't always so straightforward and ultimately it is the DAC which decides on where the boundaries lie. So, military aid is excluded, whereas the costs of using the donor's military to deliver humanitarian assistance is ODA; the enforcement aspects of peacekeeping are not ODA, but other tasks of peace operations, such as the demobilisation of soldiers after civil conflict, election monitoring, etc. are included. For a fuller list and discussion of what does and does not count as ODA see OECD (2008a).

The fourth criteria is the most tricky and worth explaining at some length. Simply put, concessional flows are those which offer reductions on market rates, i.e. they are so-called 'soft loans'. The reason this stipulation exists is because until the 1960s all types of flows were recorded as aid, including private investment (Thérien 2002). In order to defend the welfare nature of aid, in 1972 it was agreed that the grant element must be at least 25 per cent of the official loan (OECD 1972: 208). Though note that this means, contrary to popular images of aid, it is not necessarily a donation or a grant. Instead it's a loan.

Concessionality is essentially how soft the loan terms are, i.e. how far from the market rate a loan is. DAC statistics calculate this by measuring the difference between the return the lender could have expected from profitably investing it and the interest rate charged on the concessional loan (the opportunity cost). The grant element (GE) is defined as the difference between the face value of the loan and the present value of the service payments the borrower will make over the lifetime of the loan. The notional reference rate of interest used to work this out is 10 per cent. So, if a loan carries an interest rate of 10 per cent then the GE is zero, and for a grant with zero interest the GE is 100 per cent. In addition to the interest rate, the GE is also affected by the maturity of the loan, i.e. how long the loan will be paid off over, and the grace period, i.e. how long before the recipient has to begin paying off the loan. By way of example, if the maturity of a loan is less than 10 years then the interest rate will generally need to be below 5 per cent for the GE to be over 25 per cent and to qualify as ODA (OECD 2010a). If you are so inclined, take a look at Annex I of the OECD's 'Reporting Directives for the Creditor Reporting System' (OECD DDC/DAC 2007b: 54–60) for the formulas and four examples of how to work out the GE based on the type and terms of the repayment schedule.

The 10 per cent notional reference rate is notably arbitrary, but it was set in the 1970s when interest rates were rising sharply. And this has important implications for today because the much lower long-term interest rates make it much easier for donors' aid to meet the GE threshold and to qualify as ODA (OECD 2008a). Moreover if the flow meets the GE criteria for concessionality then the entire flow counts as ODA, the loan element as well as the grant element, which further inflates the amount counted as aid.

Figure 5.1 *Aid concessionality over time.*

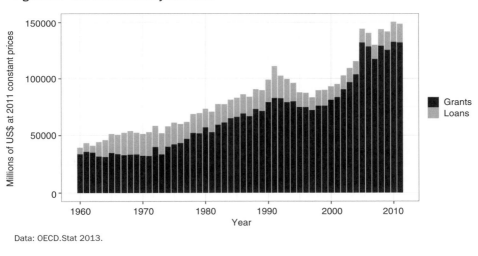

Data: OECD.Stat 2013.

How much?

In 2011 the 24 members of the DAC disbursed US$134 billion of ODA (OECD.Stat 2013). This is a net figure, i.e. it takes into account any loans repaid or grants recovered. As Figure 5.2 shows, ODA has steadily increased over time since 1970.

While the figure shows that total aid has enjoyed an upward trend over time, it is also clear that aid has experienced periods of stronger growth and periods of retraction. There is a clear peak in 1992, just after the end of the Cold War before it falls away, contrary to hopes for a peace dividend. With the end of the Cold War it was hoped that resources could be switched from military to welfare expenditure, including development. However, aid levels increased again through the late 1990s, largely because of increasing aid to new democracies and debt relief (Radelet 2006). Since

Figure 5.2 *Official and private flows from DAC countries.*

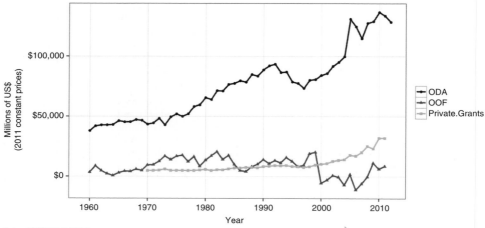

Data: OECD.Stat 2013.

then the upward trajectory has been maintained by the global response to terrorism in the wake of the attacks on the Twin Towers in New York in September 2001 and the agreements by the G8 countries at Gleaneagles in 2005 to double aid to sub-Saharan Africa by 2010 and new promises of debt relief. The recent dip since 2005 reflects a falling away in total disbursements following the unusually large amounts of debt relief granted to Iraq and Nigeria in 2005, a trend that has most recently been reinforced by economic austerity within many of the donor countries. But there has been a noticeable divergence in how donor countries have dealt with economic recession and budget pressures. The economic downturn has translated into large falls in Greece and Spain (30+ per cent from 2010 to 2011, for instance). Whereas in the UK the government committed to reaching a longstanding target of giving 0.7 per cent of the country's GNI in development aid. So political decisions and economic realities combine to create varying outcomes.

By whom? Donor volumes

There are four key types of donor. First, the DAC donor countries which account for official bilateral aid. Second, the multilateral organisations including the World Bank, the regional development banks and the UN agencies. Third, private donors such as the NGOs and philanthropic organisations. Fourth and finally, the emerging donors that have operated outside of the DAC framework, such as China and so forth. We know much more about the first two than the latter two; these latter two are becoming increasingly important but the data and research is still catching up with the new realities. Figure 5.3 shows the proportion of aid that is delivered through bilateral, multilateral and private channels from DAC countries over time.

Figure 5.3 *Net aid disbursements from DAC countries by donor type.*

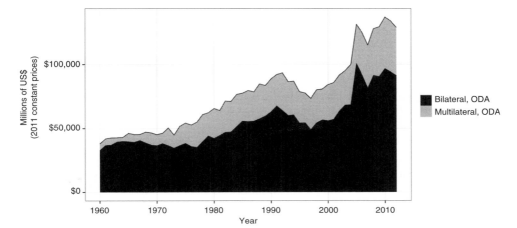

Data: OECD.Stat 2013.

Box 5.3

OECD Development Assistance Committee (DAC)

The OECD traces its roots back to its role in administrating the Marshall Plan. Hence, despite its wide-ranging role in promoting polices, cooperation and standard-setting across a variety of different issues – from tax competition to the labelling of fresh fruit and vegetables, to bribery in international business transactions – development assistance has always really been at the heart of what the OECD has done. For a wonderful book-length treatment of the OECD see Woodward (2009).

The part of the OECD which is now devoted to development issues is the DAC. The DAC was established in 1960 (called the Development Assistance Group for its first year) and became part of the OECD by Ministerial Resolution on 23 July 1961. Its current membership includes is the 24 countries with significant aid programmes: Australia, Austria, Belgium, Canada, Denmark, Finland, France, Germany, Greece, Ireland, Italy, Japan, Korea, Luxembourg, the Netherlands, New Zealand, Norway, Portugal, Spain, Sweden, Switzerland, UK, US and the European Commission (EC). The World Bank, the IMF and the UNDP participate as permanent observers. The most recent member to join the DAC was the Republic of Korea which acceded on 1 January 2010 (OECD 2010d).

DAC members agree to expand the aggregate volume of resources to transfer to developing countries for the purpose of economic development and welfare improvements. Plus, they are committed to increase the effectiveness of these flows with respect to this aim. They also agree to coordinate and to consult on such matters so as to reach these twin goals. The Development Co-operation Directorate (DCD) provides the secretariat for the DAC to help it fulfil its roles and commits to:

- be the definitive source of official development assistance statistics and mobilise the increase of ODA;

- assess members' development cooperation policies and their implementation through peer reviews;

- enhance the effectiveness of aid by making it more aligned, harmonised, results-focused and untied;

- provide analysis, guidance and good practice in key areas of development such as environment, conflict and fragility, gender equality, governance, poverty reduction, evaluation, capacity development, aid for trade and aid architecture;

- support policy coherence for development through peer reviews and collaboration with other policy communities (OCED 2010d: 5).

The DAC has been behind the recent international efforts to improve aid effectiveness as codified in the Paris Declaration on Aid Effectiveness (2005) and the Accra Agenda for Action (2008). The DAC also conducts 'peer reviews' of its members' bilateral programmes assessing *inter alia* how they meet, or not, the DAC's recommendations and guidelines, whether they are meeting aid commitments and how they manage their aid management and budget. The reviews are conducted by two other DAC members and experts from the DAC secretariat.

As well as coordinating and monitoring the traditional donor countries development assistance, the OECD also provides the most well-known and used database on international aid. The database includes comprehensive data on the volume, origin and types of aid to over 150 developing countries and territories going back to 1960 for some donors and recipients. This is because one of the key roles of the DAC is to collect statistics to help support its aims and to meet the needs of policymakers in the field of development cooperation. It provides a means by which the comparative performance of aid donors can be assessed.

As well as data on aggregate aid flows, the DAC collects more detailed data on where aid goes, what purposes it serves, and what policies it aims to implement in the Creditor Reporting System (CRS) which enables consideration of specific policy issues (e.g. tying status of aid). The statistics are available to the public in regular publications, such as the Statistical Annex to the annual *Development Cooperation Report* and the *Geographical Distribution of the Financial Flows to Developing Countries*. The completeness of CRS commitments for DAC members improved from 70 per cent in 1995 to over 90 per cent in 2000 and reached nearly 100 per cent starting from 2003 flows.

Since 1998, these statistics have been available online: www.oecd.org/dac/stats/idsonline. The DAC database covers all long-term (i.e. over one-year maturity) disbursements and commitments of official and private flows to developing countries and multilateral organisations. This is broken down into (1) ODA, (2) OOF, (3) private flows at market terms, and (4) net private grants (DCD/DAC 2007a). The 'private flows at market terms' category refers to direct and portfolio equity investments, bank loans and bond purchases by banks and so does not interest us here; they will be considered in Chapters 6 and 7. However, the other three are more genuinely considered as 'aid'.

DAC bilateral donors

Disaggregating total aid by donors, as shown in Figure 5.4, it is clear that the US is the largest aid donor, and has provided 22 per cent of all ODA over the past 10 years. Japan, which gave the next most in the past 10 years, accounted for 10 per cent. The only exception to US primacy in volumes of ODA was for the majority of the 1990s

Figure 5.4 *Net ODA in 2012.*

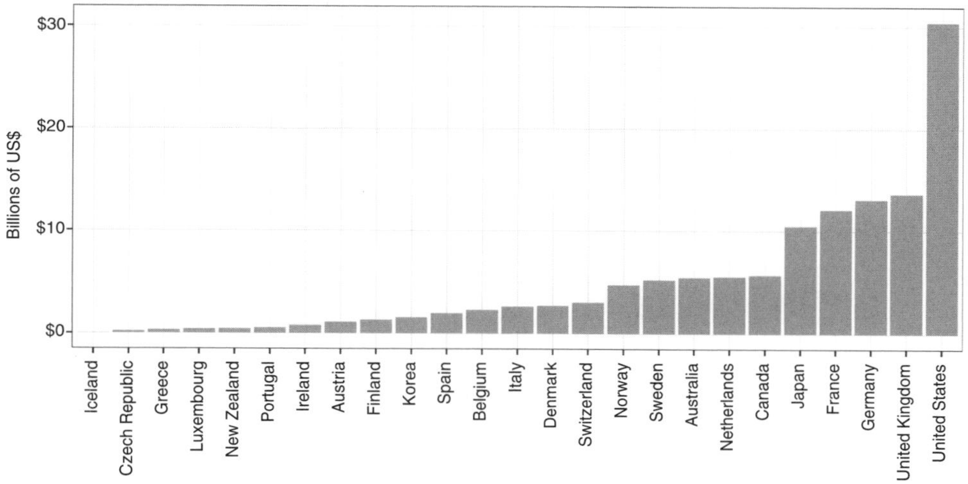

Data: OECD.Stat 2013.

when Japan was the largest donor in the world in dollar terms. This is unsurprising given that the US is the world's wealthiest country and one with many strategic interests. Along with the US and Japan, the other current largest donors are France, Germany and the UK; the five of them together have accounted for 62 per cent of all ODA in the past 10 years.

The alternative to measuring absolute levels of aid disbursed by the DAC members is to examine the amount given as a proportion of the donor country's wealth. ODA as a percentage of GNI is the common measure of donors' generosity. Importantly it is also the metric for the long-standing UN target of 0.7 per cent (see Box 5.4). The ODA/GNI measure tells a very different story about individual donor generosity as shown in Figure 5.5. Three clear groups emerge. There is a group of donors – Denmark, Luxembourg, the Netherlands, Norway and Sweden – who are already at or above the UN 0.7 per cent target. There is a second, large group which float around the donor average, and a third group of laggards scoring badly on the GNI/ODA ratio, including Japan, Greece, the US, Italy and Korea at the bottom.

Looking at ODA/GNI over time, this now tells a very different story about changes in aid levels over time (shown in Figure 5.6). The steady upward trend shown in Figure 5.2 is contrasted with a downward trend of donor ODA/GNI. So, contrary to the absolute measure, not only does the GNI/ODA ratio show that the US is less generous than other countries, but it also shows that, as a group, all donors are currently less generous than themselves in the past (and indeed the US is less generous than itself in the past). However, as shown in Figure 5.7, recently, the US and UK have been increasing their contributions, while Germany, France and Japan have been reducing theirs.

Figure 5.5 *Net ODA in 2012 as a percentage of GNI.*

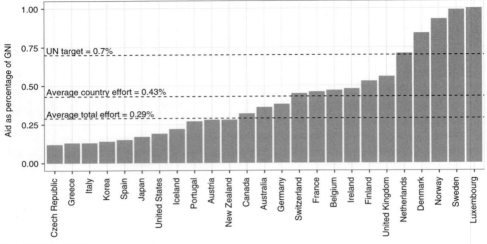

Data: OECD.Stat 2013.

Figure 5.6 *Volume of ODA and generosity of DAC donors.*

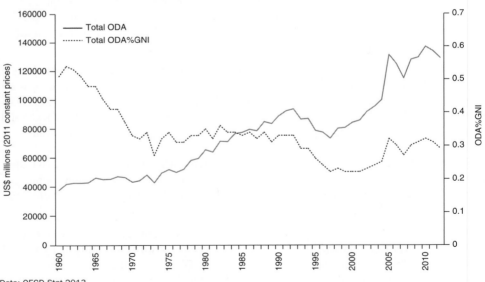

Data: OECD.Stat 2013.

Figure 5.7 *Net ODA as a percentage of GNI for the G7 countries.*

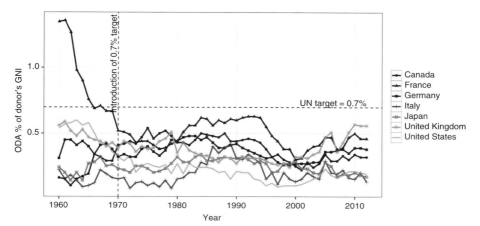

Data: OECD.Stat 2013.

Note that sometimes it is the percentage of the donor government's budget which is used rather than percentage of GNI. It is important not to confuse these figures – they often are. So, for example in the UK in 2009, where the GNI/ODA amount was 0.52 per cent, the amount of the government budget allocated to aid was 0.86 per cent. All of these different measures are useful depending on what you want to know – overall levels of aid granted by a donor and therefore maybe their influence on overall aid policy, or the amount of resources a country is prepared to commit to international development.

Box 5.4

The 0.7% aid target

The 0.7 per cent figure is probably the best-known international aid target in existence: i.e. the target for donors to give 0.7 per cent of their national income as ODA. It was established in October 1970 when the UN General Assembly adopted it:

> Each economically advanced country will progressively increase its official development assistance to the developing countries and will exert its best efforts to reach a minimum net amount of 0.7% of its gross national product at market prices by the middle of the Decade.
>
> (UNGA 1970)

The 0.7 figure was first proposed by the Commission on International Development, better known as the Pearson Commission, which had been established by the then World Bank president Robert McNamara back in August 1968. The final report stated: 'We therefore recommend that each aid-giver increase commitments of official development assistance for net disbursements to reach 0.70 per cent of its gross national product by 1975 or shortly thereafter, but in no case later than 1980' (Pearson 1969: 148–49).

Why 0.7 per cent? Back in 1958 the World Council of Churches proposed that wealthy countries should seek to transfer 1 per cent of their income to developing countries. The Council, which worked as a conduit for charitable donations to the developing world, realised that Church donations were not sufficient for stimulating development and sought to draw on rich countries' wealth. In the 1950s capital flows to poor countries were about 0.5 per cent of rich countries income: the 1 per cent most likely represented a round figure which doubled existing flows (Clemens and Moss 2005). The 1 per cent figure was supported by the UN General Assembly in the 1960 resolution, and was then accepted by the DAC members in 1968. However, the issue was that the 1 per cent figure represented *all* flows, and so it included private flows which were much more volatile and outside direct donor control. A number of other donor and development agencies were also working along similar lines:

● UNCTAD, in 1968, noted that total capital flows were falling and suggested that official assistance could form three quarters of 1 per cent.

● In 1964, the Nobel Prize economist Jan Tinbergen, who was the chairman of the UN Committee on Development Planning, had been modelling how large capital flows would need to be in order to achieve the desired growth rates in developing countries. He proposed that the combined level of ODA and OOF would need to be in the region of 0.75 per cent of donor income.

● The G-77, at the first ministerial meeting in Algiers in 1967, called for a separate minimum for ODA, net of amortisation and interest.

● The Pearson Commission was formed with the specific political end of embedding the aid target amongst donor countries as it had been noted that the public's fears of 'waste and corruption' were causing the major donors to decrease their aid levels (Kilburn 1969).

With all this as the background, the DAC suggested that the 0.7 figure was the logical amount after subtracting the relatively small amount that constituted the volume of non-concessional flows (because of the repayments going in the other direction). As Clemens and Moss (2005) suggest, the best way of looking at the figure is as a somewhat arbitrary but politically feasible compromise. Nevertheless, these origins are often forgotten and the 0.7 per cent target now enjoys broad consensus in the global development community and appears in the 2002 UN Monterrey Consensus.

Sweden was the first country to reach the 0.7 per cent target, and in 1975 the Netherlands, Norway and Denmark all reached it, and all of them have been above it since. Finland achieved it once in 1991. The UK government has been taking faltering steps to enshrine the target into law, publishing a draft bill in January 2010 requiring the UK to spend at least 0.7 per cent of GNI on development aid from 2013 onwards. And in March 2013, George Osborne, as the Chancellor of the Exchequer, announced that the UK would spend 7 pence out of every pound on aid. However, the weighted average of all DAC members' ODA has never been above 0.4 per cent of their income.

While it might sound an unproblematically 'good thing' to reach the 0.7 per cent goal, some note that while the target might be 'cause célèbre for aid activists' it is problematic that what began as a 'lobbying tool' has ended up being a 'functional target for aid budgets' (Clemens and Moss 2005: 2). The problem being that the logic for deciding on aid levels is based on rich countries' incomes – not developing countries' need.

Source: OECD (2002); Clemens and Moss (2005)

Multilateral donors

For ODA to count as multilateral it has to be given by an eligible public international organisation. Eligibility is limited to organisations which work directly on development issues (a full list is in Annex 2 of the DAC *Statistical Reporting Directives*). Organisations which are only partly involved in development and developing countries, for example the World Health Organisation (WHO) which also does a lot of its work in high-income countries, have an ODA coefficient to reflect this. So, for example, only 70 per cent of contributions to the WHO's core budget count as ODA (OECD 2008a). Furthermore, to count as multilateral ODA the aid contributions have to be disbursed at the discretion of the multilateral organisation. In the language of the OECD, individual contributions should be pooled 'so that they lose their identity'. In contrast, if the donors control the disbursement of aid, 'e.g. purpose, terms, total amount, reuse of any repayments' then it is counted as bilateral aid (DCD/DAC 2007a: 6). It's worth noting that bilateral aid can be delivered through a multilateral organisation or through an NGO when the donor retains control over where or what the money is being spent on. This might be preferred because of greater neutrality or expertise and ability to reach vulnerable or excluded populations which official actors do not or cannot cater for. But under these circumstances it remains bilateral aid.

As is clear from Figure 5.3, the greatest proportion of aid is delivered as bilateral aid. Since the mid-1970s the amount of ODA delivered through multilateral institutions has remained remarkably constant within the 26–36 per cent range (with 2005 being the exception, when it dipped to 23 per cent). Figure 5.8 shows the amount of IDA funding which originated from the DAC donors since 2010. The IDA's funding is done in three-year cycles, called 'replenishments'. The most recent replenishment (IDA16) will

Figure 5.8 *IDA disbursements and DAC contributions.*

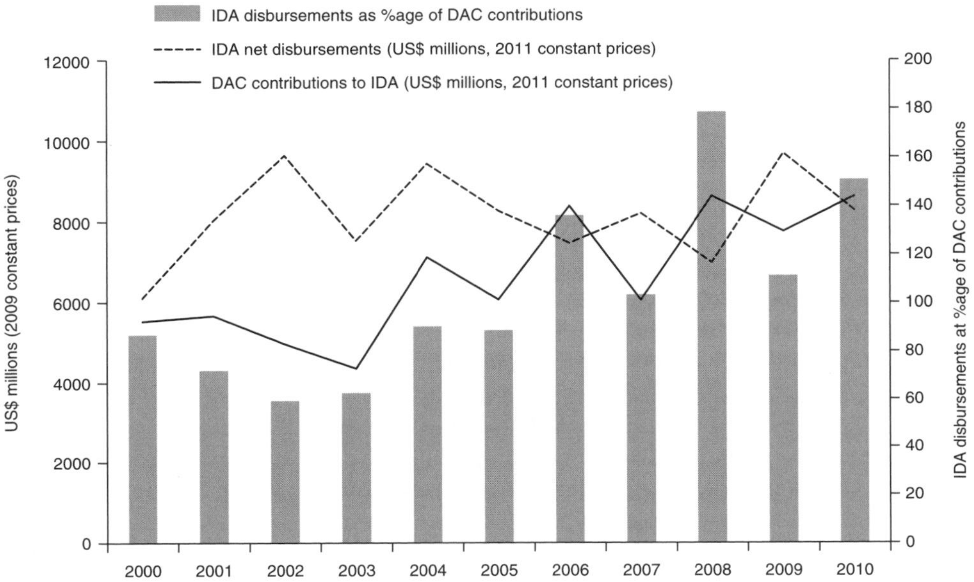

Data: OECD.Stat 2013.

Figure 5.9 *Multilateral ODA by institution.*

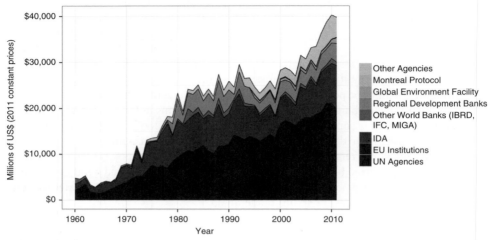

Data: OECD.Stat 2013.

last until June 2014 and totalled US$49.3 billion. Figure 5.9 displays how much DAC funding goes to the other major multilateral organisations; as is apparent, the IDA, EU and the collection of UN agencies dominate.

It is important to note that the multilateral figure represents the amount of aid given to multilateral aid agencies by the DAC donors, which is not the same thing as saying the amount of aid disbursed by the multilaterals. In reality the two differ for a number of reasons. The figure is boosted by multilateral agencies' own internal resources which come from credit reflows and invested income which they can disburse in addition to the donor contributions (see Box 5.5 on the World Bank and how the International Bank for Reconstruction and Development – IBRD – and IDA are funded). However, there are also internal administrative and research costs which account for some spending. Finally, there is often a time lag between receiving contributions and disbursement. These differences are captured in Figure 5.10.

Figure 5.10 displays the total bilateral and multilateral disbursements of ODA to developing countries for the period 1994–2005 (IDA 2008). As well as underlining the relative volumes of bilateral and multilateral ODA, it also gives a sense of the differences between what the OECD figures show as multilateral aid and what is actually disbursed. In the case of the World Bank's IDA this means US$55 million of DAC contributions compared with US$66 million of net ODA received by developing countries. The IDA is funded by a mixture of DAC donor contributions, repayments from earlier IDA credits, and transfers within the World Bank from the IBRD and IFC.

Figure 5.10 *Funding of ODA and ODA receipts by developing countries 1994–2005.*

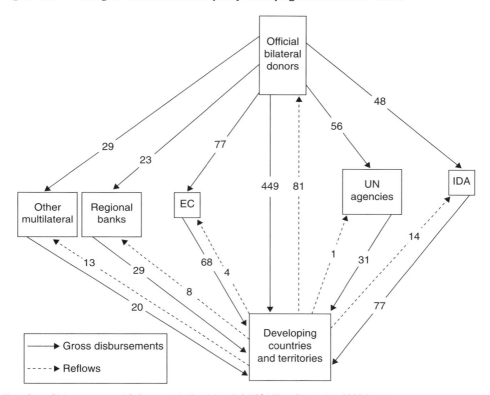

Note: Gross Disbursements and Reflows – excluding debt relief, US$ billion, Cumulative, 2005 Prices.

Source: Adapted from: IDA 2008.

Box 5.5

The World Bank's International Development Association (IDA)

The IDA provides aid to the world's 81 poorest countries. It was established in 1960 and it complements the IBRD – which was the original lending arm of the World Bank, and now operates in and with middle-income developing countries. The IDA, on the other hand, provides loans (or 'credits') which are concessional loans with either zero or very low interest charge and repayments, and have to be repaid over a period of 25 to 40 years, and include a 5- to 10-year grace period at the beginning. The IDA has also been involved in providing debt relief through the Heavily Indebted Poor Countries (HIPC) Initiative and the Multilateral Debt Relief Initiative (MDRI) (see Chapter 6).

To be eligible for IDA funds, countries must be classified as LICs (i.e. have a GNI per capita below that year's threshold, $1,175 in 2012). There are some exceptions to this, such as SIDS which face extra developmental challenges, and some countries which are classified as 'blend' countries, which are able to borrow from both the IBRD and the IDA, for example Pakistan and

China. In addition, countries must lack 'creditworthiness', i.e. not be able to borrow in the international financial markets and therefore require concessional lending. If countries meet these criteria then they are assessed to see 'how well they implement policies that promote economic growth and poverty reduction'. The outcome of this assessment is a diagnostic tool called the Country Policy and Institutional Assessment (CPIA) which has been used by the Bank since the mid-1970s. The CPIA ranks countries across 16 criteria grouped into four clusters: economic management (e.g. fiscal policy, etc.), structural policies (e.g. trade etc.), policies for social inclusion and equity (e.g. social protection), and public sector management and institutions (e.g. property rights and rule-based governance). Each criteria is given a score of 1 to 6 which are then combined to produce a Country Performance Rating (CPR) that is used to inform IDA lending decisions and the conditions and reforms that (in the World Bank's view!) the country needs to make in order to develop.

While the IBRD is self-sustaining – lending to creditworthy countries, getting loans repaid and borrowing in the international financial markets – the IDA is funded through contributions from member governments, plus extra funds from the IBRD and the International Finance Corporation (IFC) which is another arm of the World Bank. These member state commitments are collected through IDA replenishments which cover three-year periods; the last round is IDA16 and lasted until June 2014. Meetings to generate government commitments and set policy goals for IDA17 began in March 2013. The meetings involve the donor governments (over 50 now) and representatives of recipient member countries.

Source: IDA (2013)

Figure 5.11 *The World Bank Group.*

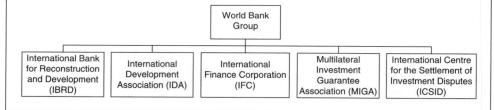

Emerging donors

Traditionally DAC members have provided around 95 per cent of bilateral aid, but many other countries have histories of aid programmes too. Non-DAC donors are beginning to change the face of the donor universe and look set to become ever more important over the coming years. Unfortunately, data and knowledge are both lagging and lacking when it comes to these 'emerging' donors. Emerging donors fall into two categories. First are the non-OECD countries which have voluntarily reported to the DAC. These include the Arab countries along with some 20 other donors including the likes of Poland, Thailand and Israel. Saudi Arabia has long been a significant donor. Taken together (for that is how the data is collected) Arab countries' aid accounted for 13.5 per cent of all aid between 1974–95 (Kragelund 2008).

However, it is the emerging economies, such as China, India, Brazil and South Africa, which are receiving the most attention at the moment and have the greatest potential to change the face of international development (Manning 2006; Kragelund 2008; Woods 2008). Since they have not reported their aid activities to the OECD thus far, it is very difficult to definitively assess their importance in the absence of official records or indeed definitions of what counts as development aid. This has invited, in equal parts, hyperbole, appropriation and perplexity. For example, Hillary Clinton's comments – as the US Secretary of State – about China plundering Africa and 'chequebook diplomacy' reflect a widespread discomfort Western donors have felt with the emergence of China and its very different way of 'doing' aid. The looser Chinese notion of development aid also reflects the non-Western view of aid, as the emerging donors are much happier explicitly blending commercial and investment activities with assistance and have long spoken in terms of 'development partners' rather than donors and recipients (Brautigam 2009).

Nevertheless, there have been a number of heroic efforts at estimating the extent of non-DAC aid. For example, one estimate put non-DAC donors' disbursements at around US$8.5 billion in 2006 (Woods 2008). The Indian Finance Ministry claims that the country distributes more than US$1billion a year (Agrawal 2007). Estimates for China's aid vary hugely, ranging from US$731 million to 8.1 billion! (Kragelund 2008) Despite this uncertainty over the precise figures there is little doubt about the direction of their influence as the two emerging countries vie for resources and soft power, especially in Africa. Chinese aid has been focused on building infrastructure – roads, ports, railways and airports – clothing and textiles factories, and securing access to mineral resources including rare earth elements. Figure 5.12 shows the relative size of selected emerging donors development assistance programmes. By way of comparison it's worth noting that in 2008 Greece's ODA was US$703 million and the UK's totalled US$11,500 million.

Figure 5.12 *Development assistance flows for selected emerging donors.*

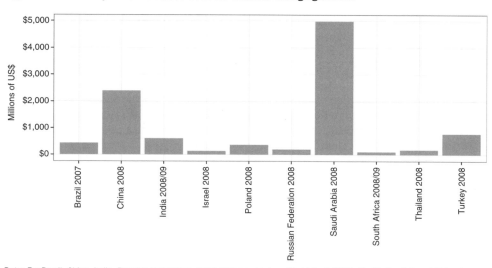

Data: For Brazil, China, India, Russian Federation, South Africa come from World Bank 2011. For Thailand, Israel, Poland, Saudi Arbia, Turkey come from OECD.Stat 2013.

Private aid

The world of private aid, officially outside of ODA, is becoming increasingly important. It includes the traditional development NGOs such as Oxfam, but also now the increasingly important activities of global philanthropists (large and small). Like emerging donors the data on private foreign aid are much more limited and patchy and a good deal less reliable. The Hudson Institute produces a report each year called *The Index of Global Philanthropy and Remittances*. It reports that overseas philanthropy from 23 OECD economies is estimated at $53 billion. Meanwhile another report suggests that philanthropic giving for international development is now equal to one-fifth of ODA (IDS 2011). The rise of so-called 'philanthrocapitalism' is, however, a highly uneven phenomenon and is centred on the US as the major source of private philanthropic activity, as the home of the Gates, Ford, Hewlett Packard, Rockefeller and MacArthur Foundations. Notably, American private philanthropy actually outstrips US official development assistance (Kharas 2007).

However, the US is the exception – philanthropy is much less prevalent in other countries. Furthermore, international development issues constitute the minority of philanthropic giving anyway, being somewhere in the region of 10 per cent of total giving, with the vast majority going to domestic causes (Sulla 2007). And as Michael Edwards (2011), a seasoned observer of these trends, has noted, the Gates Foundation is so large that it has dominated people's attention. But, once Gates is removed from the landscape, 'big' philanthropy isn't nearly as significant as trumpeted; for example, the Rockefeller Foundation's total budget is smaller than that of Oxfam GB. However, activities such as Gates's and Warren Buffet's 'The Giving Pledge', which invites wealthy individuals such as Mark Zuckerman to give the majority of their wealth to philanthropy, aim to change this.

At the other end of the scale are the small global philanthropists: the more typical members of the public who give donations to NGOs and to emergency appeals, such as those organised by the Disasters Emergency Committee (DEC) in the UK, and through NGOs such as Oxfam, Save the Children, Christian Aid, Comic Relief, the Red Cross and so forth. Some of their funding comes from private sources, such as gifts, public donations and trusts; but NGOs often receive large amounts of support from government development agencies. For example, in the UK Save the Children received around 28 per cent of its income from donations and gifts and 54 per cent as institutional grants from national development agencies (such as DFID), and multilaterals such as the UN and the EU. And, importantly, some recent and interesting research on the determinants of NGO allocations have argued that allocation decisions are driven by their 'back donors', i.e. the governments responsible for their grants (Koch 2009). The result is that NGO aid allocations mirror bilateral aid allocation. In fact, NGOs often operate in partnership with official donors. For example, DFID provides funds to MIFUMI, an organisation in Uganda and Tanzania working to protect young women and girls from domestic violence through advocacy and training (APPG DAT 2010).

And finally, in addition to traditional NGOs, new platforms have emerged, such as Kiva.org and GlobalGiving.org, which ostensibly allow for person-to-person loans

Figure 5.13 *Private flows from DAC donor countries.*

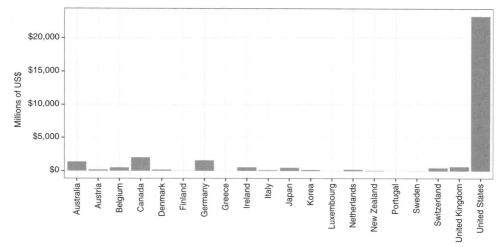

Data: OECD.Stat 2013.

(though see Roodman 2012 for how it really works). Since 2005, when it was started, Kiva has channelled loans to over 720,000 entrepreneurs worth just over US$ $282 million as at the beginning of 2012 (Kiva 2012).

To whom? The geography of aid flows

In 2011 'around 150' countries and territories were recipients of official assistance (OECD/DAC 2013). The current top recipients of bilateral aid are Afghanistan and the Democratic Republic of Congo. Whereas back in 2008 it was Iraq and Afghanistan that had dominated as the top destinations of aid for the past few years – obviously a function of the reconstruction following the Western military interventions in both of these cases – and along with China, Indonesia and India made up the top five recipients, accounting for a whopping 22 per cent of all bilateral ODA. The top five recipients in 2011 were Afghanistan (5 per cent of all ODA), Democratic Republic of Congo (3 per cent), Ethiopia (3 per cent), Vietnam (3 per cent) and Pakistan (3 per cent).

It is well established that larger countries tend to receive more aid in absolute terms, but looked at as aid per capita, larger countries receive less, i.e. less aid per person. Consider the figure below showing the top 10 recipients by total aid and per capita, and their position on the alternative measure. Many of the bottom 10 in terms of total aid are in fact receiving the most aid in terms of per person, for example Sao Tome and Principe and other small island states. Roughly a third of global ODA goes to sub-Saharan Africa. But this still came to slightly less than $35 per African in 2008. Outside Africa it works out at about $10 per head in all developing countries.

Figure 5.14 *ODA to countries by income group over time.*

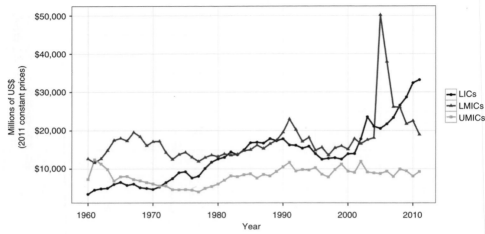

Data: OECD.Stat 2013.

Figure 5.15 *ODA to countries by region over time.*

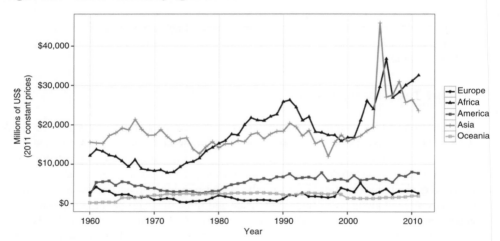

Data: OECD. Stat 2013.

The broader patterns of bilateral ODA in 2011 are that roughly: one-third went to the LICs (36 per cent), with a quarter going to the LMICs (23 per cent), and around 9 per cent going to the UMICs (a further third is indicated as unallocated) (OECD.Stat 2013). In terms of the regional distribution, around 38 per cent of all bilateral aid is accounted for by Africa, with around 28 per cent going to Asia and Oceania (see Figure 5.15). Sectorally, the majority of aid goes into the education, health and population sectors, followed by other social infrastructure and economic infrastructure.

Figure 5.16 *Largest aid recipients by total and per capita.*

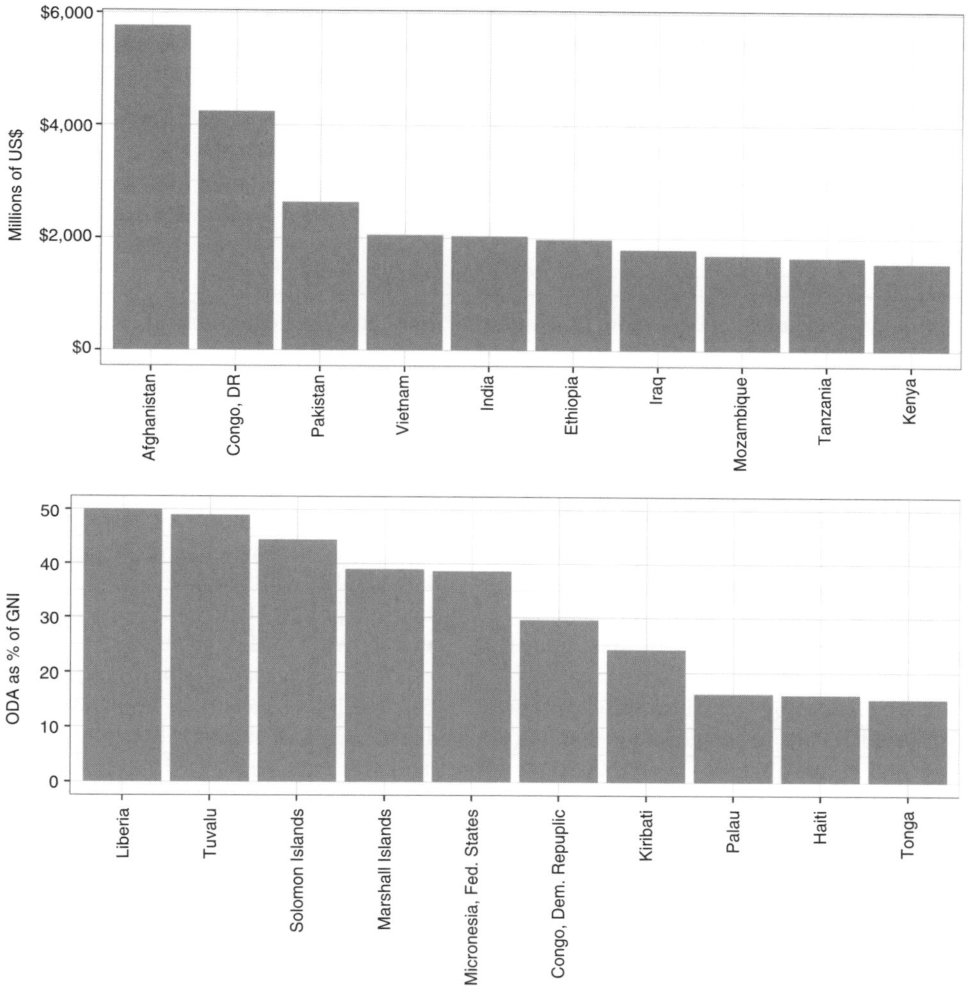

Data: OECD.Stat 2013.

Types of aid

Aid modalities

As well as distinguishing between types of donors, the other crucial specification is the form which the aid takes. In the technical parlance the different types of aid are known as 'aid modalities'. The common categories of aid modalities or aid instruments are project aid, two types of programme aid (general budget support, sector budget

Figure 5.17 *Aid modalities.*

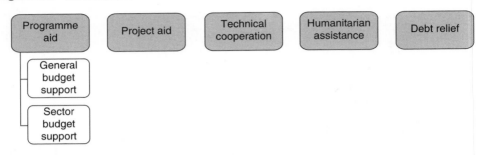

support), technical cooperation, humanitarian assistance, and debt relief (Foster and Leavy 2001).

First, project-type interventions are the type of aid which is most commonly associated with development assistance. Project aid is used to fund a specific and defined activity, for a determined period of time, with discrete objectives. For example, subsidising seeds and fertilisers to increase food security, funding an education campaign on safe drinking water and guinea worm to improve health outcomes, or legal or budgetary training for civil servants to improve governance. Project aid tends to come with its own reporting, management and budgetary systems which work in parallel with the developing country's institutions and bureaucracy. This means that projects tend to be donor-led in terms of the objectives and expenditures and generally work outside the recipient country's systems. The logic is that this improves efficiency, especially in countries characterised by weak institutions or bureaucratic capacity, or such other limitations. The advantage of project aid is that it is targeted, and therefore more likely to have visible results that can be monitored. This also increases the potential for accountability.

To make project aid work it needs to be aligned with national development strategies and coordinated. With a proliferation of projects there is the problem of duplication and even contradictions between individual projects (with each other and/or with national development objectives) as a result of a lack of coordination between donors. The proliferation of different reporting and administrative systems creates further complications and reduces efficiency (Riddell 2007; APPG DAT 2010). Moreover, because project funding does not enter the financial systems of recipient countries, beyond its immediate objectives it does little to build capacity, encourage ownership and catalyse sustainable change as well as ignoring and bypassing local knowledge and expertise. Instead, some have argued, a more sustainable approach is to build domestic capacity and country systems, such as budget systems, to solve the problems which lead to a country ownership deficit (APPG DAT 2010). And, despite all this attempt at control, there is no evidence to suggest that project aid suffers less from corruption than programme aid (Fritz and Kolstad 2008).

Programme aid, or budget support, is at the other end of the spectrum from project aid and is where large allocations of aid, earmarked for poverty reduction and development, are deposited directly into the recipient governments' budget as foreign exchange. Unlike project aid, programme aid is not linked to specific developmental activities and is instead just budget and balance-of-payments support. It's often the case that funds from a number of donors are pooled with the government's own funds. The recipient government is then free to use the finance to fund the purchase of capital goods, commodities, or to carry out macroeconomic structural or policy reforms. This means that the country is using its own systems to allocate, account, audit and report the capital flow. For example, the EU has provided Burkina Faso with extensive budget support to allow the government to proceed with governance reforms designed to strengthen the rule of law, political competition, decentralisation and internal security – reforms that would normally have required a government to take on debt to pay for.

There has been an increase in the popularity of programme aid, partly because following large-scale debt relief in the mid-2000s, donors want to work out how to keep channelling aid straight to governments. The advantages of programme aid are that it encourages administrative and governance capacity-building, it reduces transaction costs by reducing the fragmentation of donor flows, increases country ownership and alignment of aid with developing countries' development strategies, as well as providing the basis for a stronger accountability between citizens and their government. This last point is worth emphasising. As Jasmine Burnley (Oxfam 2010: 19) elaborates:

> Budget support builds the citizen-state relationship by giving the citizen[s] resources like schools and hospitals for which they can hold the state accountable. It helps this relationship to function by improving transparency. Because it pays for vital human capital such as teachers, health workers, and police employed by the state, it has the added value of putting taxes back into the system.

There are concerns that budget support may increase aid volatility as donors find it easier to stop the support abruptly. Plus there is a long-standing concern with aid's fungibility – i.e. the fact that aid can be used for purposes other than those intended by the donors (McGillivray and Morrissey 2004). And by implication, it is harder for donors to ensure that the money is spent on poverty reduction. One response has been the continuing use of conditions, such as the earmarking of budget funds and setting out medium-term budget priorities. Another response has been to use sector budget support (SBS), which earmarks aid for specific sectors.

SBS is a variation of programme aid (or general budget support). It revolves around an agreed set of objectives, which may also include specific 'basket funding' targeted at more specific types of expenditures – for example, on primary health care within the health sector (Foster and Leavy 2001). The philosophy is to build capacity jointly before graduating to developing country systems. SBS is often combined with a sector-wide approach (SWAp).

SWAps are partnerships between donors and the developing country government where a range of aid modalities, from budget support to projects, are used. Under these arrangements all funding in the sector – no matter which modality – is guided by single policy (Brown *et al.* 2001), the idea being that this helps to reduce the overlap from different donors and types of delivery, but allows a useful mix of project and

Figure 5.18 *Aid modalities in 2010.*

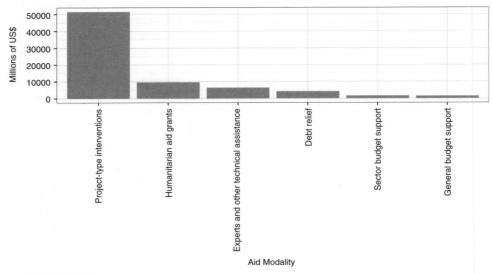

Data: OECD.Stat 2013.

programme aid. SWAps have been commonly utilised in the health, education, water and agriculture sectors.

Technical cooperation or assistance is the provision of expertise in the form of training, research, personnel and associated supplies and costs. For example, this could be providing scholarships for students; covering the costs of sending experts, teachers or volunteers; and training and teaching materials (OECD DCD/DAC 2007a). Critics charge that technical cooperation tends to ignore local expertise, is expensive, relying on Western consultants, and there is a lack of accountability to local and national development goals. Nevertheless, capacity development often demands some type of support and know-how.

Humanitarian aid is also classified as ODA. It has a much more specific and short-term aim than the bulk of ODA. It is designed to have an immediate impact, to save lives in an emergency, such as after a disaster, or be given to refugees in developing countries in order to provide relief food (often wheat and milk products), shelter, water and sanitation, and other supplies such as blankets, tents and medicines (OECD DCD/DAC 2007a, 2007b).

Finally, the other key aid modality is debt forgiveness. It is worth flagging up here that debt relief *is* counted as aid. Therefore, without an increase in the overall aid budget, when a country provides debt relief it comes at the expense of other projects or budget support. The history of indebtedness and the moves by the development community towards resolving the problem of external indebtedness will be dealt with in more detail in Chapter 6.

As is apparent, there are a range of different ways to deliver aid in terms of how it is disbursed, and the extent to which it's earmarked or has conditions attached. Understanding these differences is important for considering the reasons why development agencies provide aid and how effective aid is in promoting growth, reducing poverty or perpetuating dependence. Also note that, while these are presented as pure types, in reality many hybrid forms of aid instruments or modalities are used (Foster and Leavy 2001).

Why do states give aid? The aid allocation debate

The previous section described how much and what types of aid are disbursed by which donors and to which developing countries. But how do we explain the pattern of aid flows? There is a well-developed debate in the academic and policy literature with respect to why donors give and to whom. But, as Carol Lancaster (2007) has neatly put it, the question 'Why do states give aid?' is actually three separate questions: (1) Why do states give aid at all?, (2) Why do donors give aid to the countries they do and use the modalities they do?, and finally (3) How do donors come to their decision to disburse aid as they do? We consider each of them in turn.

Why give aid?

As the famous realist Hans Morgenthau (1960) put it, 'of the seeming and real innovation which the modern age has introduced into the practice of foreign policy, none has proven more baffling to both understanding and action than foreign aid'. If one considers that international aid is the voluntary transfer of resources from wealthy to poor countries then, given the normal behaviour of nation-states – self-interested and jealously guarded of their sovereignty and position within the world – the very existence of the aid regime is somewhat curious. But, as Lumsdaine (1993: 29) argues, 'foreign aid cannot be explained on the basis of the economic and political interests of the donor countries alone, and any satisfactory explanation must give a central place to the influence of humanitarian and egalitarian convictions upon aid donors'. Lumsdaine argues that foreign aid is driven by the same 'moral vision' and sense of justice that underpins the development of domestic welfare states. Indeed, the countries that give the most also have the strongest domestic commitment to welfare and redistribution (Noel and Therien 1995). The argument, that domestic politics and values drive aid allocation decisions, has most recently and systematically been addressed by Maurits van der Veen (2011). Van de Veen shows how Norway's aid tends to be driven by humanitarian convictions, but Belgium's, for example, is driven by a sense of obligation to its former colonies. So moral vision is important, but it does vary in its precise form across different countries.

The modern foreign aid regime, now roughly 60 years old, is commonly traced back to Point Four of Truman's (1949) Inaugural Address:

> Fourth, we must embark on a bold new program for making the benefits of our scientific advances and industrial progress available for the improvement and growth of underdeveloped areas . . . More than

half the people of the world are living in conditions approaching misery. Their food is inadequate. They are victims of disease. Their economic life is primitive and stagnant. Their poverty is a handicap and a threat both to them and to more prosperous areas.

However, there were precursors to Truman, such as Britain's relationship with its colonies. Until the 1920s no financial assistance went to the colonies except during emergencies, but this changed with Colonial Development Act 1929, which created a legal responsibility for providing financial assistance to colonies. The Colonial Development Fund was also established. Notably, while the Fund was designed to support agriculture, trade and industry in the colonies it was, crucially, also aimed to support commerce and employment in the UK (Barder 2005a).

The above discussion neatly encapsulates the major drivers behind the modern aid regime. It was developed by national governments. It was ostensibly based on the case for transferring large amounts of resources in the form of expertise and material resources from the wealthy to poor nations in the name of solidarity and justice. But it has also always been informed by a more or less enlightened or narrow sense of self-interest. As Truman put it, 'their poverty' is a 'threat' to 'more prosperous areas' (Truman 1949). In the political climate of the Cold War it was recognised that US security concerns against the spread of Communism were well served by channelling aid into the rest of the world. The notion of enlightened self-interest is a popular argument, especially among the major aid donors: states do well by doing good. This view chimes with a liberal tradition of seeing the world as a positive sum game. The act of providing aid and spurring development will foster a collectively wealthier world (Smith 1776). More recently, similar win-win arguments have been made with respect to the security of the wealthy countries. The logic being that aid offers a rational strategy to address the 'root causes' of threats to national security (which are assumed to be in the developing world) rather than their 'manifestations', such as immigration, civil conflict and terrorism. Consider the introduction to the US Development Agency's logic for international aid in Box 5.6. But, there are also commercial and economic benefits from giving aid to developing countries, especially if the benefits of that aid are likely or guaranteed to return through increased trade or through tied forms of aid.

Political scientists and scholars of international relations have long underlined the overriding importance of power politics and the self-interested logic for giving aid (Morgenthau 1962; Packenham 1966; Hayter 1971; McKinlay and Little 1979; Hook 1995; Macrae and Leader 2001; Thérien 2002; Headey 2008). The received wisdom is that donors tend to give aid to political allies to lend material support and/or to cement relationships. For example, Israel and Egypt are famously the largest beneficiaries of the US's aid programme. Between them they account for some 40 per cent of US bilateral aid and yet neither of them are LICs. As such, in contrast to Lumsdaine, critics suggest that moral rhetoric in foreign policy is purely instrumental and little more than a cover to continue with business as usual (Chandler 2003).

Box 5.6

Foreign aid in the US national interest

The new century has brought new threats to US security and new challenges and opportunities for the national interest. The terrorist attacks of September 11 tragically demonstrated the character of today's world. Globalization has sent unprecedented flows of people, ideas, goods, and services across borders, fostering growth and expanding democracy. More than ever, US security is bound up with the outside world. And as the world has become more connected, it has become more hazardous. Weapons, germs, drugs, envy, and hate cross borders at accelerating rates. Just as the tools, ideas, and resources for progress can quickly move from industrial to developing countries, many forms of risk and instability can travel in the opposite direction. When development and governance fail in a country, the consequences engulf entire regions and leap around the world. Terrorism, political violence, civil wars, organized crime, drug trafficking, human trafficking, infectious diseases, environmental crises, refugee flows, and mass migration cascade across the borders of weak states more destructively than ever before. They endanger the security and well-being of all Americans, not just those traveling abroad. Indeed, these unconventional threats may pose the greatest challenge to the national interest in coming decades.

Source: USAID (2003: 1)

Who gives aid to whom?

The second question is, given these broad moral, political, commercial reasons, how *do* donors allocate aid? To examine this further a simple and useful way of dividing up the rationale for giving aid is to compare donor-driven rationales with recipient need. Does aid go to countries based on need – i.e. high levels of poverty? Or, does the distribution of aid tend to reflect the interests of the donor country, whether those be security, economic, political or commercial interests? Or, a third possibility is whether aid is distributed according to 'merit', i.e. to those countries with a proven track record or the potential to utilise aid effectively?

First of all, geography matters. As demonstrated above, smaller countries tend to receive more aid per head of population than larger countries. It has been argued that this reflects countries' strategic aim of influencing as many countries as possible, meaning that the amount of aid needed to 'purchase' influence tends to result in smaller countries receiving disproportionate amounts of aid (Radelet 2006). It is also apparent that countries tend to give aid to geographically proximate developing countries. For example, New Zealand, Australia and Japan give more aid to Asia and Oceania than any other region (OECD 2010a).

Second, history matters. Former colonies tend to receive more aid (Alesina and Dollar 2000). For example, much more of Spain's aid goes to Latin American and Caribbean countries than to any other region (OECD 2010a). Again, this could be interpreted crudely as a tool of realpolitik. But there are also cultural reasons to support former colonies, as the example of France and French-speaking Africa demonstrates. The notion and institutions of 'Francophonie' bind French-speaking countries and imply ties of solidarity and fraternity as well as being the vehicle for France to promote the French language and culture and to maintain its own great power status. It is also the case that former colonial powers feel that they owe a debt, a moral obligation, to their former colonies. There are also good domestic reasons for politicians to show support for ex-colonies because many constituents still strongly identify with their 'mother country' – for example, the Caribbean, Somali or Pakistani diasporas in the UK.

Third, how much does geopolitics matter? Most systematic studies of the allocation of aid find that geopolitical ties tend to outweigh the importance of developmental motivations. Alesina and Dollar (2000) find that there is a significant relationship between those states that vote together in the UN and the distribution of aid. They also confirm that donors give more to ex-colonies.

However, Hoeffler and Outram (2011) come to a different conclusion. They look at the aid allocation decisions of the 22 DAC donors for the 25 years between 1980 and 2004 to see how much donor self-interest, recipient need and recipient merit explain the distribution of aid. Recipient need is measured in two ways, (1) the level of national development as income per capita and (2) how much aid (per capita) is received from other donors. Recipient merit captures the extent to which recipient countries have been able to demonstrate successful economic growth, more democratic regimes and fewer human rights violations. Finally, donor self-interest is measured as the extent to which the donor and recipient share trading links and vote together in the UN as a proxy for geopolitical allegiance.

Hoeffler and Outram's (2011) findings suggest that recipient need matters – donors *do* allocate more aid to poor countries. Recipient merit also matters, but it varies among donors. So, for example, the UK rewards economic success and democracy. France, Germany and Japan reward good human rights records. In contrast, the US actually gives more aid to countries with poorer human rights records. Finally, Hoeffler and Outram confirm that donors do tend to give more aid to countries with which they share trading relationships and which vote in line with in the UN. However, they argue that previous studies have overstated the importance of self-interest. They find that donor self-interest explains about 16 per cent of the variation in aid allocation, whereas recipient needs account for about 36 per cent. Meanwhile, recipient merit only accounted for 2 per cent of predicted aid per capita. This is important for a couple of reasons. One, it suggests that recipient countries cannot attract much more aid by demonstrating successful economic growth, becoming more democratic or protecting human rights. Two, it suggests donors' aid decisions have not (yet) been informed by liberal institutionalist insights, which have dominated the more recent aid effectiveness literature (see the next section on aid effectiveness).

Clearly, political and bureaucratic influences on aid allocation continue to matter. This takes us to Carol Lancaster's final question. For while studies such as Alesina and Dollar (2000) and Hoeffler and Outram (2011) are immensely useful for highlighting aggregate tendencies and donor patterns over time, they don't get *inside* the aid allocation process and explain why donors choose to allocate aid as they do.

How do they decide?

The reasons why Japan gives aid are different to why the US gives aid and are different to why Denmark does. That much is already clear. But the reality is that, like all aspects of politics, aid allocation decisions are usually pretty messy and contested. There is not only a mixture of reasons why a particular state or government gives aid, but these reasons vary over time and across ministries, departments and even teams. Plus, motivations are rarely singular: aid can fulfil a number of different agendas at the same time.

For example, within the formal decision-making procedures of government the executive may identify desired budgetary resources, but it is parliament which passes laws and signs off budgets, and then often a specialist development ministry or agency is in charge of delivering the aid. Not to mention the role of the media, civil society and the public in shaping parliamentarians' and governments' preferences. The different actors at work within the state do not necessarily share common objectives. And if they do, it is not necessarily the case that they share the same diagnosis of the problems and therefore the optimal way of acting to support development.

Theories of foreign policymaking point to competing organisational and bureaucratic motivations (Allison 1979). Different agencies pull in different directions trying to maximise their budgets and prestige with government. And then, within organisations, government and aid agencies, there are individuals who are not only trying to fulfil developmental objectives, but personal and professional ones too, such as promotion, or consultants trying to secure their next contract. All of which suggests that it would be wrong to read off foreign aid allocations from assumptions of a rational and unified national interest. The agonising – political, technical, and public – over whether India should remain a recipient of UK aid provides a case in point (see Box 5.7).

Box 5.7

Should India receive aid?

On the 1 March 2011, the UK Department for International Development announced the results of its Bilateral Aid Review (BAR), setting out which countries will receive UK aid for the next four years. The stated strategy behind the BAR was to reduce the number of countries the UK delivers aid to in an effort increase 'value for money' by a more efficient focusing of UK support. There were a number of controversial decisions, for example the decision to close the aid programme in Burundi and, on the other hand, continue giving aid to India.

The decision whether or not to continue giving aid to India was one which was discussed at length in the media and in the blogosphere prior to the announcement. The common refrain includes some or all of the following issues: India is growing at 8.5 per cent a year and is now the eleventh largest economy in the world, it now boasts more than 126,000 US dollar millionaires, it spends $31.5 billion on its defence budget and has launched a nuclear submarine, it is one of only six nations with satellite launch capability having a $1.25 billion space programme with a moon mission, it has over $300 billion in foreign exchange reserves, plus it has an aid programme of its own estimated at in excess of $550 million giving aid to the likes of Burma, Afghanistan and various countries in sub-Saharan Africa (as reported in Barker *et al.* 2011; Bunting 2011; Ford 2011; Lamont and

Barker 2010; Sumner 2011). And recall that this debate was during a period with a new Conservative-led coalition against a backdrop of large budget cuts and austerity elsewhere. All of which is to say that many in the media, public and indeed the government thought that the UK shouldn't be giving development aid to India.

Andrew Mitchell, the International Development Secretary, had to defend his decision. The operational plan was only released after a considerable delay in October, but it confirmed that the government announced that it would maintain the programme, spending £1.2bn in the following five years. The UK government has, for years, sold the aid programme to the public on the grounds that having a strong aid programme is both the right thing to do and in the UK's national interest. The decision to continue to give aid to India was done exactly thus (DFID 2011). As Andy Sumner noted in giving evidence to the International Development Committee (cited in Ford 2011):

> For development bods at DfID and in the UK NGOs, it's about supporting the poor. For diplomats and the Foreign Office, aid is part of bilateral ties, security and intelligence sharing. Most importantly, the balance of power has swung towards India, and the UK wouldn't want to risk a long-standing friendship with one of the world's next big powers.

In terms of development, the argument makes sense. The point that poverty and inequality are not guaranteed by growth is illustrated graphically by India. India's health indicators, especially among women and children, are among the worst in the world. Half of all Indian children are malnourished. Eight Indian states alone have more poor people than the 26 poorest African countries combined (Sumner 2011). Moreover, this poverty is concentrated in just four Indian states, which account for one-fifth of the world's poor. In one of the poorest, Bihar, only one in four people has access to a toilet (Bunting 2011; Ford 2011).

So, as Mitchell noted in an interview with the *Financial Times*, 'India has more poor people in it than the whole of sub-Saharan Africa. If you're going to achieve the [UN] Millennium Development Goals, you have to make big progress in India' (Barker et al. 2011). DFID highlighted how the aid programme was designed to eliminate poverty (rather than drive economic growth), and as such it would be working in the poorest states; spending two-thirds of its annual aid on three of the poorest states, Bihar, Madhya Pradesh and Orissa (Mitchell 2011). Andy Sumner (2011) notes that the Indian state system is decentralised, meaning that simply pointing to national wealth is to miss the point. Furthermore, the aid would be particularly targeting health and education interventions for women and children. And finally DFID would be putting more emphasis on working with the private sector (DFID 2011; Mitchell 2011). The decision to continue providing aid to India reflected the shift in seeing development as being about poor people rather than poor countries (Bidwai 2011). This observation has normative implications, but also very practical implications about to whom and why aid should be given.

In terms of the UK's national interest, as Sumner suggests, the aid programme is seen as one diplomatic tool in the toolbox. There have been reports that the aid programme was intended as part of a strategy to sell the Indian government the Eurofighter Typhoon. It didn't (Bidwai 2011; Gilligan 2012). In addition there is the colonial history which has meant – through a mixture of path dependency, liberal guilt, the Commonwealth and a sense of seeking reparatory justice – that India has been the single largest recipient of UK aid.

But in a twist in the tale, the Indian government was apparently set to reject the aid. At only 0.1–0.2 per cent of India's GNI, UK aid is a small investment relative to the size of the economy and the government's budget (Sumner 2011). But it was reported that the UK 'begged' India to take the money arguing that cancelling the programme would cause 'grave political embarrassment' to Britain (Gilligan 2012). This was after a visit to India at the end of 2010 by David Cameron where India's finance minister, Pranab Mukherjee, upon hearing about the forthcoming BAR and the possibility of the UK cutting its aid to India, described the UK's aid to India as 'peanuts' as a proportion of the country's total GDP (Lamont and Barker 2010).

The point here is that as a rising power, with a colonial history yoked to Britain, the debate on UK aid to India is just as contested in India as it is in the UK. Only vanity sees it as a UK debate. The decision to accept aid reflects the power relationship behind the tradition of aid as a transaction between a powerful giver and grateful recipient and the questioning of this reflects the growing confidence and power of the emerging donors.

In November 2012 it was announced that UK aid to India would be stopped by 2015. After this date cooperation would revolve around 'skillsharing' and trade and investment.

Carol Lancaster's (2007) book is an excellent study of *why* states allocate aid as they do. Through a series of case studies (the US, Japan, France, Germany and Denmark) she shows how four sets of key variables at the domestic level matter in determining outcomes. First, the types of ideas which are held about the world, ideas about the role of aid and ideas about what works. Second, the design of political institutions, specifically electoral rules and parliamentary versus presidential systems. Third, the access which private and commercial interests, such as agriculture and industry, and NGOs, diasporas and think-tanks have to the decision-making process. And finally, governmental organisation, i.e. bureaucratic hierarchy and the extent to which the decision-making process is concentrated in a single agency and whether it has cabinet representation and therefore influence. The particular political organisation of each country influences how decisions are made about aid. So, for example, in Denmark the advisory committee within parliament is extremely influential, whereas in France and the US its equivalent is much less powerful because of the legislative rules of the game.

An example of how and when ideas matter is illustrated by the debate about what it means to be a poor country. The World Bank's decision on which countries are eligible for soft loans from the IDA is a function of whether they fall into the low-income bracket. However, the number of countries who are classified as LICs is falling rapidly, down to 35 in 2011 from 63 in 2000 (Kenny and Sumner 2011). The established way of thinking about where aid needed to go was to the 'bottom billion' – the population of some 50 to 60 countries which were stuck in a series of development traps (Collier 2007). Thus the standard role of aid is to break these poverty traps. Yet a recent paper by Andy Sumner (2010) challenges this interpretation of the world, highlighting the fact that roughly three-quarters of the world's 1.3 billion poorest actually live in MICs; with 380 million living in LICs, mainly in Africa. The ramifications of this different view of the world are huge. To whom should donors give aid? To LICs to fill their savings-investment gaps and break the poverty trap, or to MICs who tend to have high levels of growth but still have high numbers of poor people? See Box 5.7 for this discussion in the context of UK aid to India. These are not straightforward technical questions, but are resolved politically.

An exploration of how domestic interests feed into foreign aid decision-making is provided by Milner and Tingley (2010). To do so they examine the behaviour of legislators in the US House of Representatives to see how they vote and offer amendments to US foreign aid bills over the period 1985–2003. They find that foreign aid policy is not only driven by US foreign policy interests as set by the executive, but

Table 5.1 *Levels of tied aid for selected major donors in 2007*

Country	Bilateral ODA (US$ million)	Untied Aid % of donor ODA
Australia	1,710	96
Belgium	1,587	92
Finland	661	85
Germany	9,644	73
Greece	248	13
Italy	1,440	52
Japan	12,503	75
Netherlands	4,800	78
Norway	2,898	100
UK	5,712	100
USA	24,724	63

Source: Data from Clay *et al.* 2009: 10

is also shaped by domestic support as spending decisions must be able to win Congressional approval. They find that districts with a greater proportion of low-paid and low-skilled employment and on the political right tend to oppose economic foreign aid.

It is clear that different domestic interests have to be balanced or brought on board to support aid. Commercial pressures also determine how much, to whom, and how aid is distributed. Traditionally this was done through tying aid, which is the practice of giving aid to a developing county which then *has* to be used to purchase goods and services from the donor country. Historically around 75 per cent of US aid has been tied, 70 per cent of Greek aid, and 40 per cent of Canadian and Austrian aid. On the other hand, Ireland, Norway and the UK do not tie any of their aid (Radelet 2006). Tying aid can help support certain industries and sectors as well as shore up public and political support for giving aid. Thus, the decision to tie aid reflects commercial lobbying and the political calculus of leaders about meeting broader domestic goals including employment. Yet it is widely recognised that tied aid is less efficient; recipient countries do not receive value for money because they cannot put the resources out to competitive tender. Tied aid has long been an issue the development community has tried to address. As far back as 1965 the DAC issued recommendations on 'Measures Related to Aid Tying'.

Impact: does aid work?

While the debates about who gives how much to whom are important, they are only half the story. What are the consequences of aid flows, do they tend to reduce poverty or make it worse? As some have argued, there can be a perverse obsession among donors in emphasising the amounts of dollars disbursed as though in and of itself it's an adequate measure of their commitment to development (Calderisi 2006; Easterly 2006). It is certainly easier for a donor to point to the number of dollars committed than the more tricky but significant issue of outcomes and results. Though with the emergence of the results-based agenda this is changing. The aid effectiveness debate is largely an empirical and econometric one, nevertheless the different approaches to global finance and development are significant in terms of the questions that are asked, the interpretation of results, appropriate methods and so forth. Liberal institutionalists have tended to dominate the debate in recent years, arguing that aid's success is contingent on good domestic institutions and policies. Whereas, prior to the 1990s and 2000s, the debate was dominated by neoliberals. Meanwhile, radicals have maintained a consistent but marginalised line on aid as a form of imperialism. Finally, critical reformists have tended to have the least to say about aid in recent times.

Dambiso Moyo (2009: 35) recently and provocatively highlighted that approximately US$1 trillion of aid has gone to Africa since the 1940s which, as she puts it, is the equivalent of giving everyone in the world $1,000. Moyo not only suggests this has been wasted – that aid to Africa has failed to produce growth – but that it has actually impoverished the continent. Moyo suggests that donors should cut off aid after five years and focus instead on raising investment through bond issues, FDI and microfinance. This is a popular and populist orthodox critique of aid. But is it rigorous and well founded? Critics of Moyo note that it's based on a logical fallacy (Barder 2009). Consider the following: would one conclude that hospitals are bad for people's health because that's where sick people are or that 'fire engines cause fires because you find them near burning houses' (Watkins 2009)? In basic terms this is the difficulty all assessments of aid effectiveness need to deal with.

The debates over the effectiveness of aid are long-standing. Early studies of the aid-growth relationship – although tentative and acknowledging their weaknesses such as limited data – concluded that there was a largely positive relationship between those countries which grew and those which received aid (Cassen and Associates 1986). However, over time these conclusions began to be challenged. The continuing failure of poor countries to grow and political shifts to the right against redistributive policies in key donor states fuelled the neoliberal critique. Many subsequent studies cast doubt on the apparent link between aid and development, instead concluding that aid sustained (wasteful) consumption but not investment and growth (Boon 1996).

Aid doesn't work and is possibly harmful: the neoliberal critique

The position that aid doesn't work is actually more usually the position that aid is positively harmful and produces negative outcomes. That is to say, studies that show the effect of aid is neutral and that it has no positive effect are taken as evidence that aid should be stopped, especially with the public and media (Riddell 2007). At its most simple and benign, this is an argument that the money and resources spent on aid are simply wasted; as such there is an opportunity cost to the aid and the same resources could have been spent better elsewhere, perhaps building a new hospital in the donor country or reducing taxes or the government deficit. But the ostensibly neutral position more commonly slides into a stronger statement that aid is downright harmful. Orthodox critics of aid, such as Milton Friedman, Peter Bauer and William Easterly argue that aid has sustained bad governments and created dependence; has been used to increase the personal fortunes of the elites, squirreled away in Swiss bank accounts; and has helped expand inefficient government bureaucracy and foster corruption more generally. There are a number of classic arguments offered for why aid fails.

First of all, aid is fungible. That is to say, since it is money it is easily substitutable into something else, meaning the aid given to developing country governments doesn't necessarily get spent on items which promote development. It can be used to reduce taxes or to fund military expenditure. Thus, aid dependence is often associated with rent-seeking and corruption (Knack 2001). But the real critique here is that because aid can substitute for tax revenues it creates a dependency on aid and helps to sustain bad governments and leaders who would otherwise not survive. Because foreign aid effectively sustains a government it reduces the domestic accountability of the government to its citizens because it becomes a substitute for domestic taxation (Bauer 1971; Moss *et al.* 2008). Thus governments have no incentive to reform inefficient institutions and policies (Bräutigam and Knack 2004).

Second, probably the most common criticism levelled at aid is the issue of corruption, defined as 'the abuse of public office for private gain' (Kaufmann 1997). It is easy for critics to make their point by, say, pointing to the Marcos regime in the Philippines or the Duvalier regime in Haiti where billions were embezzled into private accounts. Further examples abound. But the high-profile cases of 'grand larceny' are only part of the story. Patrimonial rule and the favouring of special interests undermines development and institutionalises unfairness (Myrdal 1989; Shleifer and Vishny 1993). In a context of weak institutions, poverty and low-paid civil servants, the everyday use of bribes can become endemic, increasing the inefficiency of getting things done and undermining trust in public institutions. Aid dependency insulates the government and leaders from these problems and there is some evidence that aid inflows actually fuel corruption as elites fight over the incoming funding (Svensson 2000).

Third, critics also highlight bottlenecks in developing countries which mean that there is a limit to how much aid can be used before it is wasted; countries are said to have an 'absorptive capacity'. Studies have estimated that, on average, the maximum growth rate is around the point at which aid hits 16–18 per cent of a country's income (Radelet *et al.* 2005). The liberal institutionalist response to these concerns is that countries' absorptive capacity varies in line with the strength of domestic institutions, infrastructure and the skills and welfare of the population. As such, increases in aid should not be limited by the existence of bottlenecks but should be targeted at relieving these bottlenecks, i.e. aid being used to improve governance, infrastructure and education, skills, human capital and bureaucratic capacity

(Sachs 2005; UN 2005). One example of this 'catalytic' view of aid is in relation to trade reforms (see Box 5.8).

Fourth, orthodox critics further argue that, even discounting deliberate abuse through corruption, aid creates economic distortions and undermines the effective price signalling mechanism of the market. It is argued that aid 'crowds out' private capital and swells inefficient bureaucracies. Ultimately this damages long-term growth because it draws investment and workers away from other productive sectors, in particular the export sector.

Fifth, at the macroeconomic level, critics highlight the so-called Dutch disease or resource curse (Younger 1992; Bulir and Lane 2002; DFID 2002; Prati *et al.* 2003; Adam 2005; Rajan and Subramanian 2005). It is called the Dutch disease after what happened to the Dutch economy after the discovery of natural gas in the North Sea in 1959 (*The Economist* 1977). Instead of bringing in new revenues and acting as an economic boost it helped to deindustrialise the Dutch economy. There is an assumed equivalence of 'any development that results in a large inflow of foreign currency' regardless of the source (Ebrahim-Zadeh 2003). Thus natural gas, oil or aid all result in a large increase in money. This means, in the short term, an increased amount of money is chasing the same amount of goods and services. The result is inflation. This in turn causes an appreciation in the real exchange rate, which makes exports relatively more expensive for foreigners and reduces export earnings. Because exports are relatively more expensive this reduces the country's competitiveness and undermines local manufacturing. The theory is clear enough, but the evidence is more mixed. Rajan and Subramanian (2005) have presented evidence of aid leading to overvaluation in the labour-intensive export sector, chiefly through the increase in the wages of skilled labour, and thus a loss in competitiveness. On the other hand Gupta *et al.* (2006) acknowledge that Dutch disease is a theoretical possibility, but find no evidence that it happens in practice.

Box 5.8

Aid for trade

Trade is a crucial source of income for developing countries, and is one of the key sources of development financing outlined in the Monterrey Consensus. While trade issues are outside the remit of this book, there is a role for development financing. The links between aid liberalisation and poverty reduction are contested and complex. Yet, there is a common recognition that for trade to be an engine for development, developing countries need to strengthen their state capacity in order to effectively integrate into the international trading system. Without strengthened capacity, for example effective policies, institutions and infrastructure, it is impossible to compete with more developed countries in international markets. This is strongly indebted to a liberal institutionalist perspective. It is, of course, in parallel with the need for developed countries to open up their markets and provide better access for developing countries, especially in agricultural sectors (a common critical reformist refrain). The fact that many developing countries depend upon a single crop, such as coffee (Ethiopia, Burundi), sugar (St Kitts Nevis, Cuba), bananas (St Lucia, Dominica), cocoa (Sao Tome and Principe, Côte d'Ivoire), cotton (Benin, Burkina Faso) or cashew nuts (Guinea-Bissau), is to a large extent due to colonialism. Monocropping was the result of colonial powers using the colony to extract physical resources and specialise in a single cash-crop; the profits and goods were repatriated back to the core with the rest of the economy being neglected. This global system has been perpetuated by the difficulty of diversifying and the way in which the global trading rules sustain protection for developed markets (Chang 2003; Pogge 2008; Wade 2010).

Nevertheless, there is the recognition that developing countries would be better off diversifying their export sectors and earnings away from single industries and especially ones that are prone to volatile or long-term depressed prices. Hence, in 2005, WTO members launched the Aid-for-Trade (AfT) Initiative at the 2005 Hong Kong WTO Ministerial Conference. There are a number of key functions that AfT fulfils:

- Technical assistance: helping countries to develop trade strategies, negotiate more effectively, and implement changes.

- Infrastructure assistance: building the roads, ports, and telecommunications that link domestic and global markets.

- Productive capacity assistance: investing in industries and sectors so countries can build on their comparative advantages to diversify and add value to their exports.

- Adjustment assistance: helping with the costs associated with tariff reductions.

For example, in the Philippines and Sri Lanka, trade officials and negotiators received training in the function, structure and rules of the trade system. Meanwhile, many African countries have received assistance to help facilitate customs reform and improve the capacity of port authorities. This has increased trade and customs revenue: 150 per cent in Angola and 58 per cent in Mozambique over the course of two years. And in Zambia, Belize and Grenada, AfT has helped strengthen new sectors and diversify their export base. Aid-for-trade stood at US$25.4 billion in 2007 (plus an additional USD 27.3 billion in non-concessional trade-related financing), growing at roughly 10 per cent per year, which the WTO and OECD report is additional aid, not at the expense of social expenditure, which means that donors are on track to meet their 2005 Hong Kong commitments.

Source: WTO/OECD (2009); OECD (2010b).

Right answer, wrong diagnosis: radical and critical reformist critiques

There is often a similarity between the criticisms levelled at aid from orthodox perspectives and those from radical critics. They also highlight that aid creates dependency, undermining the government and societies in developing countries, and that aid tends to flow to corrupt leaders. However, this is to misdiagnose the issue. The sad truth is that aid has systematically gone to corrupt regimes, but this is a function of the geopolitical and capitalist structures which drive aid. It is not only the corrupt elites within developing countries who are to blame, but the donors who give aid in the full knowledge that it serves to prop up autocratic or corrupt regimes. Radicals argue that there is no getting around this and the whole aid system needs to be junked.

Aid as imperialism

Aid, as Sandra Hayter (1971) argued in her classic *Aid as Imperialism*, is just the smooth face of capitalism. Its aim is not the liberal beneficence of donors, but instead its purpose is to extend and preserve the capitalist system in the developing world. This can be through direct political influence or tying aid for commercial gain or gaining access to new markets. But as Hayter argues, the true nature of how aid supports the capitalist system is much more nuanced than simple bribes and corruption. Through the use of conditions – privatisation, liberalisation – it is the attempt to institutionalise and embed market-based forms of social organisation and to reduce the revolutionary potential of social unrest and political alternatives in the developing world. As Hayter (1971) begins, even when financially concessional, aid has never been the unconditional transfer of resources to developing countries: it is necessary for the successful reproduction of capitalism.

As such, the real impact of aid in developing countries is to support economic exploitation as a side payment or stabiliser. Aid is a short-term solution to the outflow of private profits and the repayment of debts from developing countries which would otherwise cripple developing countries and lead to social upheaval. The consequence of this is to create a system of aid dependency. Radical critics take issue with the naïve criticisms of reformists who focus on the failure of states and aid agencies to live up to their official statements about promoting development through aid (Hayter 1971). Instead, proper critical analysis demands an examination of the underlying purpose of aid.

Saving the aid system from the donors: critical reformists

There are two particular arguments associated with the critical reformist position: aid is a small part of the solution and conditionality is part of the problem not the solution. Like radical critiques, critical reformists highlight the role of commercial and political pressures outside of developing countries in examining the impact of aid. Aid hasn't worked because it hasn't been given for the right reasons to the right governments. The historical record suggests that aid has been given to political allies, without much concern for how it was spent as long as it secured political and military alliances, or it has been given in a form which serves donor commercial interests, such as dumping food aid or tying aid. As such, why be surprised when such aid doesn't produce development outcomes?

Critical reformists tend not to focus on aid quite as much as other perspectives do. This reflects their view that aid can be positive if reformed, but ultimately it just isn't that important. Critical reformists tend to focus on the wider perspective, emphasising the role of private bank capital and unfair trading rules as more significant for development outcomes. Jonathan Glennie (2008) argues this point, saying that rather than disburse amounts of aid, the politically harder but more beneficial thing to do is for wealthy countries to reform their trading and financial dealings with developing countries. This perspective tends to be neglected. For example, take the discussion of aid for trade (Box 5.8). The emphasis in these efforts is on developing countries being 'helped' to reform; much less emphasis is put on the harms and unfairness caused by the uneven playing field of the world trade rules (Pogge 2008).

Aid and conditionality

The 1970s and 1980s witnessed a significant shift in economic and political fashions towards a strongly free-market orientation. This coincided with a significant deterioration in global economic conditions in which there was a sharp rise in the indebtedness of developing countries (see Chapter 6 for a full account). These combined to create vulnerable countries (indebted and economically unstable), in need of assistance (to be provided by the World Bank and IMF). The decided diagnosis and solution was the so-called Washington Consensus (Williamson 1990; cf Williamson 2003). The conviction of the neoliberal orthodoxy, along with the opportunity for systematic reform, meant that lending and aid from the donors went way beyond just trying to secure macroeconomic stability, budgets and balance of payments deficits: it was targeted at fundamentally changing the structure of developing countries. This 'structural adjustment' was about getting the prices rather than the policies right, seeking to remove the perceived distortions introduced by government interventions. The idea was to privatise to incentivise through the profit motive, to encourage competition to improve efficiency, to institutionalise private property rights. Structural adjustment became the goal of the World Bank and its tool was policy-based lending. That is to say, in granting development assistance to a developing country the government had to agree to a set of reforms in line with the policies espoused by the World Bank.

There are two critiques of the structural adjustment period. First, it didn't work. Despite reducing budget deficits, external indebtedness and trade protections, economic growth stagnated and poverty rates increased (Birdsall and de la Torre 2001). In short, the advantages of openness were very much oversold (Rodrik 1997; Stiglitz 2002). Moreover, with the decidedly non-neoliberal success stories of China, India, Malaysia and Taiwan the prescriptions of the Washington Consensus were severely questioned (Wade 1990). The response saw a shift from harsh neoliberalism to a more inclusive liberal institutionalism – for example, it was notable how much impact UNICEF's (1987) call to think about 'adjustment with a human face' had on mainstream thinking at the time. As we saw in Chapter 3, the shift led to a rediscovery of the importance of poverty on the one hand and institutions on the other (World Bank 1990; Naim 2000). This wider perspective was then echoed in the World Bank's 1990 *World Development Report*, which set out three key ways of securing development: (1) investing in growth, (2) investing in people – by spending on health and education the Bank would increase 'human capital' and thus productivity and reduce poverty, and (3) provide social safety nets to protect the vulnerable. This marked a partial but significant break with pure neoliberalism, at least with points two and three.

The second critique complements the first, but also stands independently. It was increasingly accepted that the conditionality associated with aid and concessional lending is damaging per se, regardless of the nature of policy prescriptions. This is because outsiders trying to 'buy' policy reforms will not be successful at bringing about sustainable change because country-level commitment will be shallow and temporary. The concerns about country ownership saw a rebranding of the SAPs in 2002 into PRSPs (Poverty Reduction Strategy Papers). Although the World Bank has presented the PRSPs as a considerable improvement in terms of flexibility, pro-poverty and the extent to which they represent the government and civil society's views on national development, existing studies suggest otherwise (Dijkstra 2005). For example, in Tanzania, agricultural market liberalisation has stalled and been reversed as local elite and bureaucratic interests, which oppose external reforms that reduce their gains from protected markets, have undermined the reforms. What has been missing is a local constituency that supported the reforms and coordinated the more dispersed winners from the reforms (Cooksey 2011). The question has therefore become, what are the political and institutional conditions that can make aid effective?

The aid effectiveness literature

Liberal institutionalists acknowledge that one can always find examples of all of the problems outlined by neoliberals, radicals and critical reformists, from the most egregious instances of personal gain and corruption, to failures caused by the institutionalised skewing of incentives. It is clear that aid has and does fail to produce developmental outcomes. But, looking at other different recipients, there is also evidence that aid does work; for every Haiti there is a Tanzania, and for every Somalia or Democratic Republic of Congo there is a Botswana, Korea and a Mozambique. As Figure 5.19 suggests, 'there is no apparent simple relationship between aid and growth' (Radelet 2006: 7). Countries that receive aid grow, countries that receive aid don't grow and countries that don't receive aid grow. There are, clearly, other factors at work. The question researchers have been trying to answer is what explains this variation in developmental outcomes? As such they have tried to go beyond individual (positive and negative) examples and anecdotal

Figure 5.19 *Foreign aid and growth, 1994–2010.*

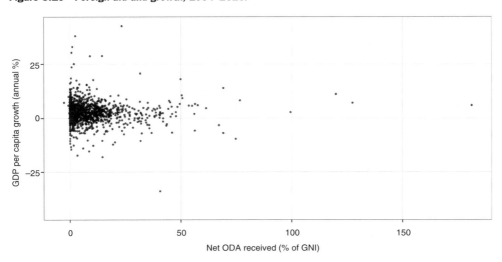

Source: WDI 2013.

evidence, and take a more systematic approach and ask 'What explains when, where, and why aid succeeds (or fails)?' And thus, if there is evidence that success happens, can it be helped through carefully designed policies?

The broad conclusion from roughly 40 years worth of cross-country aid-growth regressions is that, on average, higher aid flows are linked with higher economic growth. Note that this is different from saying aid always works; it's just that, on average, aid is *associated* with growth. There are many cases where high levels of aid have failed. But looking across all cases, aid tends to result in increased growth controlling for other explanations. This last point – controlling for all other explanations – is crucial. The models used include other plausible variables – such as institutional quality, trade policy, civil war, geography (whether a country is located in the tropics), health, budget deficits, income levels, inflation – all of which should be expected to affect growth rates and so the final result indicates aid's impact on growth in excess of what would normally be expected given these other factors.

The liberal institutionalist aid revival

The different protagonists in the debate have used many different methods, different data sets and different variables to try and identify the relationship (or not) between aid and development. So, the mainstream consensus is that, on the whole, aid works but that this finding depends upon (1) recognising that there are diminishing returns to aid, (2) the presence of an enabling environment, and (3) appropriately disaggregating aid flows. The aim for studies of aid effectiveness is to control for these factors or incorporate them as intervening variables.

First, researchers have recognised that for every extra dollar of aid given, the return on this investment falls. Economics has discussed the issue of diminishing returns since the 1950s, making it slightly odd that it wasn't until the 1990s that aid-growth regressions moved away from assuming a linear relationship between aid and growth (Clemens *et al.* 2004; Radelet 2006). Recall Figure 3.1, where the production function displays a classic diminishing returns shape – a steep initial shape before flattening out to show smaller increases in output for each additional unit of capital. It is noticeable that one of the main recent analyses that doesn't find a relationship between aid and growth, uses a linear function (Rajan and Subramanian 2005; see also Barder 2005b for a discussion). It is far more suitable to assume a non-linear relationship. Research has found that the marginal effectiveness of aid is highest when aid is delivered to the countries where there is the greatest need (Burnside and Dollar 2000; Collier and Dollar 2002).

Second, a further theoretical and methodological advance was to test for conditional relationships (Burnside and Dollar 2000; Collier and Dollar 2002; Radelet *et al.* 2005; Wood 2007). Studies show that aid effectiveness is conditional upon the coexistence of other factors. The liberal institutionalist perspective has mainstreamed the idea that liberal institutions matter for economic performance (Rodrik *et al.* 2004). In short, aid works best where secure property rights, the rule of law and an open and competitive political system is in place, and good governance is the norm. Findings confirm the conditional effect of institutional quality (Collier and Dollar 2004), democracy (Svensson 1999; Kosack 2003), political conflict (Chauvet and Guillaumont 2004), trade openness (Teboul and Moustier 2001) and so forth.

The most famous of study to look at a conditional aid and growth relationship was by Craig Burnside and David Dollar (2000). Burnside and Dollar proposed that aid only leads to growth when the recipient country has 'good polices' in place. The study has been hugely influential, but there has been an ongoing debate about the robustness of the findings. Others have rerun the analysis but have changed the data or the specification of the model slightly, for example the years covered, and have found that the relationship disappears (Easterly *et al.* 2004). Nevertheless, the position that aid can work if the right policies and institutions are in place is now the accepted wisdom.

Third, and finally, it is crucial to disaggregate aid flows and take aid heterogeneity seriously. Consider for a moment a case where a country receives aid because of a disaster. It's most likely that the country will suffer a reversal of development. If a study was to then look at the relationship between aid levels and growth it would appear that aid was related to negative development outcomes. What is noticeable is that many of the aid-growth regression analyses fall into this trap because they look at the relationship between total aid and growth (Radelet *et al.* 2005). A related problem is that a good proportion of aid is not targeted at growth, instead it is intended for investment in welfare provision. It has been estimated that over half of aid is for welfare, and only a third for productive investment (Morrissey 2002). This is a perfectly valid use of aid, in fact it may well be the strongest case for aid. However, it is unlikely that this aid will lead to sustained productive growth, especially when targeted at the poorest. Similarly, aid is increasingly intended for political institution-building or is for humanitarian reasons. Again, while aid for investment in infrastructure, agriculture and industry is clearly growth-related aid, humanitarian aid and food aid are not designed to produce growth. Aid in the form of bed nets, medicines, and democratic and judicial reforms may well produce growth but certainly not immediately. Again, most analyses fall into this trap and lump all types of aid together and assume it will impact on growth within a four year period (Radelet *et al.* 2005).

To address exactly these issues, Clemens *et al.* (2004) disaggregate aid into three categories: (1) aid for disasters, emergencies and humanitarian relief efforts (which is likely to be negatively associated with growth), (2) aid that might effect growth but indirectly and over a longer period of time (such as investment in education and health), and (3) aid directed at growth and that could be expected to work quickly (building infrastructure and investing in agriculture, for example). They examine aid flows to 67 countries for the 1974–2001 period and find a strong and positive relationship (with diminishing returns) between the third type of aid and economic growth. They estimate that for an increase in aid of around 1 per cent of GNI, the annual growth rate increases by 0.31 per cent; which is about three times greater than previous studies have found because they assumed that all aid was the same. Thinking about the impact of this over time, they calculate an additional $1 in aid will result in a cumulative increase in income of the recipient country of around $1.64. Their results also suggest that aid works *better* in countries with good policies, institutions and health, but that it still works (though not as well) in countries without these.

The aid effectiveness literature has been dominated by econometric studies which have focused on economic growth as the measure of development outcomes. However, recently, studies using a wider range of development indicators have emerged (Jensen 2008). For example, studies have looked at the impact of aid on education, health and the environment. Gomanee *et al.* (2005a) find that aid is positively associated with welfare improvements as measured by the HDI, but this effect is most pronounced (1) in countries with lower initial levels of welfare, and (2) it works indirectly through increased welfare expenditure by governments and is not a direct effect of the aid – so it is conditional on governments spending the aid money in a pro-development manner.

The area where the most effective interventions have been made and where the impact of aid on development outcomes is considered to be the strongest is health (Levine and Kinder 2004; Radelet 2006). For example, controlling river blindness, the eradication of smallpox and oral rehydration tablets to combat diarrhoea have all been achieved through development assistance programmes. See Box 5.9 for an account of the effectiveness of aid for health.

Box 5.9

Aid for health

Nicholas Kristof (2006) describes his first-hand account of seeing the difference development aid makes in an Ethiopian hospital.

> In Ethiopia, I met Catherine Hamlin, an Australian obstetrician who is overdue for a Nobel Peace Prize. She runs a hospital that repairs obstetric fistulas, one of the most awful injuries humans can sustain. Typically, a fistula occurs when a physically immature teenage girl tries to give birth, and the baby gets stuck. After a few days without a doctor around to help, the baby is stillborn, and the girl is left with perforations between her vagina and bladder or rectum. She cannot hold her wastes, which trickle constantly down her legs. She smells. Her husband divorces her, she must build a hut on her own, and sometimes she is barred from using the village well. At the age of about fifteen, her life is ruined. Some two million girls and women suffer from this affliction around the world.

> Dr Hamlin repairs the fistulas, at the cost of about $450 per operation, supplied largely by individual donors and aid groups, and she brings these teenage girls back to life. See fistulafoundation.org.

> You visit her hospital, you see the girls with fistulas, and you just want to hand over your wallet to Dr. Hamlin. I don't know what impact she has on Ethiopia's economic development, but she seems to me a saint. Easterly is absolutely right that we need to figure out ways of delivering aid more efficiently, and he has written an immensely stimulating book. But no one who sees a girl with a fistula waiting patiently for an operation could doubt that foreign aid is often the very best investment in the world.

Policy issues in development aid: a global partnership for development cooperation?

To summarise, the liberal institutionalist conclusion that aid works best in countries with the right institutions and good polices is now 'the conventional wisdom' in the global development community (Radelet *et al.* 2005). (Though critics still charge that it's a very fragile conclusion as all results from cross-country regressions are extremely sensitive to initial assumptions and model specification – Easterly 2003; Roodman 2007.) However, in addition to this 'conditional thesis', it is also increasingly recognised that it is important to disaggregate the aid which goes into hospitals, roads, schooling and institution-building. These different modalities impact on development in different ways and with different time lags.

Some have argued that it is a mistake to expect aid to work much at all. For one, much aid is experimental and so part of the benefit of aid is learning what does and doesn't work (Lancaster 1999; Riddell 2007). Second, considering the environment in which much aid is directed it would be unrealistic to expect aid to translate unproblematically into largely positive results. Nevertheless, there are clearly a number of steps which could be taken to improve the developmental impacts of aid in developing countries. There are three key areas of initiatives that come from liberal institutionalist and, to some extent, critical reformist insights; namely, first, domestic institutionalist reforms and the good governance agenda, second, moves towards the selectivity of aid, and third, global governance reforms of the aid system.

First, concerns about aid effectiveness and the failure of structural adjustment led to the rediscovery of institutions and the new language of governance. The early important statement of this was the World Bank's 1989 report *Sub-Saharan Africa: From Crisis to Sustainable Growth*. The report argued that developmental failure was not simply a result of bad (i.e. non-liberal) economic policies, but could be attributed to 'a crisis of governance' (World Bank 1989: 60; see also Mosley and Toye 1988). The understanding of 'good governance' that emerges is a domestic and technical, managerial or administrative solution to increasing developmental effectiveness. For example, the World Bank's 1992 paper on *Governance and Development* defines 'good' governance as 'synonymous with sound development management' (1992: 1). Hence, improving bureaucratic competence and reducing corruption were seen as the solution to aid failure. As a result, good governance became a central principle for donor agencies and was incorporated into the requirements they set out for developing countries in order to receive development assistance, so-called 'second generation' or political reforms (Naim 1995; Doornbos 2001). This perspective was intellectually strengthened by the liberal institutionalist emphasis on the role of institutions from Nobel Prize winner Douglass North and the rest (North 1981, 1990; Acemoglu *et al.* 2001; Easterly and Levine 2003; Rodrik *et al.* 2004).

Second, a further policy consequence has been the rise of 'aid selectivity'. Based on the insights of Burnside and Dollar (2000) some donors have moved in the direction of only providing aid to countries which have a track record of 'good polices'. This is apparent in, for example, the World Bank's allocation of IDA grants, the US's Millennium Challenge Account and some budget support from bilateral donors. The problem is, of course, that many of the countries most in need of aid are often those without a track record of 'sound policies' or even the ability to develop one. Plus, it also, rather conveniently, focuses the attention on the responsibilities of developing countries. It ignores the fact that some aid is not distributed for development purposes, and that donor countries have systematically failed to meet their own promises on aid disbursements.

Third, until recently, donors have tended to focus on increasing the effectiveness of their own aid, individually. However, there has been an increasing recognition that this is a collective action problem. With so many aid programmes and projects overlapping in developing countries, effectiveness is a function of coordination and making the aid *system* more effective, not just its constituent parts. Furthermore, donors don't stick to their commitments and the aid which does exist is made less efficient because it is fragmented and unpredictable.

Poor global governance results in poor predictability, aid orphans and the fragmentation of aid. Fragmentation was reported as a problem by the Pearson Commission as far back as 1969 (Riddell 1997). While individual donors may seek to do good things, collectively the whole is less than the sum of its parts. There is a large and growing number of donors, with varying agendas, methods and monitoring processes. Because individual donors make their decisions independently from one another, based on their own priorities, criteria and strategies, the overall pattern of aid distribution is globally sub-optimal (Rogerson and Steensen 2009). It is inefficient in that it produces large transaction costs and duplicates effort. A recent estimate by the OECD puts the direct transaction costs of fragmentation at US$5 billion or more per year, but the economic costs of sub-optimal allocation are much higher (Killen and Rogerson 2010). Furthermore, this fragmentation and lack of coordination results in inequitable outcomes in the sense that neglected countries could have benefited from the resources that are duplicated in other countries elsewhere, absent coordination. These aid holes and overlaps create what are known as 'aid darlings and 'aid orphans' (Rogerson and Steensen 2009).

Furthermore, despite well-publicised commitments, such as Gleneagles, a key problem with the aid system is that disbursing aid is voluntary and is subject to nothing more than moral sanction. This explains why foreign aid is twice as volatile as domestic resources (APPG DAT 2010). And the losses associated with aid volatility are estimated at around US$16 billion per year, which amounts to around 15–20 per cent of total ODA (Kharas 2008). Administrative bottlenecks in developing countries and a lack of political will among donors as well as their desire for flexibility and the right to withhold funding in line with conditionalities, confound attempts to make aid more predictable. Consequently, the volume of aid that flows into a country each year is the product of a range of individual, uncoordinated decisions. One response has been to try and improve alignment. It is all very well having country ownership, but it is also crucial to have donors aligning their aid with national development strategies, which should be developed with domestic stakeholder consultation and input.

These issues, along with others, have led to a series of international efforts to address the known limitations of aid effectiveness, and rethink the structural limits to aid delivery. Key to these efforts has been the OECD DAC's Working Party on Aid Effectiveness which has convened a series of high-level forums focused on aid effectiveness and MDG8's aim of 'forming a global partnership for development'. The key outcome of this process has been the principles contained within The Paris Declaration on Aid Effectiveness (OECD 2005b).

The Paris Declaration on Aid Effectiveness identifies a set of principles to improve the quality of donor aid and thereby increase its effectiveness. The Declaration was agreed in 2005 by 90 developing countries ('partner countries'), 30 donor countries and 30 development agencies (such as the UN and the World Bank) and was focused on five key principles: (1) strengthening ownership of the development agenda by partner countries, (2) alignment of donors' and partners' agendas and systems, (3) harmonisation of donor-partner work through sharing information and establishing common procedures, (4) managing for development results by using assessment frameworks which could be monitored, and (5) mutual accountability through shared assessment reviews. Similar to the MDGs the Declaration was built around a set of indicators and targets which were to be met by 2010. In September 2008 a supplemental Accra Agenda for Action was agreed which added new commitments and principles, relating to the transparency and predictability of aid flows, the use of anti-corruption measures, the untying of aid, the management of technical assistance, the use of in-country systems and the division of labour by donors at the country level (OECD 2008c).

The Declaration has been criticised for being unambitious, with watered down commitments, a result of trying to develop consensus among all donors, including the less committed. The UK All Party Parliamentary Group for Debt, Aid & Trade report into aid effectiveness (APPG DAT 2010) highlights how the issues of conditionality and tied aid are glossed over by the Paris Declaration, and ultimately the initiative remains a donor-led affair. And the Paris Declaration commits donors to no more than 'continued progress in untying aid', with no hard indicators chosen.

So where is the aid effectiveness agenda now? In 2011, the OECD published the final report on progress in implementing the Paris Declaration. The OECD called it 'sobering' reading (OECD 2011a). Only one of the 13 targets which were established had been met by 2010. The evidence clearly suggests that the most progress had been made on those indicators which required partner country effort. The least progress had been made on the areas in which donors needed to improve, specifically tied aid and donor coordination – for instance, only 8 per cent more ODA was flowing through partner country systems by 2010.

At the end of 2011 the development community gathered in Busan, South Korea, for the Fourth High-Level Forum on Aid Effectiveness. While virtually nothing was agreed by way of new monitoring standards and a tied aid agreement was torpedoed by the US, Busan made bigger strides in terms of the politics of the international aid architecture. The Busan Partnership for Effective Development Cooperation (OECD 2011b) is a watershed moment insofar as it marks a shift away from the traditional club of Western donors and explicitly recognises the role of the emerging donors, especially China. It has done so in a number of ways. First, the Declaration brings China, India, Brazil and South Africa into the fold as they signed up 'on a voluntary basis' to endorse the 'principles, commitments and actions agreed in the outcome document' which include transparency and accountability but also aid reporting standards (OECD 2011b). Second, Busan saw the launch of the Global Partnership for Effective Development Cooperation, which effectively relegates the OECD's DAC from its traditional role at the centre of the aid donor governance architecture. Notably, the new institution will have a joint OECD-UNDP secretariat, mirroring the shift from the G7 to the G20 in recognition of the new multipolar world.

Two other issues look set to shape the future of development assistance and cooperation. First, the US signed up to the International Aid Transparency Initiative (IATI). The IATI is an independent initiative which requires members to publish their aid information according to a set of agreed standards. This means that accessing aid information is much easier and more accurate, allowing for better monitoring, evaluation and coordination of donor efforts. The US signing up was crucial given the volume of aid it's responsible for and now means that the IATI covers 75 per cent of global ODA. Second, there were signs of a shift from talking about aid effectiveness to talking about development effectiveness. The recognition that aid is but one part of a series of development linkages sits well with a more critical reformist agenda. Nothing concrete has emerged from this, yet, but it is undoubtedly a key area for the future.

Conclusion

This chapter has introduced foreign aid, and more specifically ODA as a key – and for many sub-Saharan African countries and many SIDS – *the* major external source of development financing. The chapter began by defining what ODA is and the accounting debates surrounding what counts as aid: official and concessional financing administered for the promotion of the economic development and welfare of developing countries. Although seemingly technical, these definitions are intensely political as well as having important economic consequences.

The second section covered how much aid is given, by whom, how, and to whom. The third section looked at the explanations for why and where aid is given. It is clear that donors are motivated by a variety of different objectives. These sometimes conflict. In part, many of the geopolitical objectives help explain why the record on aid's effectiveness has been so patchy and why aid's critics can have a strong argument when it comes to issues such as corruption and waste. Nevertheless, recent analyses of the effectiveness of development assistance have shown that under the right conditions and with the correct measures, aid can and does have positive developmental impacts.

Over the 2000s this liberal institutionalist position came to define the mainstream collective wisdom.

Indeed, in the past 20 years there is some evidence of the international community trying to move towards a more needs-based and development focused aid agenda. However, there are still many political issues to be resolved in the international aid architecture to ensure aid's effectiveness. Attempts to increase aid's predictability, ownership and distribution hold out some hope for increasing aid's effectiveness. Finally, radical critics argue that while reformists are correct to identify the problem of donor interests, they misdiagnose the issue as a problem of sub-optimal aid delivery, rather than a case of aid working exactly as it is designed to, as a tool of imperialist domination.

Discussion questions

1 Why do states give aid? Is this different from the motives of individuals and multilateral institutions?

2 Do you think aid is a (possibly imperfect) attempt to help developing countries or do you see it as a tool of imperialism?

3 How important is aid for the external financing of developing countries; all, some, why?

4 Do you think of aid as a grant or a loan? What should it be? Why?

5 When evaluating the effectiveness of aid, should the focus be on measuring economic growth or human development?

6 How do you think the emergence of new aid donors in the coming years, such as China, will change the way international development is done?

Further reading

- The single most comprehensive book covering all of the issues related to foreign aid is Peter Riddell's *Does Foreign Aid Really Work?* (2007). It offers a comprehensive coverage of the moral, ethical and political issues to do with giving aid; the debates on whether aid works or not; as well as the different actors which inhabit the aid universe. It is a well researched, thoughtful and cautious but ultimately positive review of foreign aid.

- For a fantastically argued and critical look at the current aid system, look no further than Jonathan Glennie's *The Trouble with Aid* (2008). He takes on both the optimists and pessimists and argues that we should look beyond aid, or at least its relationship with systematic reforms such as taxation, intellectual property rights and trade rules, but also that the politics of aid and accountability need to be more central to discussions.

- To get an insight into the ferocity of some of the exchanges about whether foreign aid is a good or a bad thing (in more black and white terms) I would thoroughly recommend reading through the debate between Jeffrey Sachs, Dambisa Moyo and William Easterly conducted on the 'pages' of the

Huffington Post (www.huffingtonpost.com). It is occasionally enlightening, very revealing and quite entertaining. See: Sachs (24 May, 2009) 'Aid Ironies', Easterly (26 May, 2009) 'Sachs Ironies: Why Critics are Better for Foreign Aid than Apologists', Moyo (26 May, 2009) 'Aid Ironies: A Response to Jeffrey Sachs'.

● Finally, a strangely neglected topic: what do people on the receiving end of aid think? A recent report called *Time to Listen: Hearing People on the Receiving End of International Aid* provides a wonderful evidence base collected by talking (and listening to!) nearly 6,000 people in 20 aid-receiving countries. It's available as a free download, available at: http://bit.ly/14nB84s.

Useful websites

● The best source of aid data is the OECD's CRS. The data sets can be downloaded from http://stats.oecd.org/Index.aspx. If you select the 'Development' dataset from the left-hand column you can access data from 1960 onwards and it covers all of the donors who report to the DAC. You can download or visualise aid data in the aggregate as well as breaking it down by individual donors and recipients and by activities and sectoral information.

● The other great source of data is Aid Data (www.aiddata.org). Aid Data is a collaborative initiative that builds on the CRS. It includes the OECD verified data alongside information harvested from donor annual reports, project documents from bilateral and multilateral aid agencies, data from donor agency sources, websites and databases. So it has much more data, it's massively innovative (e.g. the data is being geocoded), and includes lots of research papers and other visualisation tools.

6 Debt financing: bank lending and bond markets

Learning outcomes

At the end of this chapter you should:

- Understand the economic logic for borrowing, but also understand the potential political and social consequences of doing so.
- Know the difference between intermediated bank lending and borrowing directly from the bond markets.
- Understand the history and geography of debt-based capital flows to the developing world, especially the rise and fall of bank lending.
- Appreciate the positions in the debates about the determinants of debt flows.
- Know the difference (and relationship) between debt flows, debt stocks, liquidity and insolvency crises.
- Be familiar with causes, consequences and responses to the Third World debt crisis, as well as a series of ongoing issues such as the threat from vulture funds.

Key concepts

Debt finance; bond markets; bank lending; banking crises; original sin; debt overhang; Third World debt crisis; heavily indebted poor countries (HIPCs); Multilateral Debt Relief Initiative (MDRI)

Introduction

The world's private capital markets constitute the largest, most liquid and most controversial element of global finance. This chapter deals with debt-creating capital flows to developing countries, the world of lending and borrowing, of credit and debt. More specifically, the chapter will focus on the two key forms of debt flows: bank loans and bonds. The following chapter deals with equity-based capital flows: FDI and portfolio equity. Just to clarify, equity-based financing is where a borrower generates investment by offering a share of the ownership in a project. On the other hand, debt-based financing is where a borrower promises to repay the lender via a series of fixed instalments such as a bank loan or a bond. As we will see, most debt-creating flows are private, but official agencies, both states and international financial institutions, play an important role too (see Box 6.2).

From a standard economic perspective, debt and debt flows allow for consumption-smoothing over time. Because governments, firms and other economic actors are faced with unpredictable income streams – as well as potentially unpredictable expenditure – they choose to borrow money in order to 'smooth' their income. Borrowing can be a short-term measure to pay for exports if there is insufficient foreign exchange from imports, or it can be for longer-term investments such as infrastructure projects or to pay for governance reforms. As outlined in Chapter 2, developing countries often face the problem of insufficient domestic savings to invest into increasing economic growth and welfare. Or, as countries develop, they often face the problem of importing more than they export, leaving domestic resources underutilised and running a current account deficit.

The politically and socially difficult solution to a foreign exchange gap is to reduce imports through slowing economic activity, which brings with it all of the associated costs of deflation, i.e. higher poverty, unemployment, lower wages and growth. On the other hand, assuming insufficient savings or taxes, the alternative is to borrow from external sources. Foreign borrowing can be used to finance a current account deficit and/or supplement domestic savings and take advantage of otherwise underutilised domestic resources and generate more foreign exchange. And as long as the rate of return on the borrowing – i.e. the extra income generated from utilising the loan – exceeds the costs of borrowing then it makes economic and developmental sense to access international credit markets. For example, in 2009 Chile was faced with the costs of reconstruction from a massive earthquake. The Chilean government chose to borrow the money from foreign lenders based on the assumption that rate of return from rebuilding its infrastructure would generate sufficient growth to repay the costs of borrowing (see Box 6.1).

There are two key ways borrowers can access credit: either through loans from a financial institution or directly from the bond markets. The first route, borrowing via a loan, is arranged through a financial intermediary such as a bank or similar institution. As we saw in Chapters 1 and 2, banks fulfil a number of economic and social functions within an economic system. Functionally speaking, banks specialise as an intermediary between those households and firms with excess capital in the form of savings on the one hand and those who wish to borrow on the other. According to standard economic theory, banks serve individual economic needs but simultaneously increase overall economic productivity. Banks are able to offer a pool of funds from which to borrow by taking deposits and aggregating these different savings streams. By specialising as an intermediary, banks reduce transaction costs, in particular the need for other economic actors to duplicate the effort of monitoring contracts, and specialise in information-gathering (Diamond 1984). Plus, by developing relationships with their clients, banks can help support long-term borrowing.

Box 6.1

Chile and debt for development

Chile, in the aftermath of the global financial crisis, suffered a massive contraction in economic activity. By April 2009, the 12-month growth rate of the economic activity index was −4.9 per cent, and by the third quarter of the year unemployment had risen 11.6 per cent (MacLaughlin and Ricaurte 2010). On 27 February, just as the prospects for economic recovery were starting to look better, a massive 8.8 Richter-scale earthquake and a subsequent tsunami hit Chile. Over 500 people lost their lives and the economy was also hit hard with some 3 per cent of the country's capital stock being destroyed, as were important industries and the country's infrastructure (IMF 2010a). The cost of Chile's planned post-quake reconstruction plans was estimated at US$30 billion. The government's share of this post-quake bill was US$8.4 billion, which is equivalent to about 4 per cent of Chile's GDP (IMF 2010a).

The problem the Chilean government faced was how to smooth out its need for income now, in effect to access its expected future income and bring it into the present. Some of the cost could be immediately financed through temporary tax increases, but to pay for the rest Chile needed to borrow. It could do this domestically or through international investors, plus it could do this by going to banks as financial intermediaries or directly to the financial markets. As it happened, Chile decided to borrow from foreign investors and directly to the markets. As such, Chile created debt by placing two bonds in international markets in July 2010 which raised US$1 billion and US$500 million respectively.

This lending by international investors, in return for scheduled repayments of the original sum plus interest, has allowed Chile to start rebuilding its devastated infrastructure and industries, and to restart economic growth. The debt will be repaid over the next 10 years with the added costs of borrowing in the form of interest being a function of the perceived creditworthiness of the borrower, i.e. the Chilean government. If things go well, as the IMF is predicting, then Chile will enjoy the economic rewards of the capital investment and the higher growth will allow it to repay the debt easily.

However, the growth could also create its own problems. There is a fear that it will result in strong inflows of private capital looking to profit from Chile's growth. Sudden influxes of capital like this can help finance investment but can equally cause price inflation and speculative bubbles in the economy. Such rapid inflows of hot money into East Asia in the late 1990s were blamed for asset bubbles which eventually popped. Policymakers have to try and benefit from private investment inflows, but also manage sudden and large inflows or outflows. Plus, there is also the possibility that the IMF will be proved wrong and growth doesn't reappear or is derailed by other events. In such circumstances, countries which have borrowed from external sources can find themselves in economic and political difficulties as they are unable to repay their debts. This was the situation many Latin American and African countries found themselves in from the 1980s onwards.

The alternative to accessing credit via a financial intermediary is to borrow directly from the international capital markets. The past 20 years have seen large increases in the levels of so-called 'financial disintermediation'. Disintermediation is the process of cutting out the middle man which, in the case of financial disintermediation, means banks. By removing the banks' need to earn a profit on the transaction governments and firms can borrow more cheaply from the capital markets as well as avoid potential conditions which come along with the loan. In order to borrow directly from the capital markets developing countries, or private actors within developing countries, issue a bond, which is essentially the equivalent of issuing an IOU or a promise to repay. The issuers of the debt or bond are therefore the borrowers. The bondholder, who buys the debt, and the right to the stream of income coming from the interest payments, is the lender. Details of bonds and how they work are given in Box 6.4 (p. 178).

The price of borrowing is determined by prevailing interest rates. It is usual for bank lending rates to be slightly lower than bond rates. This is because bank loans generally enjoy a lower default rate; a function of the relationship between lender and client that doesn't exist in the bond market. Plus, bank loans also tend to have a shorter maturity, which reduces the risk of default. The true borrowing costs that developing countries face are measured as the difference – the spread – between what is actually paid versus what is considered to be a risk-free benchmark rate. This is also called the yield, i.e. it's the profit for the investor over and above investing in the risk-free benchmark. The commonly accepted benchmarks are the 10-year US Treasury rate for bond markets and for bank lending costs the benchmark is the six-month Libor rate.

These descriptions of banks and bonds, and how they work, are the standard, functional definitions from an economic perspective. The reality of lending and borrowing entails a series of economic, political and social complications. Orthodox and reformist thinkers alike agree that the international financial markets offer enormous potential for development if successfully harnessed. However, it is widely accepted that this potential is far from being reached. Authors disagree on how likely this is as well as on how best to harness and manage the development potential of debt-creating financial flows. They disagree on how much work governments and policymakers need to do to ensure that the interests of private capital and development goals are aligned, assuming this is even possible or desirable. It is the case that capital markets, through regular and frequent financial crises, are also responsible for some of the most sudden and extreme development reversals.

Debt flows tend to be volatile in the short term. Plus, over the medium term, debt flows tend to be very pro-cyclical, meaning that they surge during periods of economic growth and stop and reverse in downturns. As such, they do not help to smooth business cycles, as they are intended to, but actually accentuate them. The problem with these sudden stops is that they are usually exactly when states, firms and individuals most need access to finance. As such, external borrowing is a major source of vulnerability for developing countries, revealed cruelly and consistently in financial crises such as the those which struck East Asia in 1997, Sri Lanka in 1989–91, Argentina in 1995 and 2001–03, Mexico in 1994–96, Brazil in 1994–98, Jamaica in 1996–98, Russia in 1998 and Turkey in 2000–01. The list could go on. Access to credit can be a wonderful thing. But the flip side is, of course, indebtedness, which at the right levels is not a problem. However, sudden changes in the economic environment, such as a jump in borrowing costs because of interest rate or exchange rate changes, can mean that borrowers quickly find themselves unable to repay the amount owed. The political and social problems of indebtedness were most severely exposed by the long-term debt crisis which engulfed many developing countries from the 1980s onwards.

Hence these risks have to be balanced against the need to borrow from foreign lenders. Even for some staunchly orthodox authors this means that the liberalisation of capital should be approached much more cautiously and with a much more critical eye than the liberalisation of trade flows (Bhagwati 1998).

Significantly, a new consensus has formed around the problem of indebtedness over the past 15 years or so with liberal institutionalists at the heart of contemporary initiatives for debt relief. This mainstream interpretation tends to emphasise the role of bad developing country politics with a technical solution of better governance. Orthodox thinkers tend to focus on issues of sustainability and access to credit. Heterodox thinkers focus on the functioning of the capitalist system and the nature of debt. This is because of their ontological groundings in the exogenous and endogenous views of money, respectively, which suggest debt is either a technical problem or a political question. Neoliberals see problems of indebtedness as the responsibility of the borrower. Moreover, they argue, well-meaning but misguided attempts to relieve indebtedness are doomed to make the problem worse as countries are taught that economic imprudence goes unpunished and they will be bailed out; the moral hazard problem. Liberal institutionalists are keener to acknowledge the problems of overindebtedness and focus on cleaning the slate and getting the governance right to avoid a next time. At this point a more sustainable system of well-run government and carefully managed debt can be established. Liberal institutionalists and critical reformists disagree as to the causes of the financial instability – whether it tends to be poor economic policies or whether it is inherent to speculative capital markets – and thus disagree on whether market-strengthening or market-regulating policies are needed. Radicals share the insights of critical reformists about how speculative markets work – or don't work – but go further in highlighting the anti-democratic nature of financial markets, demonstrating that they are not simply neutral channels of capital resources, but also operate as structures of political and social control and exploitation.

This chapter examines the role of debt in financing development. The first section describes the volume and geography of debt-creating capital flows to developing countries. The second section examines the determinants of bank and bond flows, presenting the evidence on which push and pull factors determine which countries do and don't receive flows and why. The third section asks what impact debt flows have had on developing countries. It presents the empirical studies that consider the impact of debt flows on various development outcomes before turning to the historical record on debt sustainability and analysing the causes, consequences and responses to the Third World debt crisis. The fourth and final section presents a range of key contemporary issues; namely the current status of debt cancellation, the issue of renewed indebtedness, the need for a sovereign debt mechanism and the phenomenon of vulture funds.

Box 6.2

Is it debt?

There is an important overlap between aid and debt-creating capital flows. As we saw in the previous chapter the difference between aid and other debt-creating flows are their concessional character (along with being distributed by an official agency with development as its objective). However aid *is* debt-creating. This is because despite the fact that aid must include a grant element, any loan element of ODA does have to be repaid. (IDA credits have maturities of 20, 35 or 40 years with a 10-year grace period before repayments of the principal begins. Nearly all IDA credits have no interest charge, but credits do carry a small service charge, currently 0.75 per cent on funds paid out.) As such, in examining the consequences of indebtedness the chapter has to include all borrowing, including that from official agencies such as the World Bank and concessional lending which might, at first glance, seem to belong in the chapter on aid. The important distinctions to be clear about are as follows. Official lenders as well as private actors create debt. Concessional aid flows as well as private capital flows create debt. It is only grants, as opposed to loans, that don't create a financial debt, so it is only grants which aren't included in the net financial flows data.

The volume and geography of debt flows

The section provides an overview of debt-creating financial flows and stocks. To do so it reviews the data and answers the following questions. First, what are the total levels of borrowing in the developing world? How has this changed over time and how does it compare with ODA levels? Second, who borrows? How does this vary by income group, geographical region, and is it governments or the private sector in LICs and MICs which are accessing the international credit markets? Third, who lends? Does the credit tend to come from the private sector banks and financial markets, or from public lenders, other countries or multilateral lenders such as the World Bank? Fourth, is the borrowing long-term or short-term? Fifth, what form do the debt-creating flows take: bank lending or bond issues? The data shows how these have changed over time and vary across income groups, geographical region and individual countries of interest. All the data in this section come from the World Bank's World Development Indictors which run from 1970 for most countries.

Figure 6.2 shows that private debt flows steadily increased from the early 1990s until the early 2000s and then again in 2007 where they crashed back down. Two stylised facts are apparent from all of these figures: (1) there is a long-term cyclical nature to debt flows into developing countries (although the type of flow changes) and (2) private debt flows – both portfolio and bank lending – are highly volatile. Short-term fluctuations indicate periods of a sudden interruption of financing. These sudden changes manifest themselves as financial crises.

Debt-creating capital flows come from private and public actors. The previous chapter covered the concessional element of international aid flows, ODA. Even though this chapter is focused on the role of market-based (or non-concessional) debt financing, the official public bodies such as the World Bank and IMF are still important in

Figure 6.1 *Typology of external debt.*

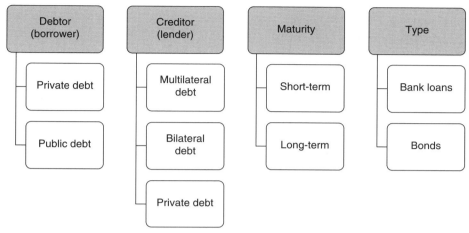

Source: Adapted from Millet and Toussaint 2004.

Figure 6.2 *Net flows on external debt.*

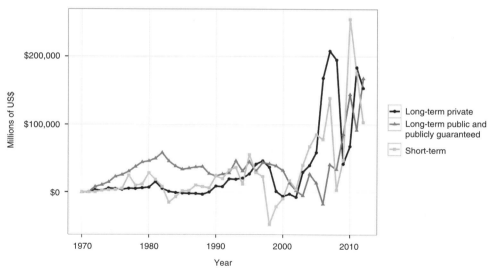

Data: WDI 2013.

understanding the role of the commercial banking system. There are two reasons for this. First, while the financial markets can be a direct source of development financing – funnelling capital into emerging and developing markets – they are also an important indirect source of development financing. This is because many rich country governments and multilateral actors, such as the World Bank and the IMF, use the bond markets to raise capital in order to provide development assistance in the form of grants and loans. Second, non-concessional lending is undertaken by official lenders such as the World

Bank (the IBRD in particular) not only commercial lenders. In 2011, the combined net debt flows to developing countries stood at $465 billion. This is considerably larger than ODA flows in 2008 ($134 billion). Moreover, because of the global financial crisis, there was a massive fall in private capital flows. In 2007, net debt flows to developing countries stood at $488 billion. So it has taken six years to get back close to pre-crisis levels.

Box 6.3

Functional short-term borrowing

While short-term bank lending is often associated with a lack of investment commitment and volatility there are good reasons for accessing particular types of short-term lending such as export credits or guarantees. These are loans provided by export credit agencies (ECAs) to allow developing countries to pay for imports. Most LICs access the international capital markets through syndicated bank loans that are used for short-term trade financing. Short-term trade financing is crucial for poorer countries that are not perceived as creditworthy and so cannot borrow longer term but still need financing to facilitate trade. Trade financing is the most widely spread and accessed form of market-based finance, much more so than traditional bank lending for investment. This is possible because the traded goods act as the necessary collateral. Trade financing is provided by banks among a range of other actors such as official export agencies and the development banks. It rose sharply in the 1990s until the East Asian financial crisis hit (UN DESA 2005). The World Bank (2004 GDF: 127–30) reports that average spreads on trade finance have fallen from over 700 basis points in the mid-1980s to 150 just before the East Asian financial crisis, lower than spreads on bank loans between 1996 and 2002. Nevertheless, critics argue that export credits are little more than export subsidies for rich countries.

Types of debt flows

The data provided by the World Bank's Debtor Reporting System (DRS) distinguishes net short-term debt flows from medium- and long-term debt flows. Short-term debt is defined as debt which has an original maturity of one year or less. Plus the DRS also divides the medium- and long-term debt flows into bond flows and flows from banks and other private creditors. These different categories of private debt flows to developing countries are shown in Figures 6.3 and 6.4.

A number of things are noticeable. Each type of flow has broadly climbed and fallen together, but it is clear that short-term flows climbed and fell the hardest, with bond flows also falling heavily. Plus, it should also be noted from the figure that bank and short-term flows were actually *negative* in 2002 and short-term lending almost turned negative again in 2008. And bank flows, after growing a lot in the seven-year period leading up to the financial crisis, actually fell the least. Finally, since the late 1990s the ratio of short-term debt to total international bank lending for all developing countries has decreased. The change is especially noticeable for East Asia. So, even though short-term debt was rising in absolute terms during the 2000s, it was falling relative to long-term debt and foreign exchange reserves, from 48 per cent in 2000 to around 28 per cent at the end of 2005 (World Bank 2006a).

Figure 6.3 *Long-term and short-term debt flows.*

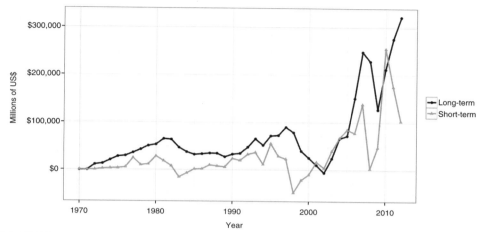

Data: WDI 2013.

Figure 6.4 *Bank and bond flows.*

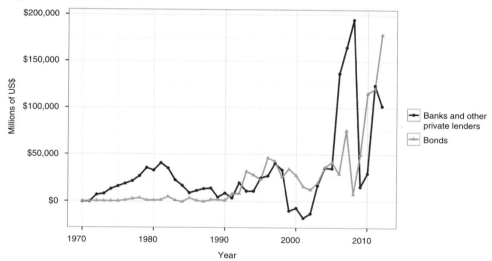

Data: WDI 2013.

Figure 6.4 also shows that, as a proportion of total flows to developing countries, banking has become less important. Part of the explanation is the consequence of the debt crisis and the way in which new bond issues were used to refinance the banking debt. This trend has since been reinforced as more emerging markets have accessed the international capital markets. Portfolio debt flows reflect this net issuance of debt in the international market (net outflows – negative figures – indicate a repayment of debt).

Despite their volatility, portfolio debt flows have become the major source of external debt financing for some countries, especially Latin America. In 2006, there were 40 developing countries which raised funds using the international bond market; this had increased from 34 in 2002–03 (World Bank 2006a). But it is also very concentrated. In 2005, sovereign borrowers accounted for 46 per cent of the total bond issuance, the average bond issue having a 12–13-year maturity.

But there is a second reason that bank flows appear to be becoming less important. There has been a noticeable change in the way in which banks are lending. Overall cross-border lending has been falling in recent years. Banks are increasingly lending via national subsidiaries, thereby funding activities domestically. This type of lending grew on average by some 29.4 per cent per year from 1996 to 2002 (UN DESA 2005). Many countries now have a large proportion of their banking sector foreign owned. In sub-Saharan Africa this is as high as 45 per cent (Spratt 2009). So what this boils down to is a shift from cross-border lending to direct investment in developing country banking sectors.

Box 6.4

What are bonds?

Bonds are a type of loan. A bond is a legal agreement between the issuer of the bond (the borrower) and bondholder who buys the bond (the lender). In return for borrowing an amount of money (the principal) the issuer agrees to pay the bondholder a fixed schedule of interest repayments and then, on a set maturity date, repay the original loan amount in full. The interest rate is also called the coupon. The name derives from the fact that historically bonds were physical notes with coupons attached to them. Bondholders were able to redeem their coupon at a bank in return for their interest repayments. The coupon is usually fixed, though it can be linked to other rates such as Libor. Maturity dates can vary from anywhere between one day and 30 years. The World Bank defines bonds with a maturity of less than one year as short-term capital.

Bonds are issued into a primary market – either through an underwriter who is then responsible for selling the debt to other investors or through an auction. Because the IOU can be sold on, bonds are traded in a secondary market and whoever buys the bond owns the income stream from the repayment of the interest.

The current yield is the return on the bond at its market price, i.e. the coupon rate divided by the current price of the bond. As such, the current yield and the price of a bond are inversely related. Before outlining some of the factors affecting bond prices a final concept to introduce is the spread. Spreads typically refer to the difference between the buying and selling price, i.e. the amount that a trade can earn from trading a financial asset. But with bonds it tends to refer to the interest rate differential in percentage points or basis points (which equal one-one hundredth of a percentage point) between a bond (e.g. an emerging market bond) and a benchmark such as a US Treasury Bond. So essentially it tells you how much an investor is earning over and above what they could have earned in a safe investment.

The price of a bond is a function of current interest rates (and expectations about future rates), the maturity of the bond and the borrower's creditworthiness. First, when interest rates rise, the price of a bond falls. The reasons are fairly straightforward. If the face value of a bond is $1,000 and the original annual coupon is $80 this means a return of 8 per cent for the bondholder. If interest rates elsewhere are 10 per cent then the bond doesn't look like a very good investment because the coupon is fixed at 8 per cent. As such the bondholder will have to sell the bond at a discount. To earn a yield of 10 per cent on an $80 coupon the bond must be sold at $8,000. Hence bondholders do badly in situations where interest rates rise and have to sell their bond at a discount. In situations where the prevailing interest rate has fallen the bondholder can sell at a premium. Of course, if the bondholder is able to and wants to hold the bond until maturity this doesn't matter.

Second, the closer a bond is to maturity the closer it will get to its par value. At the maturity date the principal amount, i.e. the original amount of the bond, is repaid to the holder of the bond (along with the final interest payment). This means that as the maturity date approaches the price tends to equal the issue or face value (referred to as trading at par). Finally, if there has been a downgrade in the credit rating of the bond issuer (see Box 6.6) this will also push the price down as the risk of default is assumed to have risen.

The advantages of bonds for investors are that they tend to enjoy better legal protection and are more likely to recoup some of their losses if the investment fails. But they are not completely protected. For example, in February 2005, Argentina completed an $80 billion debt restructuring. The deal meant bondholders losing two-thirds of their capital. This is referred to in the professional jargon as a 'hair cut'. Also, the long-term performance of bonds is slightly worse than stocks, but is much less volatile and provides a more conservative investment. Plus stocks and bonds tend to move in different directions, so bonds may be a smarter investment or a way for investors to hedge their risks if their portfolio includes stocks.

Alongside credit risk, it is inflation that bond investors fear the most. Inflation devalues the final principal repayment, because $1,000 no longer buys what it used to. To take the same example of the $1,000 bond with a coupon of $80; the nominal return is 8 per cent. If inflation is running at 5 per cent the real return is 3 per cent. But if inflation rises to 10 per cent then the real return becomes negative (–2 per cent) since inflation is outstripping the return. In addition, the inflation is eating away at the market value of the original principal. At 10 per cent inflation, by the time the bond matures it will have lost 21.8 per cent of its purchasing power.

Geography of debt flows

In terms of where the debt flows to and who it finances, the private sector is increasingly accounting for market-sourced debt with around two-thirds of the total, which is similar to the 1970s, but around double what it was in the 1990s (World Bank 2005a). More generally, investors tend to view public borrowers as more capable of servicing their debt, so in the aftermath of financial crises there tends to be a reduction in lending to private borrowers.

Looking at Figure 6.6 it is clear that the majority of debt flows accrue to East Asia, Europe and Latin America. Over time there has also been a secular fall of Latin America as well as the fall and recovery of Asia. The aggregate data hides increased flows and particularly the loss of access to the markets by individual countries, for example the East Asian crisis countries, and Argentina, Brazil, Indonesia and Turkey following their financial crises (UN DESA 2005). Interestingly, over the period of these crises, flows to other developing countries were up, helping to hide these effects. However, the Russian default in 1998 was unusual as this triggered a more generalised loss of access to bond financing.

Figure 6.5 *Net debt flows by income group.*

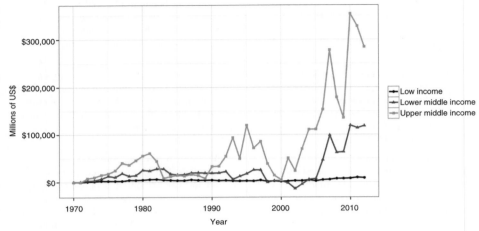

Data: WDI 2013.

Figure 6.6 *Net debt flows by region.*

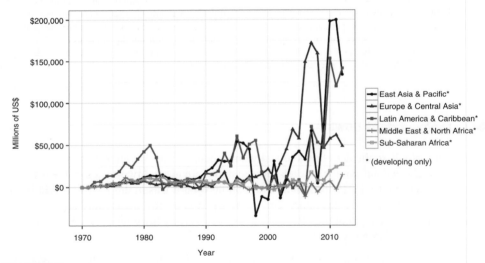

Data: WDI 2013.

Plus, the other thing the regional data hides is the fact that lending is incredibly concentrated within these regions. Between 2005 and 2011 the top 10 borrowers accounted for between 56 and 65 per cent of annual net debt inflows to developing countries. For example, China and India account for the vast proportion of inflows into Asia. The list of the top recipients makes it clear that these are the larger MICs: Brazil, Chile, China, India, Indonesia, Mexico, Poland, the Russian Federation, South Africa, Thailand and Turkey.

Figure 6.7 *Debt flows to Argentina, Brazil, Indonesia and Turkey.*

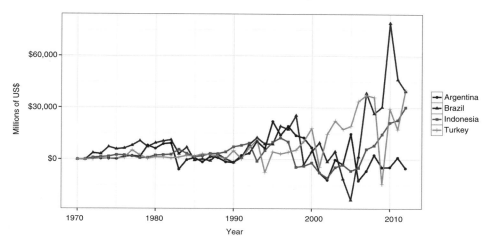

Data: WDI 2013.

Thus, the first barrier for most developing countries is not 'choosing' which type of borrowing they require, but whether they can access any type of debt financing at all. As Table 6.1 makes abundantly clear, the majority of developing countries are excluded from international borrowing. But the data also suggests that bank lending is more common for developing countries than bond financing. This is true both in the absolute levels of borrowing and also relative to the size of the economy.

The difference in access to bank lending and the bond markets is starkest in relation to the LICs. The LICs almost never borrow money through the bond markets. Poorer and more risky countries – in credit terms – enjoy greater access to bank loans because banking relationships allow for information-gathering and monitoring which allows lenders to overcome the information asymmetries which are associated with lending to LICs. The other way investors have addressed the problems of information asymmetries is through direct investment, especially by multinational corporations (see Chapter 7).

The differential access to bond markets and bank lending has effectively created three classes of developing country. This is presented in Table 6.1 which shows the 135 countries which report to the World Bank's DRS (World Bank 2006a). First, in the left-hand column there are the countries with access to both bank finance and the bond markets. All countries within this group have issued bonds in recent years, but even within this group there is a hierarchy of sorts with some countries – such as Chile, China, Mexico and Thailand – enjoying investment grade status (as indicated by the credit rating, see Box 6.6). As such, they enjoy much lower spreads and less volatility compared to developing country averages. Second, the middle column is composed of those countries which only have access to bank lending. This is because they lack macroeconomic stability and/or institutional and legal frameworks which protect investors and private property rights making them, in the eyes of international lenders,

Table 6.1 *Countries' access to international credit markets*

Countries with access to bond markets	Credit ratings[a]	Countries with access to bank lending only[b]	Credit ratings[a]	Countries with no access to private debt markets[c]	Credit ratings[a]
Argentina	B3	Albania	NR	Armenia	NR
Barbados	Baa2	Algeria	NR	Benin	B+
Belize	Caa3	Angola	NR	Bhutan	NR
Brazil	Ba3	Azerbaijan	BB	Burundi	NR
Bulgaria	Ba1	Bangladesh	NR	Cambodia	NR
Chile	Baa1	Belarus	NR	Cape Verde	NR
China	A2	Bolivia	B3	Central African Republic	NR
Colombia	Ba2	Bosnia and Herzegovina	B3	Chad	NR
Costa Rica	Ba1	Botswana	A2	Comoros	NR
Croatia	Ba3	Burkina Faso	B	Congo, Dem. Rep.	NR
Czech Republic	A1	Cameroon	B−	Côte d'Ivoire	NR
Dominican Republic	B3	Congo, Rep.	NR	Dominica	NR
Ecuador	Caa1	Djibouti	NR	Eritrea	NR
Egypt, Arab Rep.	Ba1	Equatorial Guinea	NR	Fiji	Ba2
El Salvador	Baa3	Ethiopia	NR	Gambia, The	NR
Estonia	A1	Gabon	NR	Georgia	B+
Grenada	B−	Ghana	B+	Guinea-Bissau	NR
Guatemala	Ba2	Guinea	NR	Guyana	NR
Hungary	A1	Honduras	B2	Haiti	NR
India	Baa3	Kenya	NR	Lesotho	NR
Indonesia	B2	Kyrgyz Republic	NR	Madagascar	B
Iran, Islamic Rep.	B+	Lao PDR	NR	Malawi	NR
Jamaica	B1	Liberia	NR	Mauritania	NR
Jordan	Baa3	Maldives	NR	Moldova	Caa1
Kazakhstan	Baa3	Mali	B	Mongolia	B1
Latvia	A2	Mauritius	Baa2	Myanmar	NR
Lebanon	B3	Mozambique	B	Nepal	NR
Lithuania	A3	Nicaragua	Caa1	Niger	NR
Macedonia, FYR	BB+	Nigeria	BB−	Paraguay	Caa1
Malaysia	A3	Papua New Guinea	B1	Rwanda	NR
Mexico	Baa1	Senegal	B+	Samoa	NR
Morocco	Ba1	Seychelles	NR	São Tomé and Principe	NR
Oman	Baa1	St. Lucia	NR	Sierra Leone	NR
Pakistan	B2	Sudan	NR	Solomon Islands	NR
Panama	Ba1	Tanzania	NR	Somalia	NR
Peru	Ba3	Turkmenistan	B2	St. Kitts and Nevis	NR
Philippines	B1	Uzbekistan	NR	St. Vincent and the Grenadines	NR
Poland	A2	Vanuatu	NR	Swaziland	NR
Romania	Ba1	Yemen, Rep.	NR	Syrian Arab Republic	NR
Russia	Baa2	Zambia	NR	Tajikistan	NR
Serbia and Montenegro	BB−			Togo	NR

Slovak Republic	A2		Tonga	NR
South Africa	Baa1		Uganda	NR
Sri Lanka	BB–		Zimbabwe	NR
Thailand	Baa1			
Trinidad and Tobago	Baa2			
Tunisia	Baa2			
Turkey	Ba3			
Ukraine	B1			
Uruguay	B3			
Venezuela, RB	B2			
Vietnam	Ba3			

Source: Adapted from World Bank 2006b: 50.
Note: This table classifies the 135 countries that report to the World Bank's Debtor Reporting System (DRS) by accessibility to international capital markets across bond and bank segments (based on data cover transactions on international loan syndications and bond issues reported by capital-market sources, including Dealogic Bondware and Loanware). Countries are divided into three main categories: countries with access to bond markets, including all the countries that have issued bonds between 2002 and 2005; countries with access to bank lending only; countries that have no access to either bond or bank lending, including countries that primarily rely on official financing for their financing needs.
a. Long-term sovereign foreign currency debt ratings, as of February 3, 2006. Moody's ratings were used for most of the countries. However, S&P and Fitch ratings were used for countries that are not rated by Moody's, including Benin, Ghana, Grenada, Macedonia. FYR, Mali, Senegal, and Serbia and Montenegro. NR indicates countries that are not rated by either Moody's or S&P.
b. For analytical purposes, bank lending in this table is only referred to as medium- and long-term lending (excluding short-term lending that has less than 1 year of maturity).
c. The use of the term, "no access to capital markets," is not intended to imply that all countries in this category do not have access to other types of international private capital, such as FDI and portfolio equity. International capital defined here only refers to the bond and bank segments of the market.

a high credit risk. However, because these countries generally have strong income streams from an export sector, banks and other lenders are willing to extend loans to them, confident they will earn a return on their investment. Third, and finally, there are those countries which have little or no access to any medium- and long-term debt financing. Instead these countries have to rely on other types of financing such as FDI (Chapter 7) or ODA (Chapter 5).

External debt stocks

External debt stocks – as opposed to flows – show the *accumulated* value of outstanding borrowing. This is an indication of how indebted countries are. The total stock of external debt is shown in Figure 6.8. In recent years there has been a move away from debt financing to equity financing, especially to FDI (Chapter 7). This, along with a concentred attempt to deal with developing country debt, has significantly reduced the external liabilities of developing countries. External debt ratios – as measured as a proportion of GNI and exports – have been falling over the past decade. The ratio of external debt to GNI has fallen from a high of 44 per cent in 1999 to 34 per cent in 2004 (World Bank 2006a). This is good news for the affordability of the debt and is reflected in the ratio of debt service to exports, which has halved to 9.5 per cent in the period

2000–08 (World Bank 2010e). However, these ratios do remain high in certain regions and countries as shown in Figure 6.9. Sub-Saharan Africa has seen a vast improvement on its debt indictors. This was a result of increasing commodity prices which have helped to boost incomes and export values for some sub-Saharan countries. But also because of the debt relief initiatives: the HIPCs and the MDRI. These are considered at greater length in the section on debt relief below.

Figure 6.8 *Total external debt stocks for all developing countries.*

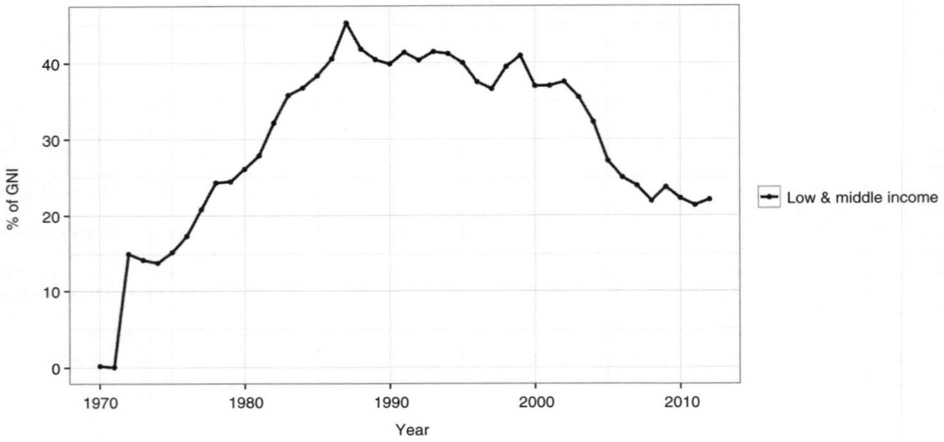

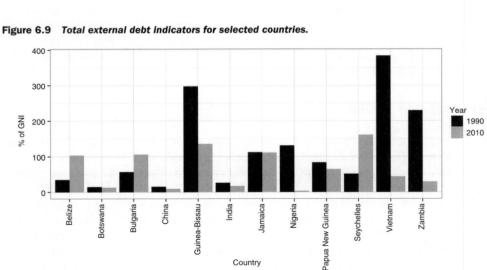

Data: WDI 2013.

Figure 6.9 *Total external debt indicators for selected countries.*

Data: WDI 2013.

Official lending

In addition to private market lenders there are also non-concessional flows from official sources, including considerable borrowing – public and publicly guaranteed debt – from the international financial institutions, such as the World Bank and IMF. The IBRD makes loans which are counted towards a developing country's public debt. The IBRD only lends to MICs, such as Brazil, Turkey and Mexico. At the end of 2008 the total amount of debt which had been disbursed and owed to the IBRD stood at $95.9 billion. And 62 per cent of this was accounted for by just eight countries: Argentina, Brazil, China, Colombia, India, Indonesia, Mexico and Turkey. The other important role public institutions such as the IBRD (and IDA) play is to provide public guarantees for private loans – see Box 6.5.

Box 6.5

The International Bank for Reconstruction and Development (IBRD)

What is the IBRD?
One of five institutions that make up the World Bank Group, the IBRD is structured something like a cooperative, owned and operated for the benefit of its member countries. Founded in 1944, it is the part of the World Bank that works with middle-income and creditworthy poorer countries to promote sustainable, equitable and job-creating growth; to reduce poverty; and to address issues of regional and global concern. The IBRD's 24-member board is made up of 5 appointed and 19 elected executive directors who represent its 187 member countries.

How does it pursue its goals?
The IBRD helps members achieve results by delivering financial products, knowledge, technical services and strategic advice, while using its capacity to call members together to discuss ways to further their specific development objectives. It strives to increase its impact in MICs by working closely with the IFC and the Multilateral Investment Guarantee Agency (MIGA); capitalizing on MICs' own accumulated knowledge and development experiences; working closely with the IMF and other multilateral development banks; and collaborating with foundations, civil society partners and donors in the development community.

Where does the IBRD get the money to finance projects in developing countries? It gets its money from the capital markets. Investors see IBRD bonds as a safe and profitable place to put their money and their cash finances projects in MICs. Annual funding volumes vary from year to year, and are currently around $10–15 billion. The World Bank has become one of the most established borrowers on the world's capital markets since issuing its first bond in 1947 to finance the reconstruction of Europe after World War II. It has had a triple A rating since 1959.

Who pays for the IBRD's operating expenses?
The IBRD covers its operating expenses primarily out of its income. The IBRD earns an income every year from the return on its equity and from the small margin it makes on lending. This pays for the IBRD's operating expenses, goes into reserves to strengthen the balance sheet and also provides an annual transfer to the IDA. The IBRD has raised the bulk of the money loaned by the World Bank to alleviate poverty around the world. This has been done at a relatively low cost to taxpayers, with governments paying in $11 billion in capital since 1946 to generate more than $400 billion in loans.

Which countries are eligible to be IBRD clients?

IBRD clients are middle-income and credit-worthy lower-income countries. The Bank classifies a country according to the wealth of its population. MICs are defined as having a per capita income of between around US$1,000 and US$10,000, which may qualify them to borrow from the IBRD. LICs with a per capita income of less than $1,000 usually do not qualify for IBRD loans unless they are creditworthy. However, LICs are eligible to receive low or no interest loans and grants from the IDA. India, Indonesia and Pakistan are examples of creditworthy LICs which are eligible for a blend of financial assistance from both the IBRD and IDA.

Why do MICs still turn to the IBRD?

MICs that are served by the IBRD have made enormous economic strides in the last few years but they still face daunting challenges to reduce poverty to meet the MDGs, which set specific targets to be met by 2016. These countries account for two-thirds of the world's population and are home to more than 70 per cent of the developing world's poor people who live on less than $2 a day. While private capital flows have risen substantially, this flow has been concentrated in a limited number of countries. Only a minority of MICs can be regarded as established bond market borrowers, able to access the market regularly at a stable cost. Other countries within the group have only sporadic access or none at all. Therefore, the majority of MICs continue to rely on the IBRD to mobilise investments in infrastructure, health, education, clean energy and the environment. The IBRD helps clients gain access to capital and financial risk management tools in larger volumes, on better terms, at longer maturities, and in a more sustainable manner than they could receive from other sources. Unlike commercial banks, the IBRD is driven by development impact rather than profit maximisation. The IBRD has also supported MICs in times of crisis when their access to capital has dried up.

How strong is the demand for IBRD services from its clients?

Some MICs no longer see the need for significant financial support from the IBRD, because they have large foreign currency reserves and are in a good budget position. However, others still have large investment needs that include funds for public infrastructure projects and social services. Increasingly, the IBRD is meeting the more sophisticated demands of its middle-income clients by providing financial services that protect them against exchange and interest rate risks and the turbulence of the commodity markets. In fiscal 2007, it carried out $5.4 billion in interest rate and currency risk management transactions on behalf of its members. In the same period, the IBRD committed $12.8 billion for 112 projects. The Bank assists client countries not only through its finance but also by providing access to its development knowledge resources. The Bank's knowledge activities range from conducting country research, to developing analytic and conceptual frameworks for country assistance, to building the capacity for sustainable development within client countries.

Source: Taken from International Bank for Reconstruction and Development (IBRD), 'Frequently Asked Questions', Online. Available: http://go.worldbank.org/YX2261GMXO

Explaining the geography of debt flows

It is useful to think of the determinants of credit flows in terms of push and pull factors. These push and pull factors operate at the international and domestic levels. On the one hand, domestic conditions and policies within developing countries are important for attracting capital flows. On the other hand, changes in the world economy as well as more specific policy decisions and economic conditions in the rich world significantly impact on these capital flows and the access which developing countries have to them.

It is important to note that the different approaches tend to place different amounts of emphasis on the different factors.

Orthodox theorists place the emphasis upon the economic fundamentals both in terms of demand and supply. To recap, the demand for credit is for economic agents to fulfil the consumption-smoothing role. Longer term, borrowing can be used to front-load investment, paving the way for developmental gains in production and welfare outcomes. The decision for a developing country to try and access external sources of borrowing is driven either by the fact that domestic resources aren't sufficient because of the gap between savings and investment, or countries wish to supplement domestic resources or avoid reducing imports. In such cases the turn to foreign sources of borrowing is also driven by the fact that it is often cheaper than domestic financing. As such, private capital flows represent the reallocation of savings across countries as the financial counterpart to trade in goods and services.

So, for neoliberals cross-border flows offer an important efficiency mechanism. From an investor's perspective they offer households and firms a way of internationally diversifying their portfolios in order to reduce the exposure to country-specific risk (Goldstein *et al.* 1991). The decision to invest in specific projects or bonds is a judgement on the risks and returns based on economic fundamentals, such as inflation risks, which are reflected in asset price changes and net capital flows.

Liberal institutionalists build on this understanding but tend to focus more on the role that political institutions play in developing countries in attracting capital but also, as we will see in the next section on developmental impact, the effective conversion of inflows into developmental outcomes. There is an emphasis upon the adequacy of institutional reforms and market-supporting policies. Liberal institutionalists also highlight the role of international factors, especially the role of rich-country government policy, in regulating the supply of credit in the OECD countries as the source of debt-creating capital flows.

Critical reformists tend to place less emphasis upon the role of domestic factors in the developing countries and focus on the operation of the financial system within the core economies and the operations of the global financial markets themselves. Critical reformists underline the role of banks, not just government policy, in creating credit and therefore creating credit booms, busts and debt crises. They also emphasise the speculative nature of financial markets, meaning that capital flows tend to become disconnected from the underlying fundamentals emphasised by neoliberals and liberal institutionalists.

Similar to critical reformists, radical critics of global finance tend to emphasise the costs of capital inflows in developing countries, especially the negative consequences of indebtedness. But, unlike liberal institutionalists who perceive indebtedness as a costly problem for developing countries which can be resolved, critics argue that indebtedness is an inevitable consequence of the logic of the capitalist system, not just a perverse or unfortunate outcome. The concentration of capital in the core of the world economy is a function of unequal international political and economic relations. Thus, while all approaches recognise that lenders lend because they have capital (or ways to leverage it) and want to earn a stream of income and profit, radicals place the most emphasis upon the internal dynamics of capitalism as a structural explanation for the geographical extension of credit flows. For Marxists especially, the need to lend capital is a function of the need to keep on earning profits. As overproduction and underconsumption characterise the industrialised core, capital has to look overseas for a spatial fix to the accumulation crisis (Harvey 1982).

Pull factors: domestic factors in developing countries

Orthodox perspectives, working with a view of rational markets, interpret capital movements as responding to the economic fundamentals in developing countries. There are a number of key issues which affect international investors' view of the credit risks associated with developing economies. These include the following: inflation, institutional and regulatory quality, amounts of short-term debt and levels of international reserves.

A track record and future promise of low inflation rates is seen as attractive because it is assumed to be a shorthand for a well run economy, but more importantly for investors it means that the value of the debt won't be eroded over time by high inflation rates – because the value of the original loan or bond is reduced. For banks or bondholders, inflation is a loss, but on the other hand the debtor gains as it is effectively paying off the loan since it will cost less in real terms to pay off the debt.

Investors care about the institutional quality and regulatory change in developing countries. Countries which have more developed legal regimes tend to be rewarded with greater debt inflows. The presence of the rule of law is especially crucial for debt-based contracts. Liberal institutionalists tend to emphasise the importance of these factors and, through econometric analysis, demonstrate that there is a correlation between countries which have a higher quality of institutions and which are more likely to attract bond financing (Faria *et al.* 2006).

The extent to which a country is reliant on short-term debt (debt with a maturity of less than one year) is another factor which affects the perceived creditworthiness of a developing country. Reliance on short-term finance makes countries more vulnerable to changes in market sentiment. To explain. Short-term debt can be taken on for a number of reasons: to finance trade, to invest in a limited project or consumption-smoothing for a cash flow problem, or because the borrower cannot access longer-term borrowing in the private markets and so has to take on shorter-term borrowing because of their perceived riskiness. In these circumstances, if the borrower needs to refinance, or 'roll over', the borrowing soon and often (which is the case with short-term debt by definition) then the country begins to look increasingly risky. Under normal conditions refinancing usually occurs routinely and without problems, however if market conditions change because interest rates rise or there is an economic or financial crisis this can result in a reduction of liquidity – i.e. lenders are less willing to make financing available. A loss of liquidity makes it increasingly difficult and expensive for borrowers to roll over their debts and ultimately increases the risk of default. On the other hand, long-term debt doesn't need constant refinancing, thereby reducing this market risk.

A country's international reserves are also a crucial factor in investors' assessment of risk. International reserves are made up of foreign exchange or gold, as well as a country's IMF reserves and SDRs (see Box 4.4), which are held by a country's central bank or central monetary authority. International reserves are the means by which countries fulfil their international payments as well as monetary policy and manage their exchange rate (see Chapter 3). A decent cushion of reserves means that a country will be able to withstand sudden reversals in sentiment within international capital markets.

All of these factors combine to form an overall perception of a country's creditworthiness and risk of investment. However, this is complicated by the fact that perceptions of creditworthiness are strongly shaped by external factors such as the judgement of rating agencies and other investors (see Box 6.6). These can be viewed as a rational reflection of the underlying fundamentals of a developing country's economy (as orthodox perspectives do), or driven by forces which are internal to the capital markets themselves (as heterodox perspectives do). Plus, regardless of the economic fundamentals of the developing countries, these can be swamped by changes in macroeconomic policies or economic conditions in the rich world. As such there are a range of push factors to also consider.

Box 6.6

Credit rating agencies

Rating agencies provide an assessment, expressed as series of letter grades, of the creditworthiness of an issuer of debt. A country's sovereign credit rating reflects a combination of factors reflecting its ability to repay its debt, such as predicted growth, existing debt burden, political risk, macroeconomic policies, and so on. The significance of a country's rating is that it acts as a shorthand or summary for all of the pull factors mentioned in the text. For the best account of credit rating agencies – how they work and the political and economic consequences of their operations – see Tim Sinclair's (2005) *The New Masters of Capital*. But in the meantime the BBC ran this useful description explaining what a ratings agency is.

> AAA, Ba3, Ca, CCC . . . they look like some kind of hyper-active school report. They are, indeed, a marking system, and one that is designed to inform interested parties. The ratings are given to large-scale borrowers, whether companies or governments, and are an indication to buyers of this debt how likely they are to be paid back. The score card can also affect the amount that companies or governments are charged to borrow money. If a country is deemed to have suffered a downturn in fortunes and its rating is lowered, investors may demand higher returns to lend to it, as it is judged a riskier bet.

> . . . credit-rating agencies [such as Standard & Poor's, Moody's, and Fitch] . . . exist to assess the creditworthiness of bond issuers – companies or, as in this case, countries who borrow money by issuing IOUs known as bonds.

> . . . A downgrade of an issuers' rating typically pushes down the value of a bond and raises its interest rate. It can mean regulated funds must now sell these bonds. But this can cause a vicious circle. If lots of funds are forced to sell, the price of the bond reduces further. That means a higher interest rate must be paid, which puts an even bigger strain on the borrower.

> . . . So how do the agencies form their judgements? Standard & Poor's says a committee of between five and eight people decides the actual rating. They base their assessment on a range

of financial and business attributes that might influence the repayment, some of which may depend on the issuer of the bond (i.e. the borrower). When asked why it changes ratings, S&P responded: 'The reasons for ratings adjustments vary, and may be broadly related to overall shifts in the economy or business environment – or more narrowly focused on circumstances affecting a specific industry, entity, or individual debt issue'. S&P gave a long list of indicators it might use, including 'economic, regulatory and geopolitical influences, management and corporate governance attributes, and competitive position'.

. . . But since the credit crisis began in 2007, these agencies have come in for heavy criticism . . . This is partly because they make their ratings available freely to investors – making their money from charging the organisations who want their bonds rated – something some believe can create a conflict of interest. As a statement from the European Commission put it: 'As a rating agency has a financial interest in generating business from the issuer that seeks the rating, this could lead to assigning a higher rating than warranted in order to encourage the issuer to contract them again in the future.'

Source: Marston (2013)

Push factors: factors external to developing countries

There are a number of factors that determine debt flows which are outside, and indeed outside the control of, developing countries. These include the collective decisions taken within the financial markets by investors, the role of credit rating agencies, and the general economic conditions in the rich world.

Economic conditions in the industrialised world are highly significant for debt flows. This is because the riskiness of any investment is relative to investments elsewhere, thus the attractiveness of lending to LICs and MICs depends upon the return available on investments within the OECD countries. Oftentimes capital is not pushed towards developing countries, but international conditions or changes in the industrialised world limit access to capital by developing countries. International liquidity is highly correlated with business cycles in developed markets (Eichengreen and Mody 1998). Downturns and recessions in the industrialised world tend to result in a reduction in capital flows to developing countries, regardless of the economic conditions and policies in those countries (Dooley *et al.* 1994).

Interest rates in the rest of the world (especially in the US) matter. Increases in interest rates typically reduce the flow of credit to developing countries (World Bank 2001c). An increase in interest rates in the industrialised world makes investing there relatively more attractive, so capital flows tend to redirect to countries with higher interest rates. Simultaneously, these higher interest rates increase the cost of repaying loans thus reducing demand for capital in the developing world. Such changes in the rich world have been shown to suddenly ration the amount of capital flows to the developing world in an 'international capital crunch' (Mody and Taylor 2002). On the flip side, when real interest rates are low in the wealthy countries international investors tend to look to emerging markets since they are prepared to take on increasing levels of risk in their search for yield. This is certainly what happened prior to the global financial crisis. In addition to short- to medium-term factors such as the business cycle and interest rate, there are longer-term push factors which suggest that the levels of capital flows to developing countries will increase steadily. For example, the ageing population in the rich world and the rise of institutional investors. Both of these mean larger portfolios and an increasing desire to diversify as well as seek high growth options.

Despite the fact that bond flows represent such a large financial flow to the developing world, only 5 per cent of global bond issues originate in developing countries (World Bank 2006a). However, prior to the financial crisis, emerging market debt was becoming an increasingly attractive investment for international investors. And it is increasingly becoming so again. This is reflected in the greater tendency for emerging market bond spreads to move with US high-yield bonds and the declining volatility in emerging market bond spreads. In turn, these improvements mean that large institutional investors can buy this debt, because they are usually prohibited from buying debt which is considered too risky. This has resulted in large increases in demand for those countries perceived creditworthy.

The final push factor is the nature of markets themselves. Efficient market perspectives emphasise the external nature of financial markets, rationally reflecting changes in economic fundamentals in the underlying economies. However, liberal institutionalists have come to accept that financial markets tend to overreact and overshoot, and the temporary mispricing of assets is a real problem that has a large impact on the levels of capital flows to developing countries. Critical reformists go further and emphasise the systematically non-equilibrating nature of markets because of the endogenous and speculative nature of market dynamics (see the discussion of Keynes in Chapter 2). Again, the volatility of asset prices and debt flows is not an aberration from the norm but is a result of the febrile and irrational nature of financial markets.

Capital flows are incredibly pro-cyclical and tend to increase following success and contract in times of crisis and need. A number of factors can combine to generate the pro-cyclical nature of debt flows. During periods of growth, spreads tend to fall, which means that borrowing is cheaper and easier. In such circumstances borrowers tend to issue more debt, often rolling over short-term debt and/or consolidating existing loans, and taking advantage of lower borrowing costs to bring forward investment projects. An increase in inflows will often result in rising asset prices and can translate into larger reserves. These changes impact positively on market sentiment, further increasing capital flows into the country.

As outlined in Chapter 3, critical reformists emphasise the fact that markets display herding tendencies. An initial decision of some or key investors to pull out of a market encourages other investors to make similar decisions, and contagion can begin to affect similar but otherwise unrelated assets and countries. The reversal in perceived creditworthiness is self-fulfilling as the social and collective psychological dynamics become economic dynamics. The financial markets' view of the deteriorated creditworthiness of the developing country is then reflected in increasing bond spreads. This makes it harder for countries to access the credit markets or at least more expensive because of the rising cost of borrowing, a reflection of increased risk premiums. The other thing which changes to reflect perceptions of increased credit risk is a shortening of lending maturities. Shorter-term, more expensive borrowing makes rolling over of existing liabilities more difficult and increases the risk of default, and so on. Furthermore, it has been shown that rating agencies tend to lag markets and change ratings *after* a crisis has struck, not prior to. Hence additional liquidity constraints, tied to ratings downgrades, rising borrowing costs and the rising insurance premiums come on the back of these downturns in flows.

The decisions taken by credit rating agencies are important in that they open up (or close) access to capital markets, plus they can change the quality of debt flows. Higher credit ratings mean that institutional investors which control large portfolios, such as pension funds, can invest. If an issuer does not receive an investment rating these institutional investors cannot buy the debt since it exposes their portfolio to greater risk than their mandate will allow. So, while banks have always been more involved in developing country lending, now institutional investors have become increasingly active in the debt markets of a handful of emerging economies with investment grade status.

In summary, private debt flows are highly concentrated in a small number of countries. This is true for both bank and bond flows, but especially so for bonds. Plus debt flows tend to be pro-cyclical and often volatile. In explaining this financial geography, neoliberal analysts focus on the role of economic fundamentals, especially in developing countries. Liberal institutionalists also highlight the economic and political quality of borrowers as driving the flow of debt, and the role of government policy in source countries. There is considerable evidence that liquidity is strongly tied to the circumstances of lenders, not the financing needs or economic condition of developing country borrowers. Reformists, meanwhile emphasise the operations of banks and the financial markets in determining credit flows in a more volatile and speculative view of the world. Finally, radicals, mostly informed by a Marxian mode of analysis, highlight the wider social structures of inequality and the logic of capital in determining the behaviour of capital flows.

Developmental impact of debt flows

The turn to external debt was driven by a lack of domestic resources in many developing countries and because foreign borrowing tends to be significantly cheaper than domestic debt. The demand for external borrowing increased with falling global aid levels and frustration with the donor driven volatility of aid flows. At least this is true for those larger and richer MICs who find it easier to access private debt financing. Yet, financing development through foreign borrowing has a long history, much longer than ODA. Despite the long history, and the recent large and increasing volume of debt flows to developing countries, there is surprisingly little research into the developmental benefits of debt flows (Reisen and Soto 2001).

Similar to the majority of aid effectiveness studies, the existing research tends to look at the impact of capital flows on economic growth and so provides some insight into the average relationship between the two. While important in highlighting general patterns it is also important to be sensitive to the diversity of developing country experiences. Institutionalist analysis highlights how the variation is a function of the different institutional mixes and absorptive capacity of different countries.

As well as variation in country characteristics there is an important heterogeneity in the composition in credit flows (World Bank 2001c). A limitation of most existing studies is that they tend to analyse all private capital inflows together (bank lending, bond flows, equity, FDI) rather than disaggregate them and analyse the impact of individual types of flows. The occasional study which has disaggregated the different types of flows shows two important things (Bosworth and Collins 1999). First, that the different types of private capital flows are not correlated with one another, meaning that they have different determinants. Second, they tend to behave and impact rather differently to one another.

The general findings from the small literature on the development impacts of debt-based financing are somewhat mixed. As such, the following sections cover the range of perspectives by focusing on (1) the developmental impacts of debt inflows, (2) the impacts of sudden stops or reversals in debt flows, i.e. crises, and (3) developmental consequences of high and ultimately unsustainable levels of indebtedness as per the Third World debt crisis.

There are two important caveats. First, any benefits from new capital do not tend to benefit the majority of developing countries, especially not the poorest. As outlined in the section above, most capital flows tend to go to MICs, the emerging markets. The empirical reality of international private capital flows diverges from the neoclassical economic theory which suggests that capital will flow to those countries which have the potential for higher marginal productivity of capital (UN 2005). Because capital flows display a tendency to reinforce a positive growth dynamic, most debt flows are highly concentrated in a small number of developing countries, generally the wealthier and larger emerging markets. Second, capital flows have been highly volatile. Volatility, the sudden reversals in capital flows, and the uncertainty this generates, result in a number of costs for developing countries.

Developmental impacts of debt inflows

What is the relationship between foreign capital flows and domestic investment? It is often difficult to say whether external capital flows provide new investment, or whether they simply finance investment that would have already been financed from domestic sources. A range of empirical studies have looked at the relationship between bank lending and different measures of economic development, for example the effects of inflows on growth, investment, savings and consumption, and financial development and deepening. On balance, existing research tends to find a weak and highly conditional positive relationship between actual capital inflows and development outcomes over time and across countries. The additional factors which condition the strength of the relationship are: the differing composition of inflows, the use that inflows are put to (investment, consumption), other country characteristics including absorptive capacity, the volatility of the flows and the pre-existing level of access to capital markets.

First, with respect to economic growth, Reisen and Soto (2001) find that foreign bank lending is negatively associated with future per capita income growth. The exception is where banks in the developing country are sufficiently capitalised. Their explanation is that banks which lack capital do not allocate capital well; either they tend to engage in risky lending, gambling on higher returns, or they take on lots more government debt. This means that the good investments – i.e. with a safe amount of risk but with decent growth prospects – remain underfinanced, whereas bad risks are overfinanced. As such future growth prospects are undermined. Under such circumstances lots of new external bank lending intensifies these pressures (McKinnon and Pill 1997). As such, external bank lending will contribute to growth only if the domestic banking system is sufficiently developed, specifically once domestic bank capital is over 21 per cent of total bank claims (Reisen and Soto 2001).

In theory, access to external capital should reduce volatility in growth. That is, access to capital in economic downturns should help smooth consumption. However in reality the opposite is true. Because capital demonstrates a decidedly pro-cyclical tendency – being plentiful in economic good times and drying up in bad – external capital flows actually increase growth volatility. Although, like everything else, volatility varies across countries and by type of capital flow. Studies have shown that it is portfolio flows and short-term debt which are the most volatile (Sarno and Taylor 1999). Crucially, volatility damages the growth impact of capital inflows. This is because the uncertainty which volatility creates undermines investors' confidence in the future causing them to prefer to hold onto their capital rather than risk investing it. Moreover, volatility has even worse effects on poverty. The poor suffer disproportionately from volatility (World Bank 2001b).

Second, with respect to domestic investment and savings, one the most well known studies to disaggregate the impacts of different types of capital flows is by Bosworth and Collins (1999). They examined the impact of the size of capital flows as a ratio of GDP on domestic investment over the period 1979–95. Their dataset and their model allows them to control for individual variations between individual countries as well as endogeneity – that is the possibility that the relationship actually goes in the opposite direction and it is domestic investment which attracts capital flows not capital flows which result in greater investment. The empirical results suggest that bank lending is positively and significantly associated with domestic investment levels. Moreover, although loans increase domestic investment they tend to be correlated with negative savings rates, which suggests that bank loans tend to displace domestic saving. However, there is almost no evidence that domestic investment is boosted by portfolio inflows (i.e. bond financing, although the study doesn't distinguish between bonds and portfolio equity). The World Bank (2001c) conducted a similar analysis using a dataset covering a longer time period and a greater number of countries; notably the results backed up the original findings from the Bosworth and Collins study.

Earlier it was shown that sub-Saharan Africa received the smallest portion of debt flows compared to other regions (Figure 6.6). However, the flip side of this is that capital flows into the region have had the strongest impact on domestic investment. This is for two reasons. First, in terms of the composition of the debt flows, East Asia and Latin America receive more portfolio flows, which enjoy a less strong association with domestic investment (Bosworth and Collins 1999). Second, the difference in the strength of the relationship between inflows and investment in sub-Saharan Africa is also a function of the pre-existing access to credit. Because other regions enjoy a greater access to credit, the marginal difference each extra dollar of inflows makes has a smaller impact than it does in sub-Saharan Africa (World Bank 2001c).

In contrast to this regional effect there is a contrary income effect. A key finding from the literature is that richer countries enjoy greater benefits from capital inflows (Blomstrom 1994). Greater economic growth from capital inflows tends to be associated with countries' income level (Eichengreen 2000). So richer countries benefit more and poorer countries benefit less or indeed not at all from capital inflows. This finding can be explained through the notion of absorptive capacity. Neoliberal economic theory suggests that developing countries will exhibit a high marginal productivity of capital because of the relative scarcity of capital – as the example of sub-Saharan Africa suggests. However, this is not automatically the case (Lucas 1990). If a country lacks sound political and institutional foundations this can undermine the relative 'benefits' of capital scarcity; crucial factors here include a skilled workforce, supportive polices and a sound infrastructure. Without these factors the ability of a country to absorb the benefits of capital inflows is seriously hampered. Thus, a country's absorptive capacity is a function of macroeconomic stability, political stability, an effective financial system, a healthy and educated workforce (human capital), infrastructure and control of corruption.

A further question is whether investment or consumption is more likely with foreign bank loans. By looking at developing countries that regained access to bank debt inflows, Cohen (1993) examines whether the renewed inflows translated into capital accumulation. The findings suggest that capital accumulation for these countries was actually less than for those that did not regain access. The author concludes that the extra capital from the renewed debt inflows is used for consumption rather than investment purposes. There is some important regional variation in this relationship. In Latin America the relationship between capital flows and investment is less strong than in Asia. The evidence suggests that this is because the inflows tend to be used more for consumption than investment in Latin America (World Bank 2001c). This is in line with the region's tendency towards lower savings rates. And it matters what type of investment the inflows are used for. For example, increases in reserve accumulation have seen the relationship between capital inflows and investment rates decrease. This is because the returns on these reserves, held in low yielding US Treasuries, are much lower than they would have been if invested domestically (Bird and Rajan 2003; Rodrik 2006b; Hudson 2010).

Finally, with respect to domestic financial development, the results are also somewhat mixed. It was noted above that there has been a shift in bank activities from cross-border lending to buying or setting up domestic subsidiaries in developing countries. For the banks this helps to minimise currency mismatch. Proponents argue that this means that the developing country banking system will benefit from the import of new techniques, ideas and business practices. Put most simply, competition results in modernisation. Local banks will learn from the new banks, become more efficient and effective, and thus more general benefits will accrue to the developing country's economy, though these benefits vary by country (Clarke *et al.* 2001b). Essentially this is the argument behind the benefits of FDI (see Chapter 7). But the trend towards subsidiaries has also raised concerns that it has actually decreased access to finance. The problem being that the new banks merely crowd out local banks and tend to focus on the best clients. The new bank lending is biased towards larger corporate clients, meaning that small and medium size enterprises (SMEs) will lose access to financing. Indeed it has been shown that foreign bank entry increases total lending, but larger banks of all nationalities are much less likely to lend to SMEs (Clarke *et al.* 2002; Spratt 2009).

In sum, there does not appear to be a strongly positive or general relationship between capital inflows and economic development. The heterogeneity of inflows is one reason for this, the ability of a country to absorb the benefits of capital is another, as is the volatility of these flows. But a further explanation emphasised by liberal institutionalists and critical reformists alike, is the damaging costs of financial crises (Prasad *et al.* 2004). A different way of approaching the development consequences of debt inflows is to look at the relationship between liberalising policies and developmental outcomes as opposed to the impact of capital inflows themselves. A range of important studies have shown that liberalisation of the capital account – i.e. opening up a country to greater inflows – is not associated with economic growth (Grilli and Milesi-Ferretti 1995; Rodrik 1998). It has since been recognised that if financial liberalisation is done too quickly, or different aspects are done in the wrong order, known as sequencing, then capital inflows can produce very negative development outcomes (Kaminsky and Reinhart 1999). Capital surges in particular can be damaging, for example, through exchange rate appreciation. This in turn undermines the competitiveness of the export sector. Within the financial sector, capital inflows are also linked to asset price bubbles as the extra credit goes into land, property and funding financial purchases. Plus, in the case of a sudden reversal, this then has further negative consequences and increases the chance of financial crisis.

Slightly differently, one way in which the bond markets operate as a form of political control is through the actions of so-called 'bond vigilantes'. Bond vigilantes, as the name suggests, are investors in the bond markets who use extra-legal punishment, acting as self-appointed economic guardians. In this case, the vigilantes use the bond market to punish bad or profligate governments. The economic logic runs as follows: the one thing bondholders fear most is inflation. Inflation directly erodes bond yields (see Box 6.4 on why this is the case). Thus, so the argument goes, investors in bonds are particularly averse to fiscally expansionary, deficit-creating government policies. Bond vigilantes punish governments who follow socially progressive welfare policies by driving up the costs of borrowing through demanding higher interest rates on government debts. This impacts directly on the government by making deficit spending increasingly expensive and forcing governments to prioritise balancing the budget instead. Furthermore, it raises interest rates across the entire economy, slowing economic activity more generally. Such pressures on the Clinton Administration in the US led James Carville, one of Bill Clinton's advisors, to famously say: 'I used to think if there was reincarnation, I wanted to come back as the president or the pope or a 400 baseball hitter. But now I want to come back as the bond market. You can intimidate everybody.'

The empirical evidence for this is mixed. It is not clear that the bond market consistently punishes governments for running deficits. Yet, the very fact that policymakers *believe* in the threat of bond vigilantes creates enough pressure to sacrifice deficit-creating budget spending, making welfare-enhancing investment and redistributive policies less feasible. Layna Mosley's (2003) work suggests that developing country governments *are* constrained by financial market pressures, but not fully. Her survey and interview data show that international investors are concerned about the risk of developing countries defaulting on their debt, which means that they tend to monitor policy rather more carefully than in developed countries. Nevertheless, the fact that there is considerable variation in developing country policies suggests that there is room for policymaking autonomy. Meanwhile Geoffrey Garrett (2001) concludes that higher levels of capital mobility in general, and growth in capital from countries opening their capital account, are not correlated with lower levels of government spending. Finally, if and when governments do default, there is only weak evidence that the credit markets punish them. For example, Gelos *et al.* (2011) found that the probability of market access is not strongly influenced by a default in the previous year, it's only about 3 per cent lower than it would have otherwise been. All of which suggests that developing country governments aren't completely beholden to the so-called bond vigilantes.

Crises

When thinking about crises, it is possible to broadly distinguish between two types of crisis: flow crises and stock crises, or liquidity and solvency crises. A liquidity crisis is the result of a sudden stop of capital flows. A solvency crisis is where the countries or companies are effectively bankrupt; this is due to a build up indebtedness to a level where the stock of debt is no longer sustainable and cannot be repaid. The distinction is useful, but it is also important to see the links between the two. A liquidity crisis creates problems for financing ongoing projects but also hits economic agents' ability to roll over, i.e. refinance, debt which is due to repaid. This in itself can lead to defaults. Bank runs are the classic example of how easily a liquidity crisis can happen and how quickly it can result in wider economic problems. Because a bank's deposits are always lent out elsewhere, a sudden generalised demand for liquidity among too many depositors at once can lead to a solvency crisis (Diamond and Dybvig 1983). It is also important to bear in mind the difference between a private banking crisis and a sovereign debt crisis, but again it's common that they tend to occur together, so-called twin crises (Kaminsky and Reinhart 1999). As Reinhart and Rogoff have shown (2011), banking crises tend to precede and precipitate currency and sovereign crises.

Flow crises of surges and sudden stops

Following the financial crises in the 1990s a lot of research was conducted into the causes and consequences of banking and currency crises. One of the key insights to come out of this was that best indicator of a coming crisis was the ratio of external short-term bank lending as a proportion of official foreign exchange reserves (Frankel and Rose 1997; Rodrik and Velasco 1999). It has also been shown that the greater the level of short-term bank borrowing, the more severe the crisis. The other significant leading indicators of crises include real exchange rate overvaluation and the ratio of the money supply to reserves (Edison 2000).

Short term debt may be the culprit in bringing about debt crises, but this may not be something which is in the hands of all developing countries. First, many LICs have little or no choice. Those countries which are perceived as less creditworthy do not have access to longer-term financing and have to rely on short-term debt (Detragiache and Spilimbergo 2002). Second, crises can be triggered regardless of what borrowing countries are doing. It has been shown that the probability of banking crises in emerging markets is more strongly tied to changes in US interest rates, and this effect is significant regardless of the economic situation in the countries in question (Eichengreen and Rose 1998).

It has been calculated that currency and banking crises have resulted in losses of 2.2 percentage points of growth per year for developing countries (Dobson and Hufbauer 2001). Over a generation, i.e. 25 years, this translates into a loss of income for the developing world of around 25 per cent (Eichengreen 2004). The banking crisis in Thailand in 1997 was a result of bad loans made by banks to the real estate and construction sectors. Too rapid financial liberalisation had triggered an asset price bubble meaning that many offices and residential properties were left vacant and banks were holding losses in their loan portfolio. These financial weaknesses set in train a series of events resulting in a speculative financial attack on the Thai baht and large capital outflows. The economic crisis resulted in millions being plunged into poverty, rising unemployment and severe social dislocation. The social costs of financial crises are disproportionality felt by the poorest. The channels through which crises impact are several. For example, the fiscal retrenchment enforced on the government meant a reduction in spending on public goods and social protection, and other redistribution or subsidisation policies. Plus, the more general slowdown in the economy increased unemployment. Price rises for imported foodstuffs, following any currency depreciation, tend to hit the poorest hardest too.

Stock crises of indebtedness

Debt crises can also result from a stock of debt that becomes unserviceable. Debt sustainability is a function of real interest rates and real income. So whether or not debt is sustainable isn't exclusively a matter of how much debt a borrower has. Most obviously it is also a function of the income of the debtor, but also an outcome of the changing costs of repaying the debt. So a debt crisis is often triggered by changes elsewhere – whether in international interest rates or in commodity or other export prices. The Third World debt crisis, beginning in the 1980s, is the best-known and most widespread example of this type of crisis. Moreover, it is the crisis which has had the most significant impact on the economic policies and development trajectories of the developing world over the past three decades. This is because many of the neoliberal economic and social policies which developing countries have followed have been a result of the conditions placed on them by the IFIs – the World Bank and the IMF – in return for helping to reschedule their debt or extend new loans.

There are strong moral arguments as to why developing countries should not continue to pay for debt they can't afford; these are detailed in the policy section below. But what of the economic effects of indebtedness? The economic and social costs of indebtedness flow through two key channels: (1) the opportunity costs of servicing the debt, (2) the costs of a debt overhang on investment and economic growth.

The most straightforward channel through which excessive indebtedness can damage development outcomes is the costs of servicing the debt. That is, the amounts of interest and principal repayments which the developing country government has to make. This reduces the amount of money which the government has to use for other budget items, such as education and health spending, and investment in infrastructure or agriculture. It has been estimated that, on average, an increase of 1 per cent in debt servicing, as a proportion of GDP, reduces public investment by about 0.2 per cent (Clements *et al.* 2003). So, if debt servicing could be reduced by 6 percentage points of GDP and about half of this saving was to go into public investment, economic growth would increase by 0.5 percentage point a year. A considerable gain when compounded over several years.

The second channel through which indebtedness damages developing countries is a disincentive effect on foreign investment. The debt overhang thesis is that the relationship between external debt and economic growth is non-linear. That is to say, there is a threshold or tipping point at which the relationship between borrowing and growth switches from being positive to negative. At first increasing debt results in positive growth as the new resources can be invested and consumed. However, after a certain point the relationship between debt and growth becomes negative as countries over-borrow. This is sometimes represented as a 'Laffer curve' as per Figure 6.10 (Cohen 1993). This is because above a certain threshold, large debt stocks – the debt overhang – act as a disincentive to new investment into the country (Sachs 1988; Krugman 1989). This is because it effectively acts as a 'tax' on new investment as some of the gains from the borrowing are used to pay off existing debtors. There are also additional problems of uncertainty for investors about potential rescheduling, or possible increases in taxation to pay for the debt, and fears that inflation will be allowed to rise to 'inflate away' the real value of the debt and make repayment less costly.

In an empirical test of the impact of external indebtedness on economic growth Pattillo *et al.* (2002) examined a large panel dataset of 93 developing countries over the period 1969–98. If a country with the average amount of indebtedness were to double their debt ratio, it would reduce its annual growth rates around a half to 1 per cent a year. Alternatively, comparing a country with external indebtedness of below 100 per cent of the value of exports with a country with external indebtedness of above 300 per cent of exports, results in a difference in annual growth rates of over 2 per cent. This is a huge difference, especially when compounded over a number of years. In terms of identifying a tipping point, the average impact of debt becomes negative at around 35–40 per cent of GDP or at around 160–70 per cent of exports (measured in net

Figure 6.10 *External debt Laffer curve.*

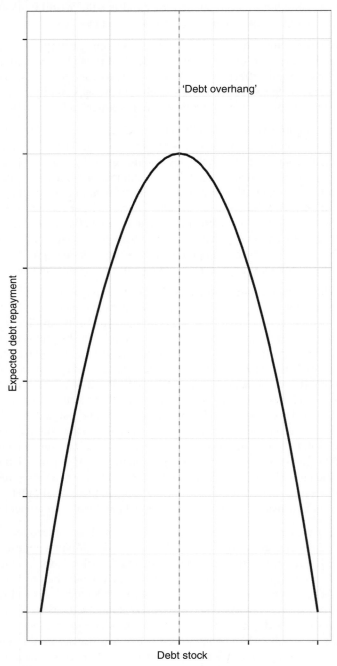

Source: Adapted from Pattillo *et al.* 2002.

present value terms rather than face value, the face value of the debt would be much higher). The marginal impact of debt becomes negative at around half of these values.

Taken together these two channels suggest that there is an economic case for reducing total stocks of debt (to increase external investment) and to reduce servicing costs (to increase resources available for public investment).

The political economy of sovereign debt

In August 1982 the Mexican government announced that it could no longer meet its debt repayments. It declared a 90-day moratorium on its repayments and requested to renegotiate the terms of its outstanding debt. This event is commonly seen as the trigger for what was to become known as the Third World debt crisis. In this section we cover four relevant aspects of the debt crisis. First, what were the causes of the debt crisis? Second, how the crisis was dealt with immediately during the 1980s. Third, what the developmental consequences of the debt crisis and its immediate resolution were. Fourth, how the ongoing problems of developing country indebtedness were dealt with in the late 1990s and 2000s.

The causes of the Third World debt crisis

A number of different factors lie behind the build up of indebtedness but it was their coming together which served to trigger the crisis. It was a toxic combination of rising lending and borrowing, poor economic management, falling commodity prices and an oil spike, and changes in international interest rates.

The historical context for the increasing reliance on bank lending was the decline of ODA from the 1960s. As a consequence aid was increasingly concentrated in the poorest countries of sub-Saharan Africa. At the same time there was a massive increase in private capital being made available to MICs that could no longer access ODA. With hindsight, the decades leading up to the crisis were characterised by what looked like excessive borrowing and excessive lending. Non-oil-exporting developing country debt went from $78.5 billion (end of 1973) to $180 billion in 1976. And, despite a round of debt renegotiations in response to this, by 1981 another surge in lending saw indebtedness soar to $600 billion. About 60 per cent of the total debt was syndicated loans from private banks, but the expansion of the eurodollar market in the 1970s was also key with borrowing in eurocurrency from private banks by non-oil-exporting developing countries going from $300 million in 1970 to $4.5 billion in 1973 (UN DESA 2005).

Domestic factors also played their part. Many countries suffered from poor public sector management. This was not helped by poor donor choices with respect to projects. The geopolitical background of the Cold War was an important explanatory factor here. It meant that the primary purpose of many loans was to maintain support among friendly regimes rather than developmental impact. This meant that corruption was rife and poor economic performance unpunished. With low growth rates and continuing external and budget deficits, developing countries took to borrowing further in order to repay existing loans. The geopolitical context meant, despite all this, lenders kept on lending.

In addition to geopolitical logics, it was also assumed that the export earnings of countries would always rise more than the cost of servicing the debt. Commodity prices rose at a steady rate of 12 per cent throughout the 1970s earning many developing countries decent export earnings. Unfortunately this did not continue. The collapse in commodities prices prior to the 1973 oil crisis created repayment problems for many developing countries. Thus, by 1974 the G77 were already calling for debt cancellation as well as rescheduling (UN DESA 2005). Yet, although many developing countries were facing falling incomes, the cost of borrowing remained low since interest rates remained low. This was because, in contrast to the falling commodity prices elsewhere, the 1973–74 oil price spike generated large export earnings for the oil-producing countries. These earnings were invested by recycling the money through the financial markets as loans to developing countries, especially to Latin America. The increase in global savings created a liquidity glut and meant that interest rates remained low.

The 1970s saw a number of countries going to the Paris Club (see Box 6.7) to seek rescheduling of their debt. But it was not until 1979 that the final jigsaw piece fell into place to trigger a widespread debt crisis; known as the Volker Shock. In 1979 the US – concerned about domestic inflation – increased its interest rates with a number of consequences. First, the interest rate increase had the desired result as economic activity slowed in the US. But this had the knock-on effect of reducing imports, meaning many developing countries' export earnings fell yet further. Second, the interest rate increases also saw repayment rates increase on borrowing. Faced with little option the first response of developing countries was to increase borrowing further to meet the shortfall. A tactic that ultimately proved disastrous.

In 1982, Mexico was the first country to announce it could no longer service its debt obligations. At this point lenders' confidence was finally broken and lending went from being expensive to non-existent. This sudden drying up of liquidity meant that other developing countries could no longer service their debt repayments either, as they needed new loans to service their old ones and roll the debt over.

Box 6.7

The Paris and London Clubs

Debt rescheduling was arranged through the Paris or London Club, for official and commercial debt respectively. The Paris Club was established in 1956 in order to deal with debt servicing problems in Argentina. It is an informal group of 19 permanent members which are all official lenders. The Paris Club has offered different types of debt treatments depending on the situation of the country facing debt repayment difficulties – i.e. the so-called Classic, Houston, Naples and Cologne terms. The Paris Club has no legal status and has always agreed debt treatments on a case-by-case basis. Meanwhile the London Club is an informal group of private creditors – the commercial bank lenders – which was established in 1976 to deal with Zaire's debt problems (now the Democratic Republic of the Congo).

The policy response in the 1980s

The initial response to the widespread threat of defaults by the financial and international community was to avoid a bigger financial meltdown by restructuring the outstanding debts. Many banks would be in serious difficulties if developing country governments defaulted en masse (Sachs and Williamson 1986). The IMF stepped in, providing loans, with conditions, for the developing countries to repay the creditors.

The consequences of the debt crisis were threefold. First, it caused extreme socioeconomic dislocation in the indebted countries with rising poverty, unemployment and falling economic growth. Austerity hit many countries extremely hard, leading to large reversals in development gains. For example, Peru struggled with falling living standards, deteriorating terms of trade, negative growth, falling real wages which by 1985 were half of what they were in the previous decade and a crumbling social fabric, making murder and terrorism 'the daily fare of Lima' (Sachs and Williamson 1986: 405).

Second, it also shrunk the policy space available to developing countries and the restructuring of their economies. This was because of the way in which the IMF reshaped its role within the international system in response to the increasing debt problems and renegotiations. Through conditions and structural adjustment loans the IMF became more directly involved in developing countries' economic policymaking (see Box 3.5, p. 60). To play its new role the IMF developed indicators of countries' need for rescheduling and when countries were again able to continue debt servicing. These included the ratio of debt to exports and GDP, debt servicing as a proportion of exports, GDP, and reserves, and the ratio of amortization to debt (Lee 1993). As we will see these indicators were and remain important.

Figure 6.11 *GNI to debt ratios by region.*

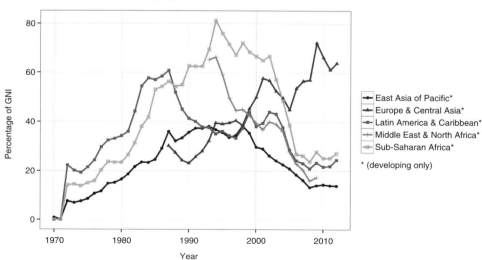

Data: WDI 2013.

Third, developing countries' indebtedness continued to rise. Despite the debt restructuring and structural adjustment the 1980s and 1990s saw a continued build-up of external indebtedness: from $500 billion in 1980, to $1 trillion in 1985, to $2 trillion in 2000. The debt levels of the very poorest countries – HIPCs – rose from $60 billion in 1980 to $190 billion in 1990. It was increasingly recognised that although the Brady Plan (see Box 6.8) had resolved the debt problem for many MICs, there needed to be an alternative solution for LICs. The 1990s were a period of rising debt distress amongst the poorest countries. Continued dependence on commodities – which continued to suffer from volatile and falling prices – led to low growth. This was accentuated by poor revenue management in many countries.

Box 6.8

The Brady Plan

The Brady Plan takes its name from the then US Treasury Secretary, Nicolas F. Brady, who in March 1989 outlined the strategy to break the continuous cycle of rescheduling and restructuring the debt. Although the immediate threat of financial crisis had been averted, and the immediate interests of the commercial banks protected, it was slowly recognised that the economies of the developing countries were stuck with low growth levels, increasing poverty, and were unable to access the international capital markets on account of the debt. Ultimately, if economic growth didn't return many of loans would never be repaid. This was enough to galvanise action.

The Plan has several objectives. First, to reduce the debt servicing levels and make the debt service affordable. Second, to help the developing countries affected regain access to the international credit markets. Third, to diversify the credit more widely throughout the financial system and away from just the badly exposed US commercial banks. Fourth, to force through market-based reforms in the participating countries in return for the debt relief. Assessed on its own terms, the Plan was largely successful.

The basic operation of the Plan was for debtor governments to swap their bank loans for 30-year collateralised bonds. Two general options were available: Par bonds or discount bonds. Par bonds were where the face value of the debt remained the same (hence par), but the government faced a concessionary fixed rate of interest, guaranteeing lower debt service repayments. Discount bonds were where the government swapped its loans for a bond which still paid the market-based rate of interest, but the value of the debt was reduced, in the region of 30–50 per cent. Each country negotiated debt relief terms on an individual basis, so there was no set formula. Mexico was the first country to restructure its debt under the Brady Plan over the course of 1989–90. Mexico was followed by Argentina, Brazil, Bulgaria, Costa Rica, the Dominican Republic, Ecuador, Ivory Coast (Cote d'Ivoire), Jordan, Nigeria, Panama, Peru, the Philippines, Poland, Russia, Uruguay, Venezuela and Vietnam.

The Brady Plan was successful insofar as it helped reduce the debt servicing for participating countries and allowed them to regain access to the international capital markets. But the LICs – especially in sub-Saharan Africa – as opposed to the middle-income, emerging markets that were involved in the Brady Plan, remained indebted until the HIPC and MDRI Initiatives in the late 1990s and 2000s. The Brady Plan was also successful in diversifying the debt away from the banks and creating a secondary market in 'Brady bonds', hence helping to encourage the shift from bank to bond flows. Most Brady bonds have now been bought back by the original issuing countries.

Source: EMTA (2009)

Debt relief

The debt restructuring of the London Club (the equivalent of the Paris Club for commercial banks) and the Brady Plan had proved a partial success: financial meltdown was avoided but it left many of the poorer countries with continuing debt problems. There was also a shift in the nature of the outstanding debt, with an increasing proportion of it being multilateral debt. It was increasingly recognised that these were not temporary payments problems, but had in fact become structurally unmanageable debt burdens with many individual countries facing debt servicing levels which were over half of their export earnings. As such, a more systematic approach was called for, and one which would be focused on dealing with the multilateral debt. In October 1996 the HIPC Initiative was launched by the IMF and the World Bank to address the problems of the HIPCs. This was following in 1999 by the enhanced HIPC Initiative. And in 2005 HIPC was supplemented by the MDRI. The logic behind the HIPC Initiative was to reduce external indebtedness in order to increase pro-poor government spending.

The HIPC Initiative follows a two-step process (IMF 2010b). First, to receive interim debt relief, the IMF and World Bank formally decide whether a country is eligible for debt relief. This is called the decision point. To reach the decision point a country must (1) qualify for the World Bank's IDA lending and IMF's extended credit facility, (2) face an unsustainable debt burden which cannot be solved through traditional debt rescheduling, (3) have established a track record of 'reform and sound policies'; and (4) have established a PRSP. To move from decision point to completion point, where full and irrevocable debt relief is granted, a country must (1) continue with its track record of good performance under programmes supported by IMF and World Bank lending, (2) implement reforms agreed at the decision point and (3) implement the PRSP for at least one year.

As of October 2012, 39 countries are (potentially) eligible for HIPC Initiative assistance, 34 countries have reached completion point, a further 2 have reached decision point and are receiving interim debt relief, and another 3 are eligible but have not reached the decision point.

The outcomes of debt relief

Two economic logics for reducing the debt burden – the costs of debt servicing and overhang – were outlined above. But, as well as being an economic argument, the debt servicing logic is also an inescapably moral argument too. The World Bank (2001b) reports figures from UNICEF and Oxfam that in 1999 six heavily indebted countries in sub-Saharan African were spending over a third of their national budgets on debt servicing. At the same time they were spending only a tenth on basic social services. This is both a moral and an economic case for forgiving the debt of heavily indebted countries. As Jeffrey Sachs has stated: 'No civilised country should try to collect the debts of people that are dying of hunger and disease and poverty' (cited in England 2004).

Meanwhile, in addition to focusing on the consequences of indebtedness, Noreena Hertz has written about the immorality of how the debt has been accrued. She argues that the debts are 'odious debts'. Odious debts are illegitimate debts. The term was defined by Alexander Sack in 1927. Odious debts were not used for the purposes they were intended (often in the full knowledge of the lender) and didn't benefit the population of the country. The South African debt incurred under the apartheid regime 1985–2002 is a classic example (Howse 2007). Private banks continued to lend money to the apartheid regime, despite it being under sanctions from UN, and used this money to shore up the government and oppress many of its citizens in a racial dictatorship.

The moral case for debt relief is often well made. But what have the consequences of debt relief been? Does the removal of the high debt burdens provide a potential route out of permanent debt rescheduling by indebted countries, placing them onto a path of growth and development, being able to spend freed-up monies on welfare provision?

Nguyen *et al.* (2005) find that there are some gains to be made by reducing the debt stock (and therefore overhang) but only after its face value exceeds 50 per cent of GDP. They suggest that improvements in growth rates in the 2000s can probably be attributed to debt relief. Furthermore, they argue that to reap the real benefits of debt relief, reducing the stocks has to go hand in hand with reducing the service payments and getting governments to channel these savings into public investments. Without both of these the full benefits of debt relief won't be realised. Meanwhile, Depetris *et al.* (2005) provide a considerably less rosy picture. In their own words, they 'find little evidence' that debt relief has produced positive outcomes in developing countries' public spending or raised growth, investment rates or the quality of policies and institutions. They don't argue that debt relief is useless, but just that we shouldn't expect huge gains.

In the best book length-treatment of the issue, Geske Dijkstra (2008) provides a great organising framework for unpacking the question. She differentiates between stock, flow and conditionality outcomes. The stock effects of debt relief work through removing the consequences of debt overhang, increasing private investment and the creditworthiness of the country. The flow effect relates to the increase in resources available to governments to spend on development goals such as increasing welfare. Finally, conditionality outcomes relate to the impact of conditions of debt relief on achieving policy change. Dijkstra's conclusions are largely pessimistic. She argues that too much emphasis was placed on resolving the debt service flows rather than underlying stock. Furthermore, the initial levels of debt relief were too small to see stock effects, the wrong modalities were used, and the debt relief was not additional to existing aid flows. Moreover, debt relief was given alongside new loans, creating new indebtedness. She also highlights the negative consequences of the conditionalities attached to new lending and debt relief. Her conclusions, therefore, are more a criticism of the way in which debt relief has been carried out than of debt relief per se.

Ongoing issues with sovereign debt

More recently, critics have suggested that current lending by the IMF in the wake of the financial crisis contains the seeds of a new debt crisis. The lending was necessary given the global liquidity crisis. But some observers argue that the IMF needs to be more careful about lending to poor debt-sensitive countries (Leo 2010). Debt sustainability is key here. The IMF and the World Bank have a Debt Sustainability Framework (DSF) which dictates whether a country should receive loans or grants. The DSF determines up front the balance of grants and loans depending on a country's ratios and on its institutional strength. A traffic-light system is used and so-called red-light countries – those with high debt levels, in debt distress, i.e. in arrears or default – should not receive grants. However nearly half of IMF lending in 2008 and 2009 went to red-light countries and the amount the IMF lent to HIPCs in new loans in 2008 and 2009 was equal to $2.6 billion (the amount of debt relief agreed at Gleaneagles in 2005!). For example, the IMF disbursed US$276 million in new loans to the Democratic Republic of the Congo in 2009, yet its debt-to-GDP ratio is almost 120 per cent. Meanwhile, through the HIPC Initiative the IMF is also committed to roughly US$320 million in future debt relief to the Democratic Republic of the Congo. Hence we see an effective cancelling out of debt relief.

Finally, debt prevention is clearly important, but so is debt resolution. Given the re-emergence of possible sovereign defaults on the international political agenda following the global financial crisis and the evidence that increased indebtedness is again on the rise, the case for some sort of mechanism to manage debt restructuring in an orderly fashion is due. Collective action clauses have tended to be inserted into bond contracts which have been issued since 2002 (World Bank 2005a). But there is no systematic process available. The Monterrey Consensus contained a reference to developing a 'sovereign debt workout mechanism'. In 2001 the IMF published just such a proposal: the Sovereign Debt Restructuring Mechanism (SDRM) (Kreuger 2002). The SDRM was designed to be a kind of voluntary 'Chapter 11' for countries, similar to the bankruptcy provisions in domestic law (Chapter 11 is a US reference). However, due to US opposition to the idea it never got beyond the proposal stage. New sovereign debt crises may well force the issue back onto the table.

The other threat to developing countries in terms of the consequences of indebtedness is the phenomenon of 'vulture funds'. Essentially these are companies – often private equity or hedge funds – that have bought up old distressed debt. So this is debt that in all likelihood won't ever be repaid because the debtor can't – see Box 6.9 for the case of Zambia. Vulture funds take advantage of the situation by paying a heavily discounted price for the debt and then aggressively pursuing the debtor, often through the courts, to get back more than they paid for the debt. Politicians, NGOs, the IMF and the World Bank have all spoken out against the practice, not least because it undermines the gains from debt relief, i.e. interfering with the 'orderly restructuring of sovereign debt'. Plus, most people find the practice of profiting from financially distressed developing countries morally repugnant. Gordon Brown, back in 2002, said at the UN, 'We particularly condemn the perversity where Vulture Funds purchase debt at a reduced price and make a profit from suing the debtor country to recover the full amount owed – a morally outrageous outcome'. Other examples of vulture funds going after debt they have bought include Peru, the Democratic Republic of Congo, and most recently, Argentina in 2013. Both the UK and US moved to address the problem through introducing legislation.

Box 6.9

Donegal International and Zambia

Where did the debt come from?

In 1979, Zambia was given credit of $15 million by Romania towards agricultural machinery and vehicles (not all of which arrived in usable condition). By the 1990s, because of crushing poverty, it was unable to repay its external debts and became eligible for the process to qualify for debt relief. It began negotiating with its creditors about partial payment of old, bad debts.

How did Donegal get hold of the debt?

In 1999, as Zambia was trying to negotiate clearing the debt it owed to Romania, Donegal International swooped in and bought up the debt – then valued at around $30 million with accrued interest – for a knockdown price of $3.3 million. It then sued Zambia for the full amount of the debt, plus compound interest, demanding a staggering $55 million in total!

Why didn't Donegal get their full claim?

The judge ruled Donegal's claim of the full worth of the debt was not justified. But in the end he ruled that, legally, it was entitled to something from Zambia, which he judged as being $15.5 million. These rich pickings might be just business to Donegal. But for Zambia, it is money that could train doctors and nurses, pay teachers and build hospitals.

What did Zambia say about the debt?

Zambia claimed in court that Donegal had got hold of the debt – and an earlier agreement by Zambia to pay – dishonestly, by bribing officials. The judge found that there was not enough evidence of this, but he did state that the businessmen controlling Donegal had been dishonest in court. In the final hearing, he said, 'I do regard the dishonesty with which I was confronted as quite serious . . . it is not what you might call individual fibs popping up in the witness box here and there.'

Shouldn't Zambia just pay up?

Zambia is a very poor country, which needs its money to meet the needs of its own people. Eighty per cent of Zambians live on less than $1 a day.

Moreover, it is not even as if this is a debt that Zambia incurred towards Donegal: Donegal bought the debt at a vastly reduced price at a time when it knew perfectly well that Zambia was desperately poor and considered so indebted that it had to have debt relief. In fact, Donegal even used this fact in its attempts to persuade Romania to sell it the debt.

Other creditors have, in good faith, cancelled debts that are in some cases much larger. Rich country creditors, including the UK, have agreed to cancel Zambia's debt to them on the understanding that Zambia should then have extra funds available for poverty reduction, rather than extra funds to repay other creditors in full. Zambia had to promise these countries that it wouldn't give better terms to other creditors: Donegal is now forcing Zambia to break this promise by demanding a huge sum that could otherwise go to poverty reduction.

Source: Taken from Jubilee Debt Campaign, 'End the Culture Vulture, Case study: Zambia', Online. Available:
http://www.jubileedebtcampaign.org.uk/Case3720study373A3720Zambia+2968.twl

Conclusion

Private capital flows in the form of debt – whether through bank lending or bond issues – offer developing countries an important source of development financing. Access to credit from external sources of borrowing offers the potential to smooth consumption and boost future growth through investment. The growth of bond lending to developing countries – even with the caveat that it is highly concentrated – is important but certainly not unprecedented. Historically, bond issues accounted for a large proportion of flows to developing countries in order to develop infrastructure – British and German investors buying up, for example, Argentine bonds used to finance the building of the railroads. However, after World War I and in particular the Great Depression, international capital flows were largely restricted to FDI. In time bank lending grew again in the postwar period but it was the 1980s debt crisis that was crucial in determining the re-emergence of bond flows. The twin responses by the international community of first issuing Brady bonds and secondly the policy conditionalities of the SAPs, including the liberalisation of capital accounts, resulted in the re-emergence of bond issues (portfolio debt flows).

However, despite increased flows of debt-creating finance to developing countries there remain problems. First, these private flows are incredibly concentrated in a few large MICs and do not offer much for lower-middle income, low-income, and small poor countries. Second, there is much evidence to suggest that debt flows are volatile, pro-cyclical and are crisis-prone. Plus, there is a good deal of evidence to suggest that these negative aspects of debt flows are endemic to the financial markets and linked to the economic conditions in the wealthy countries, and measures taken in developing countries can be swamped by these factors. The success of attempts to deal with financial crises and the debt crisis have been mixed. Many countries have benefited from having their debt levels reduced, but not as much as was hoped. The structural causes of international poverty clearly lie beyond just dealing with indebtedness.

The recognition that external short-term borrowing, especially through the banking system, has been a key factor in developing country financial crises has resulted in support for measures which help limit developing countries' reliance on this kind of financing. For LICs this is through alternatives such as ODA and long-term official lending, and for MICs this is through FDI, and increasingly international remittances for all developing countries. It is to these sources of financing that the book turns next.

Discussion questions

1   In your view, are debt crises an inevitable part of development? And ultimately do the potential benefits of borrowing outweigh the costs?

2   Are international or domestic factors the primary driver behind developing country indebtedness?

3   Is debt relief a radical initiative, or an orthodox measure to maintain the continued smooth operation of capitalist finance?

4   How much policy space do developing countries have with respect to the demands and expectations of international investors and bank conditions?

5   Are vulture funds morally repugnant or are they a justified business model and help to minimise systemic moral hazard?

Further reading

- For a really nice, critical and comprehensive coverage of the debt crisis – its nuts and bolts and the consequences for the relationship between developing countries and the international financial institutions, take a look at *Debt, the IMF, and the World Bank, Sixty Questions, Sixty Answers* by Éric Toussaint and Damien Millet (2010). Ultimately they advocate complete forgiveness. It uses a very accessible Q&A format.

- David Graeber is an anarchist anthropologist, but don't let that put you off reading his *Debt: The First 5,000 Years* (2011), it's brilliant. Graeber provides a historically grounded and politically attuned analysis of how debt is related to money, the economy and community. It serves to strip away a lot of myths, plus is really well-written. Highly recommended.

- Although it's a little dated now *I.O.U.: The Debt Threat and Why We Must Defuse It* is a very readable account of how developing world debt came to be and what the author – Noreena Hertz – argues we need to do about it. Lots of praise on the jacket from rock stars, but it's sound stuff and comes complete with some considered solutions.

Useful websites

- The World Bank's websites on debt relief and debt management are both worth taking a look at. They provide details of how the Bank is dealing with developing country indebtedness as well as containing lots of useful links to the latest publications and the latest data too. Debt Management http://go.worldbank.org/W7V1F1A6S0. Debt Relief http://go.worldbank.org/N8RWGKBOA0.

- The Jubilee Debt Campaign was behind the 24 million person petition back in 2000 and tirelessly campaigned to get debt relief on the political and economic agenda. They succeeded. But they continue to campaign because they feel that many unpayable poor country debts remain and work on issues like vulture funds. The website is a great resource with lots of information and – perhaps most usefully – provides the most succinct (critical) evaluation of where the debates about developing country indebtedness are now: http://www.jubileedebtcampaign.org.uk/.

Equity finance: foreign direct investment and portfolio equity

Learning outcomes

At the end of this chapter you should:

- Be able to distinguish between FDI and portfolio equity (or indirect investment), and how they are defined and measured.
- Appreciate the increasing importance of FDI to many developing countries as the largest element of global financial flows.
- Understand the history and geography of equity flows to the developing world, especially their concentrated nature.
- Understand the key drivers and developmental impact of equity, the role of transnational corporations (TNCs) and financial markets, and the importance of spillovers, complementarities and linkages.
- Be aware of the controversial nature of foreign investment – despite the much-touted benefits of equity flows – in terms of the debates around exploitation, extraction, dependency and their pro-cyclical nature.
- Understand the contemporary policy and political debates around regulation and taxation.

Key concepts

Equity finance; portfolio equity; foreign direct investment (FDI); transnational corporations (TNCs); determinants; development impact; growth; inequality; export processing zones (EPZs); spillovers; linkages.

Introduction

The previous chapter looked at the two main debt-based types of development finance. Equity finance is the second major category of private capital flows to the developing world. This chapter will turn to the two equity-based forms of financial flows to developing countries: FDI and portfolio equity flows. In contrast to the last big rise in private capital flows to the developing world in the 1970s which was bank-led, the more recent peak has been portfolio equity and direct investment flows. The commonly cited advantage of equity financing is that it does not incur a debt on behalf of the actor seeking capital. Instead the investor makes the capital available in return for a share of ownership in the enterprise, which is where the notion of shareholders comes from.

This means that a larger proportion of the risk is borne by investors. However, it also raises very sensitive issues about foreign ownership of companies, the balance of power between large transnational corporations and developing country governments, and exploitation within the international division of labour.

As noted, equity financing can be further split into FDI and portfolio equity. However, the distinction between the two is often clearer conceptually than it is in reality. But the distinction remains intuitive nevertheless. The difference between portfolio equity and direct equity investment is that in the former the investment is limited to purchasing shares or equity in an enterprise with the purpose of making a financial profit; perhaps a return from a revenue stream such as dividend payments and/or an increase in the value of the shares. These investments can be done privately and directly or through stock markets where companies are publicly listed for investors to buy shares in them in order to raise financing. The purchasers of shares can be private investors, or more likely large institutional investors (such as pension, insurance or mutual funds) which often invest their funds in equities to hold in their portfolio. In order to diversify their exposure to credit risk, investors often hold lots of small investments across different firms, industries and countries.

In contrast, direct investment is made for the purposes of management control over the physical assets of an enterprise. Another way of thinking about it is that the investment is made for business rather than financial purposes. As such, the share of the investment is often much larger. Instead of trying to control their risk by diversifying their portfolio across many small investments, direct investors seek to control risk by actively guiding the management of the firm and making sure that it is a success. The common vehicle for FDI is the TNC. According to orthodox proponents, because FDI goes to the private sector, rather than the public sector, and supports investment rather than consumption, it is viewed as having a larger catalytic effect on overall economic growth than, say, ODA. Moreover, because it comes from the private sector it is assumed that it's driven by stronger efficiency concerns than public sector aid.

Box 7.1

When is it direct and not portfolio investment?

For accounting purposes distinguishing between portfolio and direct investment is done by asking how much the investment is as a proportion of total ownership: the larger the share of the investment the more likely it is a direct rather than portfolio investment. This distinction is far from watertight, but the desire to make a distinction for accounting purposes means that many countries have settled on a percentage figure, often quite arbitrary, to indicate when an investment becomes a direct investment. International standards have settled on 10 per cent (see below). But this isn't exclusively used. For example, Korea and the Netherlands will ignore investments of greater than 10 per cent of ownership if the non-resident investor isn't deemed to have an effective voice in management. On the other hand Turkey doesn't use any percentage of ownership criterion – all enterprises with foreign ownership are treated as FDI, regardless of the percentage of ownership by non-residents (IMF 2003).

At the international level the OECD provides the 'benchmark' definition of FDI (OECD 2008b: 17).

> Direct investment is a category of cross-border investment made by a resident in one economy (the direct investor) with the objective of establishing a lasting interest in an enterprise (the direct investment enterprise) that is resident in an economy other than that of the direct investor. The motivation of the direct investor is a strategic long-term relationship with the direct investment enterprise to ensure a significant degree of influence by the direct investor in the management of the direct investment enterprise. The 'lasting interest' is evidenced when the direct investor owns at least 10% of the voting power of the direct investment enterprise.

> The emphasis on control and lasting interest are the significant conceptual points, the proxy of 10 per cent of voting shares is the practical response. In contrast, the OECD defines portfolio equity investment as that which is 'not made to acquire a lasting interest in an enterprise' (OECD DCD/DAC 2007a) and where 'investors do not generally expect to influence the management of the enterprise' (OECD 2008b). The OECD alongside the IMF has been trying to standardise the way in which countries compile and report their balance of payments statistics and international investment positions. This has been done through the snappily named Survey of the Implementation of Methodological Standards for Direct Investment (or SIMSDI for short) which was launched in May 1997 and is revised every five years or so.

While both forms of equity finance avoid generating indebtedness, there are benefits and disadvantages that are specific to each. In general, certainly in recent years, FDI has tended be viewed in a more positive light; indeed as the best form of capital for sustainable development. Liberal proponents make a number of arguments in its favour. First, and most basically, FDI offers an important, and debt-free, way of adding to a developing country's capital stock. Second, theoretically the interests of FDI and TNCs in creating wealth are more aligned with the interests of the host society. FDI offers greater risk-sharing characteristics than debt, so if things go wrong it isn't exclusively the responsibility of the borrower. Although debt can involve lenders taking a hair cut (absorbing the losses), as we saw in the previous chapter this isn't usually the first port of call. Moreover, unlike lending, investors tend to take more interest in the success of the investment, as the return is proportional to success as opposed to the fixed income stream of debt repayments. Third, proponents of FDI emphasise a series of indirect benefits, so called spillovers. In the words of Paul Romer (1993), FDI helps fill 'ideas gaps' as well as 'object gaps' (i.e. physical capital and investment). Romer (1993: 543) argues that developing countries do not just suffer a lack of 'valuable objects like factories, roads and raw materials' but the other reason they are poor is 'because they do not have access to the ideas that are used in industrial nations to generate economic value'. FDI and the activities of TNCs help foster development through importing know-how and advantages in technology, management skills and export expertise. Finally, FDI flows have tended to be much more stable than debt flows, as well as portfolio equity. This is because of the sunk costs of direct investment; that is to say, the investment is often made into physical capital such as machinery, buildings, etc. As such, FDI is less likely to pull out en masse with short-term changes in economic sentiment (World Bank 2006a).

The benefits of portfolio equity investment, most especially from a neoliberal perspective, but in a qualified form from liberal institutionalists too, are less direct, but potentially no less important. Like FDI, equity investment can be beneficial because of its risk-sharing characteristics (Claessens 1995). But in addition, inflows of portfolio equity boost and develop domestic stock markets. As outlined in Chapter 2, larger and deeper stock markets help to increase firms' ability to raise capital to fund enterprise, partly because this increases the amount of capital available through a more efficient and greater reallocation of capital from savers to investors, and partly because it helps reduce capital costs. In addition to creating a deeper pool of liquidity, stock markets are held to play an important signalling mechanism for the economy, helping to allocate capital more efficiently and offering investors the ability to diversify their holdings over a wider range of assets and therefore reduce the risk they bear, further reducing the cost of capital. Liberal proponents also argue that deeper stock markets, with greater investment opportunities, unleash the forces of competition and allow investors to reallocate away from poorly performing enterprises encouraging better corporate management (Diamond and Verrecchia 1982).

Because of these perceived advantages, recent years have seen a surge in private capital flows to developing countries, peaking in 2007. Much of this increase is specifically due to increases in equity finance flows as there has been a move away from debt financing and towards equity financing, particularly FDI (see Figure 2.4, p. 40). This has been especially pronounced in East Asia and Latin America, the two regions that suffered the most from financial crises in the 1990s (see Figure 7.2). The shift to equity financing has been a deliberate strategy on behalf of developing countries to reduce their dependence on external debt (World Bank 2006a).

However, in contrast, critical reformists and radicals (and some liberal institutionalists) are much more antagonistic or downright hostile towards private equity flows to developing countries. Critics charge that the way in which developing countries seek to attract these private capital flows as well as the manner in which TNCs bargain and conduct their business in poor countries results in a 'race to the bottom' (Tonelson 2002). That is to say, the pressure on developing countries to win a 'beauty contest' to attract potential investors means that wages, labour and environmental standards are under constant downward pressure. The playing field between under-resourced developing country governments and mobile capital is an unlevel one, forcing governments to offer sweeter deals at the cost of protecting a country's labour laws, environment and tax receipts. Competition is detrimental to development, not necessarily an ally. And when it comes to portfolio equity flows, critics argue that far from encouraging a more efficient allocation of resources, stock markets encourage speculation and more closely resemble casinos than rational pricing and allocation mechanisms. When 'hot money' rushes into a developing country it causes inflation and price bubbles as assets become valued at many times their true value. This results in a *worse* rather than improved allocation of resources in the economy, for example pushing up land and property prices and diverting investment into building high-rise luxury apartments, as happened in Thailand. And these flows often reverse quickly when the bubble bursts, resulting in high costs. Bangkok's skyline is dominated by unfinished empty towers. Hot money resulted in asset inflation, bad loans, followed by bankruptcies and collapse.

Liberal institutionalists acknowledge that managing foreign investment successfully is incredibly important for developing countries since FDI represents the largest capital flow for the developing world as a whole. Institutionalist research suggests that deriving benefits from FDI depends as much on how the host country manages the inflows as it does on the type of assets bought and the nature of the firm. The consistent lesson from this research suggests that developing country governments must seek to embed the TNC into the local economy and create economic links with local enterprises such as the local sourcing of inputs. Critical reformists go further than liberals, noting that it is the larger and more activist governments which tend to do better, with South Korea and Taiwan often cited as successful cases. It is not just a case of developing a pro-investment and market-friendly environment, reinforcing private property rights and limiting the risk of expropriation, but that governments can and should actively manage the market.

This chapter is structured as follows. Having introduced the difference between portfolio equity and FDI, the next section details the increasing importance of both of these flows for developing countries, but again highlighting their concentrated geography. The next section explores the literature which deals with the determinants of equity flows: essentially why do investors invest where they do, why do TNCs locate where they do? This provides some insights for government policymaking. The subsequent section evaluates the developmental impacts of inflows, detailing the evidence on capital formation, economic growth, poverty and inequality. Plus we review the mechanisms through which impacts travel. The evidence is mixed, but liberal institutionalist insights about good institutions and absorptive capacity as well as critical reformist insights about managing the market both appear to suggest why some countries have been successful and others could be.

Volume and geography of equity finance

There have been considerable surges in investment flows to developing countries, and in the aggregate they now dominate all other forms of capital flows. The notable increases occurred in the 1990s and in the mid-2000s, before the East Asian and the global financial crisis broke, respectively. These surges were in part driven by a move away from debt financing to equity financing by developing countries. Figure 7.1 illustrates a number of things: the secular growth in equity flows, the relative composition of portfolio equity and FDI, and the shifting balance away from debt towards equity for developing countries as a whole.

Portfolio equity flows hit a high of $135.4 billion in 2007. But after the crisis broke they reversed sharply to turn negative, recording an outflow from developing countries

Figure 7.1 *FDI and portfolio equity flows to developing countries.*

Data: WDI 2013.

of $57.1 billion (World Bank 2010a). FDI flows, while slowing slightly, continued to grow through 2008 reaching $593.6 billion. Moving forwards, in the view of organisations such as the neoliberal Institute of International Finance, emerging markets offer the greatest growth potential for the world economy and a good deal for international investors. Indeed, in 2011, FDI to developing countries reached a new record of $684 billion (UNCTAD 2012).

FDI

FDI has grown massively as a proportion of all resource transfers to developing countries. FDI predates World War II and was part of colonial financial transfers and indeed political control. But it was after World War II that modern FDI took off (Sumner 2005). FDI grew twice as fast as economic output in the 1950s and 1960s and then really took off in the early 1990s as shown in Figures 2.4 (p. 40) and 7.1. As a proportion of all capital flows to developing countries it grew from very low levels in the early 1990s to a massive 83 per cent in 2001 before dropping off again. Indeed, the growth in the volume of FDI and the recognition of the increasing importance of 'non-debt-creating-flows' meant that in 1996 the World Bank's World Debt Tables were changed to the now familiar Global Development Finance to include FDI and equity portfolio flows (Woodward 2001).

FDI flows, like the other forms of private capital, have been dented by the global financial crisis. Global inflows of FDI declined by 16 per cent in 2008 and 37 per cent in 2009 to $1.1 trillion. FDI inflows into the developing world fell by 27 per cent to $548 billion in 2009 (UNCTAD 2010). This was a smaller decline than for the developed countries, for which it was 44 per cent. There has been a growing trend of more and more FDI going to developing countries and it represents nearly half of global flows (UNCTAD 2012). China is now the second largest recipient of FDI flows, after the US.

Measuring FDI and the composition of FDI

The key source for getting inside FDI is the United Nations Conference on Trade and Development, or UNCTAD, as it is more commonly known (data also comes from the IMF's balance of payments database, see the links at the end of the chapter). The latest data and extended commentary and analysis is published in July each year in UNCTAD's *World Investment Report*. A word of warning: data on FDI is notoriously shaky. A common complaint is the discrepancies between total inflows and outflows of global FDI. In principle the two should be equal: outflows from one country being inflows into another, but some years ouflows are higher than inflows (e.g. 2010 there was $80 billion difference) and some years inflows are higher than outflows (e.g. 2009 there was a $15 billion difference) (UNCTAD 2011). Partly this is due to differences in national reporting schemes, partly because of blurred definitions between FDI and portfolio capital, and partly because of errors.

UNCTAD breaks the FDI data into various categories which are important to clarify; for, similar to ODA, the heterogeneity of FDI is important – not all flows are created equally, as it were. There are five key accounting distinctions worth flagging up: (1) the components of flows, (2) the distinction between stocks and flows, (3) the distinction between mergers and acquisitions (M&A) and greenfield investment, (4) the sectoral composition of FDI flows and (5) the activities of TNCs.

First, an FDI flow is composed of three different components; the major element is the equity investment component, accounting for around two-thirds of the total, and the other two components are intra-company loans and reinvested earnings, which are the larger and smaller elements of the remaining third (Sumner 2005). These distinctions matter in terms of the stability of FDI flows. Indeed, one of the key benefits and characteristics of FDI is that it has characteristically been more stable than other forms of private capital. FDI has been demonstrated to be less volatile than bank lending, or equity portfolio or bond flows (Mody and Taylor 2002; Prasad *et al.* 2004). The World Bank (2006a) calculated a measure of stability for 72 developing countries with access to international capital markets over the period 1980–2004 by using the correlation between current and previous levels of capital flows. If the flows were perfectly stable the measure would equal 1. FDI on average is 0.61, while debt is less stable at 0.52. This relative in/stability is graphically reflected in the contrast with portfolio flows to developing countries since the global financial crisis (Figure 7.1). Moreover, a regional focus shows that FDI flows even held up over the 1998–2002 period when the Asian and Russian crises occurred. However, it is important to highlight that it is the equity capital component of FDI which tends to be stable. The intra-company loan and reinvested earnings elements are as volatile as portfolio equity and debt flows.

Second, FDI figures are reported by UNCTAD as both stocks and flows. Most basically, as the labels suggest, inflows and outflows capture the value of foreign investments over a

Figure 7.2 *FDI flows by region before, during and after the Asian financial crisis.*

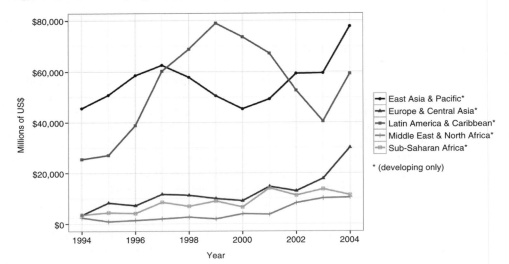

Data: WDI 2013.

defined period of time. In turn, these flows add to or subtract from the total amount of foreign investment in a country, and this amount is the stock. In reality the way in which TNCs deal with their profits complicates this picture slightly (Woodward 2011).

Third, there are two broad types of FDI: (1) M&A, and (2) greenfield. M&A is the form of direct investment which purchases existing productive capacity, i.e. investing in a firm which is already trading. On the other hand, FDI that results in the creation of new productive capacity through the building of a new warehouse, factory or quarry is classified as greenfield investment. Most M&A activity occurs in the richer world, accounting for just over two-thirds of the global total. But the developing world receives over half of greenfield investment flows. Most of the decline in FDI following the financial crisis is explained by a decrease in M&As which contracted by 34 per cent in 2009, whereas greenfield investment fell by only 15 per cent (UNCTAD 2010). This reflects the fact that M&A is usually a shorter-term investment than greenfield projects and tends to rely on price signals from stock markets, which are more volatile. As such, private equity funds that tend to fundraise through the financial markets for leveraged buy-outs were hit particularly hard (UNCTAD 2010).

Fourth, UNCTAD also reports on FDI across different industries. The broad categories are the primary, manufacturing and service sectors. Originally most FDI was concentrated in the extractive industries such as mining and primary commodities. But a couple of important shifts have taken place since 1980 (Woodward 2001). First, manufacturing FDI used to be driven by market access concerns: TNCs looked to invest in companies that sold to domestic consumers. But more recently it has shifted away from financing import substitution to enabling export-led growth. In part this has been driven by the processes of globalisation and liberalisation resulting in falling

Figure 7.3 *Primary manufacturing services as proportions of M&A FDI.*

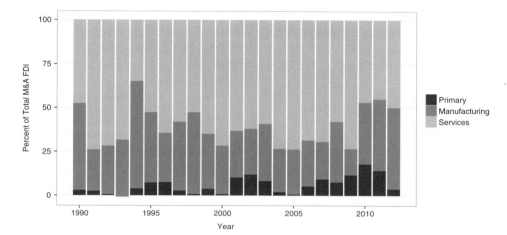

Data: OECD World Investment Report 2013.

transport and communications costs, falling tariffs and trade barriers, and a rise in the number of TNCs operating across several territories and engaging in intra-firm trade. Second, and more recently, there has been another sectoral shift with an increase in the amount of FDI accounted for by services (Sumner 2005). Again, this has been driven by the globalisation of information services, liberalisation and, crucially, privatisation, with financial services, consumer care and data storage being off-shored from OECD economies to the likes of India.

Finally, the manifestation of FDI is the TNC. In 2009 there were 82,000 TNCs worldwide. The foreign affiliates of TNCs account for 11 per cent of global GDP and employ 80 million workers (UNCTAD 2010). UNCTAD (2007) has an interesting set of transnationality measures and indexes to assess the extent to which TNCs are embedded in their home or host countries. The transnationality index (TI), for example, examines the ratios of foreign assets–total assets, foreign sales–total sales and foreign employment–total employment. The newer internationalisation index (II), also from UNCTAD, measures the ratio of the number of foreign to the total number of affiliates and how many countries the company's operations and interests are spread over. Nestlé comes out as the most transnational company.

Geography of FDI

The geography of FDI is very uneven, and like other private capital flows it tends to be very concentrated. Figure 7.2 shows the regional distribution of FDI over a particular period; note that Africa received just $53 billion in 2009 and this had fallen to $43 billion by 2011. Moreover, like other capital flows it tends to be very concentrated in just a few countries, generally the larger emerging markets and other

Figure 7.4 *FDI inflows as a percentage of GDP across regions.*

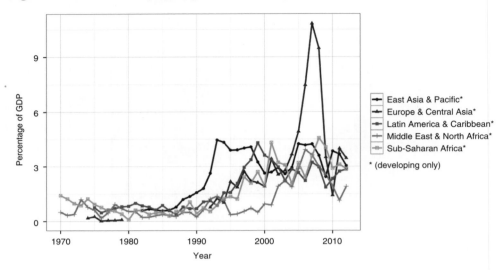

Data: WDI 2013.

higher income developing countries. Figure 7.4 shows how *relatively* important FDI inflows are as a proportion of the overall size of the economy.

There has been a small but steady decrease in the concentration of FDI in recent years (which was accelerated slightly by the global financial crisis), but it remains very high. In the mid-1990s the top 10 recipients of FDI – China, Russian Federation, Brazil, Mexico, Czech Republic, Poland, Chile, South Africa, India and Malaysia – accounted for 75 per cent of the total going to developing countries. This now stands at around 65 per cent. The LICs now account for around 10 per cent of inflows to all developing countries. While this has been increasing over time, it is mostly accounted for by resource-seeking FDI. Increases in commodity prices – from the late 2000s – served to drive more FDI investment into resource-rich countries. Like aid and remittance flows, this geographical concentration of flows changes when measuring FDI as a proportion of GDP (see Figures 7.5 and 7.6). Like aid (Chapter 5) and remittances (Chapter 8) this alternative measure catapults many poor and small countries into the top 10.

This suggests it is very hard for developing countries that are not emerging markets, namely the poorest and most vulnerable countries, to attract FDI. But this is not entirely true. It's important to note that despite the concentration of FDI in a few larger, wealthier developing countries, it remains a very significant source of external finance for the majority of developing countries. This is not appreciated enough and is a point well made by Andy Sumner (2005). This is partly because of the concurrent concentration of different capital flows, meaning that FDI is large as a proportion of total inflows for individual poor countries, but this can be because all other inflows are also/even smaller. This matters for both thinking about development financing from a

Figure 7.5 *FDI inflows, top 10 countries in 2011.*

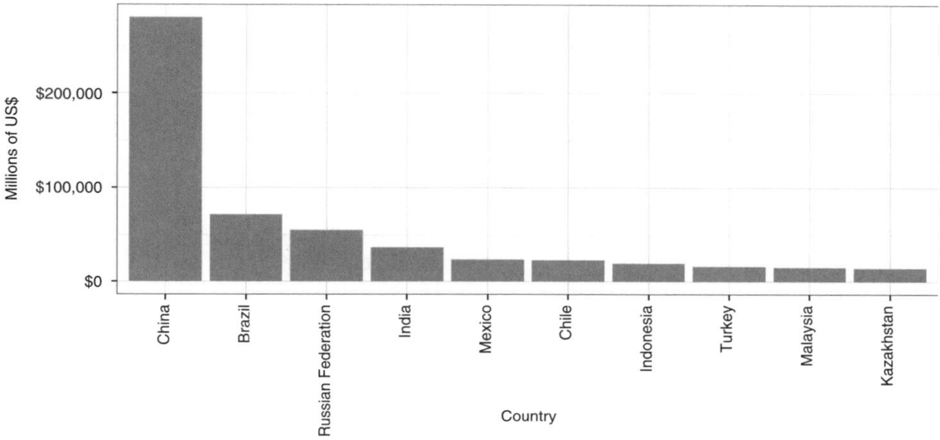

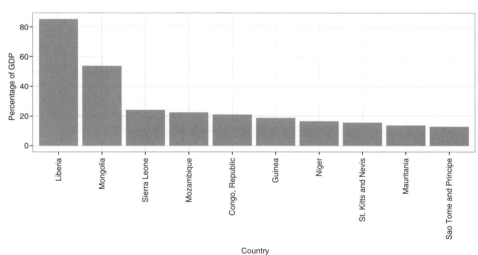

Data: WDI 2013.

Figure 7.6 *FDI inflows as a percentage of GDP, top 10 countries in 2011.*

Data: WDI 2013.

global perspective, but also in terms of the policy implications for developing countries from a domestic perspective in the difficulties of trying to attract FDI, trying too hard to do so, and the vulnerability and dependency on external finance.

Some countries, often smaller, are noticeably dependent on FDI. For example Hong Kong and Singapore, which receive more than 10 and 20 per cent of investment from overseas. Moreover, in terms of the dependence of poor countries on FDI for gross fixed capital formation, it appears to be higher than in the wealthier countries – both developed and MICs. During the 1990s, FDI flows to sub-Saharan Africa accounted for 11 per cent of gross fixed capital formation. This was twice the percentage for developed countries (Griffith-Jones and Leape 2002).

FDI is simultaneously elusive but also fundamentally important to the most vulnerable poor countries – the LDCs, the LLDCs and the SIDS). The 49 LDCs received 3 per cent of global FDI inflows and only 6 per cent of the flows to developing countries. And, again, within this group the bulk of FDI is concentrated in those countries with natural resources. But, as a group, these countries depend on external investment for capital formation. For this group FDI was equivalent to 25–40 per cent of their gross fixed capital formation in 2009. There are 31 LLDCs which have inherent geographical disadvantage in attracting FDI because of higher transportation costs. Kazakhstan accounted for 58 per cent of the total FDI to the 31 LLDCs in 2009! There are 29 SIDS. Again, because of their small markets, limited resources and high transportation costs, they face an inherent geographical disadvantage and tend to rely on tourism. Over half of this group's total FDI flows go to Jamaica, Trinidad and Tobago, and the Bahamas.

A final important trend has been the growth of *outward* FDI *from* developing countries, both in general terms, but also to other developing countries (i.e. south-south flows). And importantly, this was during a period where north-south FDI flows were declining. China, in particular, has become an important source of FDI outflows, mainly into natural resources, but the Russian Federation is also in the top 20 largest investors in the world (UNCTAD 2010). The World Bank has estimated that south-south flows increased from $14 billion in 1995 to $47 billion in 2003 (World Bank 2006a). To give a sense of proportion, this is comparable to ODA flows (Sumner 2005). So south-south flows are becoming an increasingly significant part of the global financial landscape.

Portfolio equity

Measuring portfolio equity

There are two broad ways of measuring the importance and significance of portfolio equity activity. First, equity flows into (and out of) developing countries. Second, the performance of stock markets.

Portfolio equity flows emerge from one of two types of transaction: (1) through the initial public offerings (IPOs) a.k.a. equity placements, which is where shares are first made publicly available, and (2) through sales in the secondary market. The World Bank (2006a) reports that in 2005, 63 per cent of all emerging market transactions were IPOs, up from 47 per cent in 2004.

Again, it is China which has dominated these figures, accounting for 61 per cent of all emerging market placements (and 21 per cent of the world's total).

The geography of portfolio equity

The World Bank (2001b) has described the growth in equity market flows as one of the most important long-term trends in development financing. At the turn of the century, equity flows were booming, increasing from $8.6 billion to $34.8 billion in the two years between 1998–2000. And in 2000, equity flows accounted for roughly 15 per cent of all capital market flows to developing countries (up from just 5 per cent in 1998). During the mid-2000s portfolio equity flows surged and the stock issuance of emerging market economies rose significantly. It was the privatisation process in many developing markets that drove this boom as state-owned companies were publicly listed. And yet, after the initial purchases, the growth in portfolio equity declined as international investors switched to direct investment in order to take control of these companies (UN DESA 2005).

Between 2002 and 2005 the capitalisation of East Asian stock markets increased by 300 per cent (World Bank 2006a). And share prices rose. Standard & Poor's/IFCI index rose from $1.7 trillion in 2002 to $4.4 trillion in 2005. The procyclical nature of the markets means that 'success' begets 'success', or at least inflows beget inflows. The gains in portfolio equity were also driven by increased interest in emerging markets from international institutional investors – i.e. pension, insurance and mutual funds – in the US as emerging markets outperformed the US market. US-based investors accounted for $20.7 trillion of the global total of $46 trillion in 2004 (World Bank 2006a). The reason why US investors' behaviour makes such a difference is partly because of the large sums involved, but also because most of their portfolio is invested domestically, with only 13 per cent of equity investment of the major US pension funds being invested overseas. In contrast, UK pension funds are 28 per cent international and those in the Netherlands 40 per cent. Hence small changes in the portfolio of US investors account for large changes in the absolute volume of flows to

Figure 7.7   *Net equity flows by income group.*

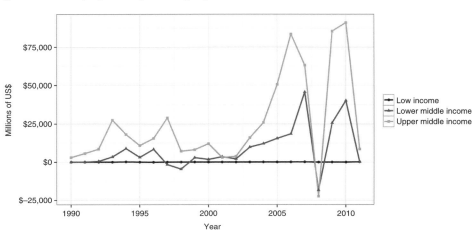

Data: WDI 2013.

developing countries. We return to this point when discussing the determinants
of portfolio flows and their implications for policy.

Portfolio equity flows are smaller in size than other private flows, representing about
7 per cent of all aggregate net resources in 2010. Like the other private capital flows,
portfolio equity flows tend to be concentrated, and are even more concentrated than FDI.
However, for a few individual countries they have been an important source of external
financing. Ten countries typically account for around 90 per cent of all portfolio equity
flows (World Bank 2010a). In 2005 almost two-thirds of all portfolio equity flows went
to just three countries: South Africa, India and China. China accounts for around a third
of all portfolio equity flows to developing countries, and around half of all flows in Asia.
Around 94 per cent of all the flows to East Asia went to China, India and Thailand. And
back in 1991–92 portfolio equity flows accounted for about a quarter of all external
financing for Mexico (Claessens 1995). Of course, Mexico was to suffer a financial crisis
two years later, reiterating the argument set out in Chapter 3 about the correlation
between capital account liberalisation, capital inflows and financial crises.

The consequence of this concentration was such that during the 1990s although the
overall levels of private capital rose, the LICs lost out yet further as their share of
private capital flows fell from their existing low level (Mishra *et al.* 2001). However, for
exactly the same reason, when the global financial crisis hit in 2008, the fall in portfolio
equity flows only hit a small number of countries, for example India and Russia were hit
hard. China saw a slow-down, but still recorded positive flows. In theory, equity
investors should have longer-term horizons than debt-holders in a crisis as liquidating
their position will incur large losses (World Bank 2001a). However, this is not the case
in reality. Equity flows often turn negative during crises reflecting a withdrawal of

Figure 7.8 *Top 10 equity flow destinations in 2011.*

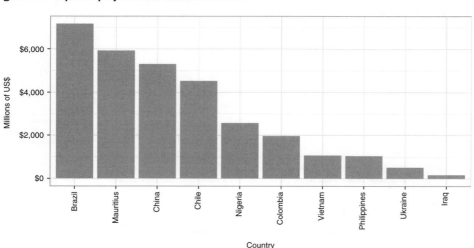

Country

Data: WDI 2013.

money leading to a loss in liquidity. To illustrate this point, portfolio equity flows registered as negative US$57.1 billion in 2008 (i.e outflows), way down from their peak of $135 billion of inflows into developing countries in 2007. Equity prices also fell, reflecting the fall in flows following the financial crisis. The MSCI global emerging market index was down by 55 per cent in dollar terms in 2008 (World Bank 2010a).

Why do investors invest where they do?

How can we explain these patterns of investment flows? Why do investors seem to prefer some countries over others? Does this suggest that developing country governments might be able to do something in order to attract (the right kind of) investment? Why do investment flows seem to vary over time, in particular their cyclical quality and sudden surges and reverses? Two very different literatures have grown up to explain the patterns of portfolio equity and direct investment.

Portfolio equity

There are a number of 'push' and 'pull' factors which explain the flows of portfolio equity, although there is an important interaction between the two because investors' decisions are guided by the *relative* risk and rate of return between a foreign investment and a domestic one.

The pull factors – the conditions in the developing countries that serve to attract investment flows – are the most studied. This reflects an orthodox preoccupation with getting the policies and governance right. Changes in policies, especially to more business-friendly policies, are often the most cited ways of securing investment inflows. Liberals argue that policy decisions such as liberalising the capital account and/or the stock market not only send out an important signal, but in some instances actually make it possible to invest in the first place.

The increases in cross-border equity flows have been driven by liberalisation, specifically the removal of capital controls by developing countries, for example, the removal of restrictions on foreign ownership and liberalising capital account transactions (Claessens 1995). Research that has examined the impact of capital controls on portfolio equity flows, i.e. restrictions on foreign ownership of domestic equities, shows that liberalisation (according to this measure) has proceeded rapidly (Edison and Warnock 2001; Prasad *et al.* 2004). The adoption of international accounting and information standards has made developing countries more attractive to investors. Empirical work in the orthodox tradition has found that portfolio equity favours countries with good corporate governance and management accountability, plus an independent and transparent legal system to protect investors' rights (Doidge *et al.* 2004; World Bank 2006a). In sum, market-friendly.

Further barriers to equity flows into developing countries include macroeconomic instability (i.e. high and variable inflation rates), an unfavourable credit rating, an underdeveloped domestic stock market (in terms of its size – Portes and Rey 2003 – and its legal protections), the presence of a good standard of regulatory and accounting frameworks, and a lack of *perceived* openness and efficiency (Bekaert 1995; Claessens *et al.* 1995). *Perceptions* are clearly important. Surprisingly, formal ownership restrictions do not seem to matter as much as might be expected (Bekaert 1995); others have argued that this might be because they are simply circumvented (Claessens 1995).

In terms of 'push' factors, we are thinking here of policy decisions as well as business cycles in the developed countries. Investors make their decision based on the trade-off between the expected rate of return against the risk. This involves an assessment of the prospects of the asset itself, but also broader issues to do with the liquidity and efficiency of the stock market, the legal and accounting standards and political risk, such as the possibility that capital controls could be reimposed (Claessens 1995).

The average returns on equity in emerging markets tend to outperform the global and industrialised world's averages. However, they also tend to be more volatile. Nevertheless, large international institutional investors have an incentive to diversify their portfolio across different countries as stock performance tends to be less correlated across countries than different assets within a domestic stock market tend to be (Claessens 1995). In addition, high growth rates in China, India, Brazil and Indonesia etc. have undoubtedly attracted investors. But the other factor in the 2000s was the decline in interest rates, meaning that investors were looking to emerging and developing markets in search of yield. In addition to the low interest rates in the West, regulatory changes in the US and Europe in the 1990s made it easier and possible for funds to invest in emerging markets. So, it's important to note that the surges in capital flows in the 1970s, 1990s and 2000s all had as much to do with factors in the rich world as the developing countries (Calvo *et al.* 1996). It's not just a case of developing countries getting their house in order and the capital will come.

The importance of wealthy country conditions in limiting the amount of capital available has been demonstrated (Mody and Taylor 2002). The business cycle in industrialised countries is key in rationing capital to developing countries. Unlike FDI, portfolio flows tend to be negatively related to the business cycle. Edison and Warnock (2001) have found that portfolio equity flows from the US to developing countries tend to be negatively correlated with output growth and interest rates in the US. They also find that this relationship is particularly strong for Latin American countries (less so for Asia). The findings suggest that flows surge and reverse for reasons independent of domestic factors within developing countries. In an attempt to parse out the relative importance of the different factors, Lane and Milesi-Ferretti (2004) examined the relative contribution of (1) bilateral factors, (2) source-country factors, and (3) host-country factors over the period 2001–08. They found that a country's' overall volume of equity investment is most closely correlated with the depth of its domestic stock market and its income level. They also found that geography matters: doubling the physical distance reduces equity holdings by 61 per cent. Shared institutions also matter. Common legal origins and a common language matter, the latter increasing equity holdings by around 40 per cent. And patterns of international trade are also an important determinant of equity investment flows.

FDI flows

Why does FDI flow where it does? Why do TNCs locate where they do? While the popular perception of TNCs, such as Nike, Apple and The Gap, locating in poor countries to exploit low wages and lower labour standards has some truth to it, it can also be seriously misleading when thinking about FDI more generally. FDI is also attracted by countries with good infrastructure: a highly skilled labour force, and economic and political stability are also crucial. Thus it is again important to take the heterogeneity of capital flows seriously – there are a range of factors which vary in importance across different types of FDI, and the cost of labour is but one.

Box 7.2

Apple and Foxconn

'Foxconn City' is the informal name for the large factory owned and operated by Foxconn Technology in Longhua, Shenzen, China. It has also been called 'iPod City'. To say that it's a large factory is an understatement: it employs 230,000 workers, working 12-hour shifts. Specialised guards are required to manage the flow of people coming to and leaving from shifts. Many of these workers live in company dormitories and earn less than $17 a day. The reasons are pretty straightforward. As reported in a *New York Times* piece:

> One former executive described how the company relied upon a Chinese factory to revamp iPhone manufacturing just weeks before the device was due on shelves. Apple had redesigned the iPhone's screen at the last minute, forcing an assembly line overhaul. New screens began arriving at the plant near midnight.

> A foreman immediately roused 8,000 workers inside the company's dormitories . . . Each employee was given a biscuit and a cup of tea, guided to a workstation and within half an hour started a 12-hour shift fitting glass screens into beveled frames. Within 96 hours, the plant was producing over 10,000 iPhones a day.

> 'The speed and flexibility is breathtaking,' the executive said. 'There's no American plant that can match that.'

But there are human costs built into Apple's products. The factory runs an interesting line in 'motivational posters' with banners displaying the message 'Work hard on the job today or work hard to find a job tomorrow.' Labour abuses at the Foxconn installation – according to the *New York Times* feature – are widely known about within Apple, but are tolerated. They are seen as the price to be paid for producing high-quality products, at low prices, with a high degree of flexibility, as reported above. Workers live in overcrowded dormitories. On occasion, the dormitories would have to fit up to 20 people into a three-room apartment.

Apple does have a Code of Conduct for its suppliers. And between 2007–10 it carried out around 312 audits every year of suppliers and its suppliers' suppliers. About half were guilty of violations, and there were 70 core violations – including 'involuntary labour, under-age workers, record falsifications, improper disposal of hazardous waste and over a hundred workers injured by toxic chemical exposures'. In October 2012, Foxconn admitted to using underage interns in its factories in China, employing children as young as 14, breaking Chinese law.

Sources: Duhigg and Barboza (2012), Duhigg and Bradsher (2012)

There are a number of pioneers in the study of FDI and TNCs, including Caves, Hymer, Vernon and Dunning. Stephen Hymer (1976) must be one of the few Marxists read in business schools (Strange 1997). He argued that direct investment wasn't affected by interest rates in the same way as portfolio equity. Instead direct investment was to do with the firm-specific characteristics, such as economies of scale, technology and managerial expertise, which meant that it could out-compete domestic firms in the host country. And then, once invested overseas, access to and control over the overseas productive assets (whether labour, capital or natural resources) gave it a further competitive advantage. Meanwhile Raymond Vernon (1966) argued that the 'product life cycle' explained why firms looked overseas in the first place. The product life cycle evolves over four phases – firstly a producer enjoys a competitive advantage over other domestic producers for, say, a vacuum cleaner, and also exports its products abroad. But then, other domestic competitors start to develop their technologies to compete more directly and often undercut the original producer and exporter. This then stimulates the original producer of the new technology to move production overseas to save on costs.

John Dunning (2008) built on the work of these pioneers in his 'eclectic' model. Dunning brings together three key factors: (1) ownership-specific advantages, (2) the internalisation of these, and (3) location-specific factors. Ownership-specific advantages directly build on Hymer's insights about firm advantages such as size, expertise, patents, technology or access to credit. These advantages mean TNCs can compete in other markets even against domestic firms with better local knowledge and established suppliers and markets. Second, the internalisation of these advantages refers to the fact that it would, in a world of textbook markets, be more efficient to just export overseas rather than produce there. However, because of transaction costs and imperfect markets there is an advantage to internalising these market transactions (Coase 1937). Crucially, the ability to internalise these costs and retain ownership advantages is a prerequisite of a firm transnationalising. Finally, the location-specific factors refer to the advantages that accrue from locating in a specific area. For example, access to markets (especially within tax or tariff areas), access to natural resources, lower production costs or access to a labour force that is suitably productive, low-cost or high-skilled.

The balance between these different objectives varies between different industries. For example, the extraction of physical resources such as oil, minerals and forests is unavoidably all about locational issues. Manufacturing, especially garments, traditionally tends to favour locating in low-wage economies. Whereas access to markets, especially the large domestic markets like Brazil and now China and India, and Canada and Mexico as part of NAFTA, is becoming increasingly important for consumer goods.

As is evident from the concentration of FDI in a few countries, and specifically those countries that are not the poorest, FDI tends to flow where economic growth is already taking place. Since the 1990s most FDI has been market-seeking (World Bank 2006a). Therefore it is attracted by growth and its potential. The exceptions to this – as evidenced by the sectoral disaggregation of the data – are in natural resources and primary commodities. Resource-seeking FDI is slightly different. Countries such as Nigeria and Angola receive large amounts of FDI because of their oil reserves. The value and non-negotiable geographical nature of natural resources mean that other factors are much less important for such investments.

Increases in FDI can also be traced to significant political changes, such as membership of a multilateral organisation or signing up to international agreements. For example, since China joined the WTO in 2001 there has been an increase in FDI as foreign investors have been moving in as the economy is liberalised. For MICs on the border of the EU, potential membership offers similar benefits as FDI inflows increase in candidate countries in anticipation of accession. Joining means countries generally become more investor-friendly in line with EU law, plus membership offers market access to the rest of Europe, reduces currency risk (at least in theory!) and EU membership gives states access to EU structural funds to improve their infrastructure as well as other financial incentives, such as tax breaks.

There are a number of measures that developing country governments can take to increase their attractiveness to FDI inflows. For example, investment in infrastructure and investment in education in order to increase the skill base and entrepreneurial capacity of the host country. Political stability and an investor-friendly business environment also help. Hence the recent increases in FDI were in part driven by strong global growth, but also reflect shifts in developing country policies towards foreign investment. Many countries have put in place investor-friendly policies, improving corporate governance and reducing restrictions. Investors like transparent, consistent and liberal rules and regulations. But it is also clear that a lot of FDI flows have been driven by large-scale sell-offs of state-owned assets, which is part of the liberalisation trend but not reducible to 'good' policies. In the 1990s privatisations in Latin America and Eastern Europe were driving increases in FDI, as well as explaining more recent large FDI inflows into countries like Turkey and Indonesia (World Bank 2006a).

In sum, we can see that both portfolio equity and FDI tend to favour the MICs. The reasons being that the LICs tend to be the ones which suffer from a lack of effective domestic demand, weak legal systems, political and economic instability, poor infrastructure and a poorly educated workforce (Woodward 2001).

But, just like portfolio equity flows and debt flows, FDI flows are also conditioned by economic conditions in the rich world (Prasad *et al.* 2004). Research has demonstrated a significant positive correlation between business cycles in the rich world and FDI flows to the developing world (Reinhart and Reinhart 2001). Recall, though, that this is the opposite to portfolio equity flows, which tend to be negatively related to the business cycle (Edison and Warnock 2001).

Developmental impact

As highlighted above, domestic growth attracts private capital inflows. However, do private capital flows increase growth further, and with it increase employment, reduce poverty and expand freedoms? Like other private financial flows dealt with in the previous chapter, the aim of equity and direct investments is for the investor to profit from their investment. The returns on investment comprise two key elements: (1) profits to the owners of capital (for equity investors this is in the form of regular payments, referred to as dividends, which are made to shareholders), and (2) any capital gain from the increase in an asset's value. This capital gain is only realised upon reselling the asset at a higher value (or, obviously, losses if sold at a lower price). So it is important to note, corporate social responsibility put aside, that the developmental consequences of private investment arise as either a beneficial side-effect of an investor's self-interest or as the negative consequence of extracting maximum profitability. The question, then, is to what extent does this occur and what are the channels through which it works?

Portfolio equity inflows

The liberal orthodoxy is that increases in private capital flows should increase the volume of investment, but also the efficiency of investment, and with it growth and employment, thus reducing poverty. With respect to the impact on investment rates, the World Bank (2006a) finds that private capital flows are, on average, positively correlated with domestic investment in developing countries. It compared aggregate domestic investment as a percentage of GDP across two periods 1999–2001 and 2002–04 for a sample of 72 developing countries with access to international capital markets accounting for 95 per cent of private capital flows to developing countries. The analysis suggests that private capital flows have a positive and significant effect on domestic investment as long as countries have already reached a minimum threshold of financial development and capital account openness. Out of the different types of capital, FDI was found to have the strongest correlation with domestic investment. Bosworth and Collins's (1999) study was similar in that they found a significant positive impact on domestic investment, with FDI and bank lending having the stronger impact and portfolio equity and debt being weaker.

Second, are equity flows positively correlated with growth? Reisen and Soto (2001) compared the effect on growth rates of FDI, portfolio equity flows, bond flows, long-term bank credits, short-term bank credits and official flows (i.e. aid). Their panel included 44 developing countries and covered the period 1986–97. Out of the six different kinds of flows only FDI and portfolio equity flows were positively associated with economic growth. Meanwhile, in a separate study, Marcelo Soto (2000) confirms the importance of the threshold effect of financial development as well as the advantages of equity financing over debt financing. His study covered the period 1986–97 and included 44 developing countries. He found that debt inflows actually display a negative correlation with growth, whereas both FDI and portfolio equity flows have a positive and significant relationship. A 1 percentage point increase in the FDI to GNP ratio is associated with a rise in income of 0.6 per cent in the steady state level. Moreover, portfolio equity flows are also strongly positive; a 1 percentage point increase in the portfolio to GNP ratio is accompanied by a 0.68 per cent increase in short-run economic growth. Soto suggests that the surprisingly high coefficient for portfolio equity can be theoretically explained through the impact of portfolio flows on stock market liquidity and then the subsequent consequence of that for growth.

Indeed, in a world of efficient markets, increased integration of developing countries with the international capital markets should bring down the cost of capital. Portfolio equity flows do tend to be associated with the development of domestic capital markets in the sense of growing them and increasing liquidity (Mishra *et al.* 2001; World Bank 2001a). However, as we saw in Chapter 3, increased capital account liberalisation and larger equity inflows also make an economy more vulnerable; both to asset price inflation, bubbles and misallocation of productive resources, but also to crises. Overheating can arise from a too rapid increase in private flows. This manifests itself as rapid increases in investment, inflation, currency appreciation and consumer goods imports. Moreover, because developing countries' stock markets are small or 'thin' it only takes moderate increases in portfolio flows to create sudden and significant asset price inflation. For example, Standard & Poor's/IFCI index more than doubled in a staggering three years from $1.7 trillion in 2002 to $4.4 trillion in 2005. During this same period the capitalisation of East Asian stock markets increased by 300 per cent (World Bank 2006a). Such rises do not usually augur well for future economic stability.

How stable or volatile are equity flows? In theory, equity investors should have longer-term investment horizons than debt-based flows. Indeed bond flows represent the most volatile capital flow. In contrast to bond investors who are concerned about keeping their capital, equity investors are more likely to stick it out through more challenging economic times. However, there is evidence for similar sudden and financial crises caused by sudden turnarounds in equity flows, leading to equally damaging declines in economic output (World Bank 2001a). But, say neoliberal defenders of portfolio flows, the flip-side to the cost of portfolio volatility is that portfolio equity flows to emerging economies tend to recover quickly. For example, they displayed a 'remarkable rebound' from negative $85 billion in 2008 to $119 billion in 2009 (IFF 2010). Often this is a consequence of investors taking advantage of some good buys. A similar pattern was evident in East Asia after the financial crisis there where Western investors cashed in on 'fire sale' prices of distressed companies (Krugman 1998; Wade and Veneroso 1998). So, as critical reformists, such as Wade and Veneroso, point out, what appears to be a return to good conditions of market liquidity should really be interrogated as a massive transfer of 'ownership and power' (Wade and Veneroso 1998: 20). This was not exclusive to the East Asian crisis, but characterises all financial crises. As such, many developing country governments have sought to minimise short-term flows through capital controls and encourage longer-term flows such as direct investment.

FDI, growth and poverty reduction

As suggested earlier, FDI is feted as the most pro-development private capital flow. The logic is that the additional capital inflows will increase total investment and fuel economic growth. Assuming, that is, that the capital flows *are* additional, i.e. the FDI inflow needs to exceed profit repatriation, royalties, intra-company loans and transfer pricing; the dollar value of FDI exports needs to exceed the dollar value of FDI imports to keep the current account transfer positive; and the taxes paid by the TNC need to exceed the subsidies paid by the government to attract the FDI for it to benefit the public purse (Sumner 2005).

The evidence suggests that FDI can boost investment and growth and there is a more mixed assessment on whether it can reduce poverty. Again, any positive effects are conditional on host country conditions on the one hand (the stock arguments of liberal institutionalists) and appropriate government policies to extract the maximum benefits from foreign investors on the other (the main insights of critical reformists). Furthermore it is also conditional on the type of FDI (as per the accounting distinctions between M&A and greenfield sector and so forth, above). Again, it is important to take the heterogeneity of FDI flows seriously.

In terms of investment, the evidence suggests that FDI can increase domestic investment in excess of the original investment, i.e. a 1 per cent increase in inflows is associated with an increase in investment of over 1 per cent, a phenomenon called 'crowding in' (Borensztein *et al.* 1998; Mishra *et al.* 2001), or at least raises domestic investment on a one-to-one basis (Bosworth and Collins 1999). But empirical work also suggests (1) diminishing marginal returns to external investment, and (2) interaction effects between FDI type and/or the policy regime in explaining whether a particular country experiences crowding in or out.

With respect to diminishing returns to external investment, Mishra *et al.* (2001) and the World Bank (2001a) present a couple of explanations which are of interest here. First,

Figure 7.9   FDI as a percentage of gross fixed capital formation by region.

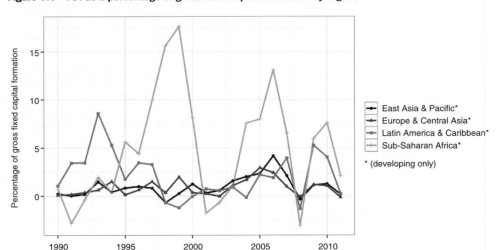

Data: WDI 2013.

as capital flows increase there is a change in the composition of these flows; they tend to contain less greenfield investment flows and more M&A FDI. Simultaneously, there is a greater shift away from FDI in general and towards portfolio equity and bond flows. All these flows represent less bang for the buck when it comes to increasing investment. Second, as outlined in Chapter 3, countries have increasingly been using capital to build up reserves as insurance against sudden reverses in capital flows. These reserves tend to be saved in a relatively liquid form and therefore contribute less to productive investment. With respect to the interaction effects between FDI and regime types, crowding in is less likely in liberal policy regimes (Sumner 2005). It appears to be the developmental policy regimes, such as in East Asia, which have fared better (UNCTAD 1999).

In terms of growth the overall story is also positive, but highly conditional. Though note that one influential study found that FDI is not positively associated with growth at all (Carkovic and Levine 2000). But the weight of the evidence is that just as investments in, say, human capital – through education and up-skilling the workforce – help attract investment, this also increases the productivity of FDI (Mishra *et al.* 2001). Countries with a more skilled and better educated workforce (Borensztein *et al.* 1998) and with a better physical infrastructure tend to enjoy greater productivity benefits from investment. Similarly, another study found that only countries with a sufficiently high level of financial sector development could benefit from FDI inflows (Alfaro *et al.* 2004). Another important condition for generating a positive relationship between FDI inflows and economic growth is to have a sufficiently large market for the FDI inflows to serve (Balasubramanyam *et al.* 1999), which points towards the importance of an export-oriented growth model (Balasubramanyam *et al.* 1996).

In terms of FDI's potential for reducing poverty the evidence is generally much more negative, suggesting that economic liberalisation and growth fails to lift all boats (Sumner 2005). And when it comes to inequality the evidence almost uniformly points to FDI inflows being associated with increases in domestic inequality (Tsai 1995; Beer 1999; Kentor 2001; Beer and Boswell 2002). These findings corroborate the radical critique of capital dependency theorists that dependency on foreign investment tends to result in undederdevelopment, in the sense of lower rates of growth and higher levels of inequality. In other words, capital-poor countries are poor precisely *because* they rely on imported capital (Chase-Dunn 1975).

The seminal work in the capital dependency tradition is Bornschier and Chase-Dunn's (1985) *Transnational Corporations and Underdevelopment*. Their central thesis is that TNC investment in developing countries is characterised by unequal power relations. And the unequal international division of labour is the primary mechanism of capitalist exploitation and serves to maintain the unequal exchange at the heart of the world economy's core/periphery hierarchy. This is because TNCs tend to exacerbate existing income inequalities by displacing labour-intensive, low-wage industry, repatriating profits and unduly benefiting from tax breaks and other favourable government policies (Feenstra and Hanson 1997; Beer 1999). Bornschier (1980) argues that, over time,

TNCs will tend to 'decapitalise' the host economy as opposed to adding new capital via the processes of profit repatriation and transfer pricing.

Further studies, looking at foreign capital penetration, measured as the ratio of foreign capital stock to total stock, find that the countries which are more dependent on FDI experience lower economic growth, crowding out of domestic investment and higher income inequality (Dixon and Boswell 1996). Others have argued that what the results really show is that domestic capital is better than foreign capital (which can still be beneficial and doesn't necessarily crowd out foreign investment) (Firebaugh 1996). A further specification of the argument finds that it is actually high investment concentration that matters (Kentor and Boswell 2003). That is to say, where a country relies on most of its capital from a small number of sources these foreign investors enjoy higher levels of control over the government, and economic and social policies, and the domestic elites tend to become captured and their interests aligned more with the interests of foreign capital. This is obviously to the detriment of a government's autonomy to act in the long-term interests of the national economy (Evans 1995). In a further piece, Herkenrath and Bornschier (2003) find that the negative effect of FDI has somewhat diminished over time – in contrast to earlier results, they find no effect on growth, just rising inequality. They argue that this ostensibly mixed record might actually be a conditional one – conditional on the way in which governments deal with TNCs – suggesting that there are opportunities for intelligent government policy to create 'outlier' states that buck an otherwise negative trend.

Taken all together, the empirical record suggests the importance of a country being more rather than less developed in order to take advantage of inflows and/or a strong and interventionist government to manage the foreign investors in the interests of national development. Such conclusions lend much greater credence to the vision of critical reformists and an interventionist developmental state as set out in Chapter 2. The insights of liberal institutionalists into the importance of good governance, private property rights and high levels of human capital and so forth are clearly important, but it is less clear how useful these insights are to developing countries who don't yet enjoy these conditions. In order to understand a little better the ways in which FDI *may* translate into development outcomes we must look at the mechanisms involved.

Mechanisms of FDI

In one well known study of Moroccan FDI, following a period of trade and investment liberalisation, Haddad and Harrison (1993) failed to identify higher productivity growth in domestic firms in the sectors where FDI was coming in, which one would expect if there were positive spillovers. Why might this be?

A good way of thinking about the developmental impact of FDI is to distinguish between the quantitative and qualitative effects of FDI (Cypher and Dietz 2009). In addition to the quantitative effects on economic growth and employment the literature tends to emphasise the indirect resource transfer effects. Recalling Romer (1993), FDI not only transfers capital but also *ideas*, which increase recipient country productivity.

The positive externalities of FDI are called 'spillover effects' which can operate through the movement of labour from TNCs to domestic firms or through 'demonstration effects' that enable the transfer of new technology and ways of doing things, such as new production and management techniques. Others have argued that there are important competition effects too (Aitken and Harrison 1999). For a country to successful harness these spillovers it is particularly important to develop the linkages between the TNC and other firms in the local economy (Lall 1978).

It has been shown that the spillover benefits only accrue to those countries with a sufficient absorptive capacity. That is to say, if the domestic workforce has sufficient skills and education to benefit from FDI, or if the existing industrial structure matches the incoming FDI, hence creating synergies, or if the financial system is sufficiently developed to allow domestic firms to upgrade and expand to act as potential suppliers to the new foreign firms (Borensztein et al. 1998; Reisen and Soto 2001). Some relevant issues flow from this and are discussed below.

First, M&As versus greenfield investment. It is necessary to distinguish between FDI that directly leads to new capital formation, and those flows that don't necessarily do so. M&A FDI is primarily about transferring the ownership of capital from home to overseas. UNCTAD documents that 20 per cent of FDI flows into developing countries are M&A and just buy out existing companies. However, this proportion can be much higher for individual countries, for example in Mexico in 2001: it received a large amount of FDI, but 71 per cent was just buying up existing companies (Cypher and Dietz 2009). This may well have positive spillover effects, greater and more efficient investment, but it does not result in new capital formation. The empirical record shows that there was an increase in M&A during the 1990s as a proportion of overall FDI inflows and this was accompanied by a fall in the relationship between inflows and domestic investment and productivity (UNCTAD 2000; World Bank 2001a). While M&A *can*, over time, lead to increases in productivity in contrast to greenfield investment, there is no immediate additional new productive capacity.

Second, the investment purposes FDI is put to. FDI does not necessarily go into productive investment that directly supports economic growth: this could be – for a number of different reasons – oil companies, luxury hotels or apartment projects (Cypher and Dietz 2009). 'Enclaves' of investment don't necessarily create sufficient ripple effects by creating extra employment or boosting domestic demand. Research has highlighted the importance of backward linkages into the economy to facilitate the spillover, or qualitative impacts, of TNC investment to occur. Examples of backward linkages include local sourcing and procurement. Without these linkages much of the production will happen in-house and the intangible benefits from the investment will not flow to local firms. Firms that are market-seeking and target the domestic economy tend to have greater backward linkages into the economy, and as such tend to rely on local skills, knowledge and materials. Meanwhile, export-oriented TNCs are less embedded in the domestic economy. Likewise, production systems which are capital intensive, not labour intensive, fail to maximise employment. Governments have an important role to play in discriminating between FDI which will boost the domestic economy and create synergies, and facilitate the transfer of skills and fill ideas gaps, and that which won't.

Henderson's (1989) study of the electronics industry in South-East Asia confirms that the extensive backward linkages into the domestic economies through the widespread use of sub-contracting served to build up domestic industry and employment. The existence of electronics industrial clusters in places such as Penang and Johor in Malaysia, around Bangkok in Thailand, and in Singapore, is no accident. Industries tend to show strong natural agglomerative tendencies as producers seek to benefit from proximity in order to share services and supply chains, but this is also a process which is encouraged by governments through creating special economic zones and/or directing credit and protecting infant industries. But creating linkages between TNCs and the economy is the crucial issue.

Third, in relation to the arguments about decapitalisation there are the processes of transfer pricing and profit repatriation. Repatriation is an open process of decapitalistaion, whereas transfer pricing is much more hidden. Sumner (2005) reports that for every dollar of net FDI inflow, roughly a third leaves in profits back to the TNC's home country. In 2001 profit repatriation was roughly equal to total aid flows to developing countries. However, there are important and significant cross-regional differences: in sub-Saharan Africa profit repatriation is 70 per cent or higher, while in South Asia it is only around 10 per cent. The differences likely reflect policy differences between governments and the openness of the capital account (Sumner 2005).

Transfer pricing is the accounting process of allocating a price to an input or unfinished good as it is traded within a firm across divisions. This is a necessary and perfectly normal part of business, allowing a firm to keep track of costs and productivity. However, when a firm is transnational, with different divisions in different tax regimes, there is an incentive to overstate value-added in low-tax jurisdictions and vice versa, reducing the overall tax burden. Because the firm is on both sides of the transaction – as the buyer and the seller – it is free to set or manipulate the price to a level different from the price which would be set in the market. There is an incentive to understate returns in countries with higher tax regimes; weak tax regimes and liberalised reporting and registration requirements accentuate this.

In contrast to the mainstream assumptions about how beneficial FDI is, David Woodward (2001) has calculated the costs of FDI for financing development as opposed to official or commercial borrowing. The costs of official loans are below commercial rates for LICs, around 3–4 per cent and 1–3 per cent for sub-Saharan Africa, and around commercial rates, i.e. around 6 per cent, for MICs. Private loans to governments are slightly higher – at around 5.5 per cent for low-income and sub-Saharan countries. Commercial loans to the private sector were a bit more expensive to take into account the risk premium in developing countries, another 2 per cent over lending to governments. In contrast, the estimated rate of return for FDI is in the magnitude of 16–18 per cent; twice the cost of private lending (World Bank 1997).

And for sub-Saharan Africa, this rises to around 24–30 per cent, which is four times the cost of private lending and 16–10 times the average costs of official lending. Meanwhile, the rate of return on equity investment is about 29 per cent per year for developing countries. The message here is that the costs (and profits for investors) tend to increase as a country switches towards FDI and equity.

Policy debates

At the national level it is useful to think of (1) policy initiatives to attract portfolio equity and FDI, and (2) policy initiatives to capture and guide the benefits of investment. And at the international and global levels the debate is about regulating private capital flows and corporate social responsibility.

If developing country governments are concerned with stability of the inflow, based on evidence from Sarno and Taylor (1999), they would be well advised to focus on attracting FDI flows rather than bond, portfolio equity or even ODA. They find that commercial bank flows contain a good deal of permanent flows and FDI flows are almost completely permanent. However, as Woodward (2001) nicely summarises, the fact that FDI does not flow to the poorest countries has a number of significant and largely negative consequences: it obviously limits the amount of capital these countries can access, especially given how important FDI flows are as a proportion of overall development financing; it means that TNCs require a far larger return on capital to risk investing in LICs; and it increases the competition among LICs for available FDI – all of which results in a range of negative outcomes for the poorer developing countries as they look to compete against one another and to attract investment inflows.

In terms of attracting equity inflows, it is significant that equity investors are interested in quite different indicators and factors than debt-based investors. The change from debt to equity financing has thus required a change in orientation in developing countries' policy goals. For example, equity investors prefer floating exchange rates which can help increase the competitiveness of firms based in that country's economy. However, bond investors much prefer fixed or at least stable exchange rates because this protects the value of their investment. Moreover, there is now more emphasis on the underlying base of the economy and its productivity, education and skills levels, the depth and liquidity of the stock market, whereas bond investors are much more interested in aggregate indicators relating to currencies and the current account.

One of the characteristics of LICs, by definition, is a low per capita income. This translates into low consumption levels. Plus, not only are many developing countries poor, but they are often small too, both in a population and in an economic market sense. These factors all work against these countries attracting inward investment (UN DESA 2005). They cannot take advantage of economies of scale. One solution is for official aid flows to be specifically geared towards supporting inward investment either through providing public goods or sharing the risk, for example through the MIGA. Another solution is for countries to engage in regional integration. For example, the Southern African Development Community (SADC) has enabled investment and trade from South Africa as a larger member to the smaller member states (UN DESA 2005).

The decision for national policymakers in developing countries is often characterised as striking a balance between liberalising their economies sufficiently to attract investment on the one hand, with social and environmental public policy objectives on the other (UNCTAD 2010). UNCTAD (2010) characterises the 1950s–70s as a period of state-led growth and the 1980s–2000s as a period of market-led growth. Since 1992 UNCTAD has been reporting new national policy measures taken by governments which affect foreign investment. So, for example, in 2009 UNCTAD (2010) identified 102 new policy measures towards foreign investment. Of these, 71 were liberalising measures. Liberalising measures included, for example, privatising state-owned enterprises, opening up sectors to foreign investment, liberalising land acquisition laws, offering tax incentives, speeding up approval and licensing procedures and reducing corporate tax rates. On the other hand, 31 new national policy measures were tightening up regulations for FDI. These included, for example, placing limits on foreign ownership, tightening procedures for project approvals and expropriating investment. The case for such measures is to increase social or environmental protections, to protect strategic industries or because of national security concerns. Plus, in 2009, this included measures taken to facilitate the large bailouts of firms following the financial crisis which were classified as de-liberalising.

As we have seen, attracting investment is not even half of the story. Developing country governments must seek to regulate or harness FDI and TNCs as best they can. Cypher and Dietz (2009) suggest that three conditions must be in place for developing countries to benefit from the spillover effects of FDI. First, countries must actively select TNCs. Second, the developing country has to have the capacity to absorb the benefits. Third, TNCs need to see knowledge transfers as beneficial to them, otherwise they will guard proprietary advantages.

EPZs (export processing zones) or special economic zones (SEZs) have been the traditional way in which developing countries have sought to attract new investment. They are special areas where different – i.e. business friendly – regulations and rules apply; for example, some combination of import and export tax exemption, corporate tax holidays, subsidised building leasing, services such as electricity, water and gas, plus infrastructure. Sometimes they exclude the usual minimum wage and labour laws as well as trade unions. They often host manufacturing or assembly processes that are labour intensive, which in theory should lead to high levels of employment. However, they often have very few forward or backward linkages into the host economy. One study found that for every five jobs created in an EPZ only one extra job is created in the domestic economy (ILO 1998). This is a very low jobs multiplier, meaning that 'EPZs can create the appearance of industrialisation and development in a country, without the substance' (Cypher and Dietz 2009: 481). An irony of the shift towards pro-FDI policy positions by developing country governments has resulted in a bias against domestic firms because many of the tax incentives and subsidies to attract FDI are often not offered to local companies (Sumner 2005).

Meanwhile, at the international level, the investment regime is constituted by international investment agreements (IIAs) such as bilateral investment treaties (BITs) and double taxation treaties (DTTs). Arbitrations are settled under a number of different rules including the International Centre for Settlement of International Disputes (ICSID), the International Chamber of Commerce (ICC), and the UN Commission on International Trade Law (UNCITRAL). There were 357 known state-investor dispute settlements filed by the end of 2009. The world of IIAs, BITs and DTTs is incredibly dynamic; in 2009 there were, on average, four new agreements per week, making a total of 5,988 by mid 2010 (UNCTAD 2010).

Like the domestic regulation of FDI, the international investment regime seeks to strike a balance, but in this case between protecting investors' rights on the one hand and TNCs' obligations to states and their populations on the other (UNCTAD 2010). Heterodox approaches highlight the costs for developing countries in tying their hands through agreeing to such arrangements. The logic of developing countries signing up is that in exchange for giving up some of their sovereignty it acts as a substitute for good governance reforms and provides a legal security for investors; a kind of developmental short-cut. The evidence on whether investment treaties do anything to attract new investment is mixed (Hallward-Driemeier 2003; Neumayer and Spess 2005; Sumner 2005). Neumayer and Spess (2005) tested to see whether the logic for BITs holds up and found that the more BITs a developing country had the greater the FDI inflows. However, Hallward-Driemeier (2003) suggests that BITs act as complements rather than substitutes for good institutional quality and local property rights. She goes on to argue that the legal cases which have been brought recently have served to highlight how foreign investors enjoy many rights which are not available to domestic investors and, crucially, that the agreements severely curtail the policy space available to governments in regulating and trying to extract benefits from FDI. In UNCTAD's words: 'Making IIAs work effectively for development remains a challenge' (2010: xxvi).

Conclusion

There has been an important and significant shift from debt-based financing to equity-based financing for developing countries. Many commentators, policymakers and some researchers see this as a broadly positive step. The crucial advantage of equity financing is that it fundamentally requires investors to take a greater stake in the investment, its success and risks. Thus this also, proponents argue, creates a longer-term investment ethic in equity flows. Orthodox proponents argue that the advantages of equity are a result of two mechanisms: a boost to capital accumulation through augmenting the capital stock and, second, providing efficiency gains from higher quality capital (human and ideas as well as funding). With respect to portfolio equity the additional gains are through the deepening of capital markets allowing them to serve the price discovery role more effectively. With respect to FDI, in addition to the debates about whether new investment crowds domestic investment in or out, there is the important debate about positive spillovers.

As both this and the previous chapter highlighted, large surges of portfolio equity often result in asset bubbles, such as the stock market and land or real estate bubbles. This misallocation of capital is costly. And the sudden bursting of the bubble is *extremely* costly. As such, developing country governments have sought to lengthen the maturity profile of debt and increase longer-term investment.

Assessments of the developmental impact of equity flows have been just about positive at the macroeconomic level with inflows potentially positively associated with economic investment and growth. However, MICs with a deeper financial system and higher levels of education seem to do far better. Moreover, both forms of equity flows – portfolio and FDI – display the same geographical and middle-income, large-country concentration as debt flows do. The liberal institutionalist insights about the importance of a stable and market/investor-friendly environment do seem to be borne out to a certain extent. FDI has an aggregately positive effect in those studies that take account of various institutional or human capital threshold effects. That said, it is the decidedly non-liberal policy regimes that appear to have extracted the most developmental benefits from FDI inflows, for example the developmental state policies of South-East Asia.

More radical critics point out that the shift from borrowing to investment, especially FDI, represents a transfer of ownership and productive potential from domestic capital to foreign hands. Loans to governments and firms in developing countries mean that they are beholden to their creditors, but the profits earned remain within the country. However, with investment, both profits and ownership are transferred overseas.

Discussion questions

1 Given that the foreign affiliates of TNCs are only responsible for 7.9 per cent of all employment in developing countries (UNCTAD 2012), what is their importance? Are they important for development?

2 Do the benefits of not building up indebtedness outweigh the costs of devolving sovereignty or ownership to foreign economic actors?

3 How should countries best attract and harness the potential of equity flows? Liberalise? Strengthen institutions and rules? Operate with strict seclusions and manage firms? Minimise these flows?

4 Given that equity portfolio flows are extremely concentrated in emerging markets, are there any advantages of LICs trying to attract this type of investment?

5 What do you make of Joan Robinson's (1962: 45) notion that 'the misery of being exploited by capitalists is nothing compared to the misery of not being exploited at all'?

Further reading

- Ted Moran's *Foreign Direct Investment and Development: Launching a Second Generation of Policy Research: Avoiding the Mistakes of the First, Reevaluating Policies for Developed and Developing Countries* is an interesting statement of the liberal institutionalist worldview. Arguing against econometric studies that claim to find an on average positive or negative relationship, he disaggregates the sectors' FDI flows into and then focuses on the policy and institutional changes that governments can and should make to maximise the developmental benefits of FDI. There are plenty of studies and plenty of case studies in here – for example, covering Thailand, Costa Rica, Mexico. He concludes by arguing that FDI has the potential to reduce poverty, governments should liberalise, but that it's no panacea.

- John Gerrard Ruggie, a professor in human rights and international affairs at the Kennedy School of Government, and since 2005 the UN Secretary-General's Special Representative for Business and Human Rights, was responsible for developing a set of principles on business and human rights, which have recently been adopted by the OECD, the International Standards Organization, the IFC and the EU. In *Just Business* (2013) he has written a fascinating insider's account of navigating and rethinking the relationship between global corporate business and human rights and development.

- Madeley's *Big Business, Poor Peoples* (2008) is a radical critique of the impact of TNCs on poor people in developing countries. With an explicit focus on the social costs of FDI and telling the stories of the poor using detailed case studies, such as Mitsubishi in the Philippines and Shell in Nigeria, Madeley writes as a journalist determined to provide a counterpoint to the mainstream academic analysis and myopia around TNCs and poverty.

Useful websites

- UNCTAD's Annual *World Investment Report* provides details on FDI trends at the global, regional and country levels with a special emphasis on the developmental consequences. It also provides a ranking of the biggest TNCs, an in-depth analysis on a particular topic and policy recommendations. The *Report* also comes with a statistical annex with data on FDI flows and stocks for 196 economies. You can also download individual *Country Fact Sheets* with the most relevant indicators about FDI in a country, including data on FDI flows and stocks, mergers and acquisitions, the largest TNCs and regulatory changes. Available at: http://unctad.org/en/pages/DIAE/World%20Investment%20Report/WIR-Series. aspx.

- CorpWatch has been operating since 1996, from San Francisco in the US. It is a non-profit organisation which describes its mission as 'investigative research and journalism to expose corporate malfeasance and to advocate for multinational corporate accountability and transparency. We work to foster global justice, independent media activism and democratic control over corporations'. The website contains constantly updated news and analysis on corporations, as well as a substantial archive of investigations, exposés and first-person accounts. A great place to start critical work on an industry or company. Available at: http://www.corpwatch.org/.

8 Remittances

Learning outcomes

At the end of this chapter you should:

- Know the history and emergence of international remittances within the development agenda, so-called 'remittance euphoria'.
- Be able to explain the individual motives and the aggregate drivers of remittance flows.
- Appreciate the importance and occasional invisibility of migrant labour as the very possibility of remittance flows.
- Understand the impact of remittances on poverty, inequality, welfare outcomes and financial development in and between developing countries.
- See and explain the influence of the liberal institutionalist paradigm in shaping the public policy attempts to harness the developmental potential of remittances.
- Be aware of the radical and critical reformist critiques of the orthodox mainstream logic and agenda.

Key concepts

Remittances, migration, development, growth, poverty, inequality, leveraging remittances, public policy

Remittances and development

Remittances are the portion of migrant workers' earnings which they send home to their country of origin, often to their family (see Box 8.2). While the act of sending money home is not new, the size of the phenomenon and its recognition is. The World Bank estimates that in 1990 remittances to developing countries were around US$31 billion. The latest estimates suggest that remittance flows to developing countries reached $351 billion in 2011 (Mohapatra *et al.* 2011). Remittances dwarf aid flows; remittance flows were almost two and a half times the volume of global ODA, which stood at US$128.5 billion in 2010 (see Chapter 5). In 2008, remittances were equal to over 40 per cent of *all* capital flows to developing countries: remittances added an extra US$327 billion worth of inflows to the combined net official and private flows to developing countries (i.e. all the flows covered in Chapters 5, 6 and 7 combined) which stood at US$780 billion (World Bank 2010a) (see Figure 2.4, p. 40). And in many countries and some regions remittances are larger even than FDI flows (see Figure 8.6).

Precisely because of the size of this financial transfer to developing countries, policymakers have seized on remittances' potential for promoting development. At the macro-level remittances provide developing countries with a very significant flow of foreign exchange and source of hard currency, and at the micro-level provide what is often an essential source of extra income that households can utilise.

Technically speaking, remittances are not really a capital flow. They are actually unrequited transfers and so get counted as an above the line item and get recorded in the current account in the balance of payments, not the capital account. But because they are one of the largest flows of money and savings internationally, and constitute a significant resource for poor countries and their populations, it seems entirely appropriate to dedicate a chapter to them. Furthermore, as we will see, some of the key public policy debates – about how best to manage and leverage their impact – are debates about financial intermediation.

Plus, it is worth flagging up that although remittances are usually equated with financial transfers they can also be transfers in kind or indeed flows of social, technical or political resources. These are all significant and have important impacts on diaspora communities and their home towns (Levitt 2001). Our attention in this book will be limited to financial transfers.

Box 8.1

'Hidden in plain view'

In the early 1990s most development economists viewed remittances as small change. This seemed to be borne out by official statistics, for example the IMF had estimated that the Philippines received some $122 million in remittances. However, Dilip Ratha, now a World Bank economist, influenced by his own experiences of migration and sending money back to his family and his home village of Sindhekela in Eastern India, became interested in the overall volume of remittance flows to developing countries (De Parle 2008). He had in the past been toying with the idea of setting up a money transfer service from the US to India. Following more extensive and careful research into the volumes of remittances, Ratha published the first global tally of remittance flows in 2003 which included a new figure for the Philippines of US$6.2 billion; 51 times higher than the previous estimates. Moreover, global remittance flows were shown to be almost three times greater than global aid flows. Ratha confesses that he was surprised, but 'the finding actually stunned experts in the field' (Banerjee 2010: 14). It's fair to say that from this point forward the international development community took notice; took notice, that is, of something which many migrants and households had long known about. Ratha's insights have been built on in the intervening years and a series of studies have now overturned the previous view where 'most economists saw remittances as small private sums that were irrelevant to development. There was also a feeling that remittances fuelled so-called conspicuous consumption, and therefore, were not good for development' (Banerjee 2010: 15). As Ratha goes on to point out, we no longer see consumption as necessarily a bad thing – sometimes that is exactly what poor households need to do with their income to survive, sometimes consumption is actually an investment, and who exactly are we to tell other people what to do with their money? Ratha has probably done more than any other individual in ensuring that remittances are now taken seriously and are one of the key issues at the top of the international development agenda. He is now a lead economist and the manager of the Migration and Remittances Unit at the World Bank and continues to work on issues around migration, remittances and innovative financing.

The global development community's sudden and significant interest in the potential of remittances has been quite remarkable (see Box 8.1). It is safe to say that recent levels of attention are unprecedented. Until recently, in the words of the IADB, remittances were 'hidden in plain view' (IDB 2006: 1). Now there is an enormous interest 'on the part of academics, donors, international financial institutions, commercial banks, money transfer operators, microfinance institutions, and policy makers.' (Perry 2007: v). For example, the World Bank and the IMF have both dedicated recent editions of their flagship publications to the development potential of migration and remittances: the World Bank in 2006 with its *Global Economic Prospects*, and in 2005 the IMF used Chapter 2 of its *World Economic Outlook* to focus on the determinants and implications of remittances, plus more recently the UNDP's *Human Development Report 2009* addressed the benefits from a human mobility perspective. The OECD (2005a), and the International Fund for Agriculture (IFAD 2007) have also been involved in setting the agenda and publishing influential reports.

The G8 has also moved on the issue of remittances with important Declarations at Sea Island, 2004, Heiligendamm, 2007 and Toyako 2008. The Seven Recommendations of the G8 Outreach Meeting on Remittances in Berlin in November 2007 has proved crucial. The G8 made the following recommendations: (1) improve the data on remittances, (2) leverage their impact by building on the research into the development impact of remittances, (3) implement some General Principles on International Remittance Services, (4) improve international coordination to facilitate remittance flows, (5) attract remittances into the formal financial system, (6) encourage the use of technology advances for payment instruments, and (7) establish a Global Remittance Working Group to forward this agenda (Federal Ministry of Finance 2007). This chapter will address these issues; that is to say: what does the available data tell us about remittance flows? What does the existing research into the developmental impact of remittances suggest? And what are the principles and policies being followed by the international community to advance recommendations 3 through to 7?

Academic research into the drivers of remittances and their impact has started to shed some light on whether they meet their development promise and under what conditions. The consensus is well summarised by the IADB and World Bank and runs as follows (World Bank 2006c; Fajnzylber and López 2007; Perry 2007): remittance inflows tend to be associated with higher and less volatile growth and investment rates, as well as reducing poverty and improving welfare indicators and outcomes in education and health. However, the impact can be modest and varies greatly from country to country with the governance capacity and policy environment being crucial. And some care should be paid to the potential downsides, including the large costs remittances incur on individuals and households with the break-up of families, and the losses for developing countries of what are often the young, entrepreneurial and educated.

Remittances sent home, by migrants like Nancy (see Box 8.2), clearly change people's lives, reducing poverty, raising incomes, improving health outcomes and providing investment for the future. They also provide much-needed foreign exchange for developing countries. This chapter examines the development impact of remittances, assessing how remittance money tends to be spent. There is an important debate as to whether it is spent productively or unproductively, and whether this distinction even holds up. Plus, the difference between its role in reducing household poverty versus broader development. The chapter also considers what lies behind these flows, why people migrate, why they send money home and how much it costs. Finally, how has the international community sought to increase and leverage the impact of remittances?

The very possibility of remittances: migration

The UN Population Division (UNDP) reports that there were an estimated 214 million international migrants in 2010, which is equivalent to about 3.1 per cent of the world population (UNPD 2009). Most of these migrants live in Europe (70 million in 2010), Asia (61 million) and Northern America (50 million). The US had an estimated 43 million migrants in 2010, the most in the world. Roughly 60 per cent of the world's total migrants live in HICs; roughly 1 in every 10 people, which compares to 1 in every 70 people being a migrant in developing countries. However, perhaps contrary to sensationalist media reporting, only a third of all migrants moved from a developing to a developed country; the majority of migrants move from a developing country to another developing country, so-called south-south migration (UNDP 2009). Recent work is starting to compile the first estimates of bilateral migration data which looks set to reveal further insights into migration and remittances and overturn yet more comfortable assumptions, for example Ratha believes that the 'largest migration corridor in the world may not be Mexico to US as commonly believed, but that it is Bangladesh to India' (Banerjee 2010: 16). Moreover, most migration is actually internal migration with people remaining within their country. The UNDP (2009) estimates that internal migration stands at about six times greater than those who migrate internationally, so roughly some 740 million internal migrants. It is fair to say that there are many, many more unknowns than knowns with respect to migration and development.

There are a number of studies which have shown when, where, how and why migrants remit and how much, and these are surveyed below. But it is even more important to first reflect on 'the very possibility of remittances': the ability for some migrants to somehow make their way from their home to the rich world, and often in extremely testing circumstances. This is partly an issue of politics with respect to the barriers placed in the way of would-be migrants. It is partly an issue of social relationships – often the strength of family or community bonds – and historical ties. But, ultimately the logic of remittances rests on the existence of employment opportunities in the rich world and the ability to send money back in the opposite direction (see Box 8.2). All of this reflects international and economic inequalities.

Box 8.2

The affective economies of migration and remittances

It is important before, and indeed after, thinking about remittances that we do not lose sight of the human lives and labour which makes them possible. The abstractions of statistics and bourgeois economics both have a tendency to erase the labouring subject from view (Davies and Ryner 2006; Shelley 2007).

Nancy Wambui Itotia works as a nurse in England in order to support her family back home in Kenya. She sends money home to her sons, daughter and wider family. In 2005 there were 427,324 Kenyans living overseas from a total population of 35 million, and the UK was the top destination (World Bank 2008). In 2006 Kenya received over US$1,128 million of remittances, a staggering 5.3 per cent of Kenya's GDP that year. And Kenya is by no means unusual in this; many developing countries receive in excess of 5, 10, 15, even 25 per cent of GDP in remittances from migrant workers living overseas. They send remittances by bank transfers, through money transfer agents like Moneygram or Western Union, or deliver them by hand when visiting. It is salutary to note that the World Bank numbers are only the official *recorded* figure; the true figure, including unrecorded, informal remittances, is certainly much larger. Vanessa Baird (2005) describes the global, economic, and emotional economies of migration through the case of Nancy.

Nancy's reason for migrating was simple. 'In Kenya I could not make ends meet on a nurse's salary.' She has three children who she has to support without the help of an estranged husband. As a senior and experienced nurse in Kenya she was earning just a sixteenth of what she earns in England in a nursing home in Oxfordshire. If she works overtime this can amount to 30 times what she was earning in Kenya. But it's a difficult life. There are large cultural differences and she misses her home and family; Nancy left her daughter Ruth and two sons Ayub and Joel as well as her elderly mother (they were 15, 19, 22 and 86 years old respectively when she left home). But four years on the salary which Nancy can now save and send home means she stays.

Nancy sent money home to her cousin Florence so that she could buy a taxi to provide a livelihood. Tragically, Florence had breast cancer and sold the taxi to pay for treatment including a mastectomy. Nancy sent more money for cancer drugs. Florence lived thanks to the money Nancy could send home. Nancy has also been saving up in order to buy a small plot in the poor Nairobi neighbourhood of Githurai so that when she returns to Kenya she can set up her own clinic.

Nancy has also bought a home back in Kenya for $25,000 – impossible on her previous salary. Her eldest son Joel currently lives there. On one visit home Nancy brought a Sony video camera back for him. It is an important investment, not simply a gift. He is a communications studies graduate who has done some filming for the Kenyan Television Network, but is limited in his work opportunities because of his lack of equipment. When asked what difference it will make Joel replied that 'It will have an immediate effect on my income. I can use it for filming news footage and for hiring out.' But supporting her sons is not entirely altruistic, it is an implicit investment in her future, or a social contract. When Nancy grows old, she will go to live with one of her sons and be looked after, as is customary in Kenya. It is in her interest to make sure they are well set for the future.

But what of the Kenyan health system? The World Bank (2008) reports that over 8 per cent of Kenyan nurses have left Kenya. Baird (2005) reports that 'there are more doctors from Benin working in France than there are in Benin; more Ethiopian doctors in Washington DC than in the whole of Ethiopia'. Meanwhile, around 40 per cent of nurses entering the British health system have been trained overseas. Yet, both the UK government and the opposition try to outdo one another in who had best got 'the problem' of immigration 'under control'. The truth of the matter is that the hospitals and nursing homes in countries like the UK, not to mention many other sectors of the economy, would struggle to function if the likes of Nancy hadn't migrated.

The IDB (2006), in its survey of Latin American immigrants to the US, found that 56 per cent of migrants in the US did not have a job in the year before deciding to migrate. Why is this striking? First, it is significant that the perceived benefits outweigh the potential risks for a good proportion of migrants. And second, after migrating to the US, over half of these migrants without ready waiting jobs had found employment in their first month (and 38 per cent were successful in finding work within just two weeks). The survey also reveals that the average salary was $900 per month; six times the amount they were earning back home. The prospect of getting a job, and earning more, is clearly a powerful one.

On the other side of the equation, this supply of migrant labour has a demand. Many rich countries have come to depend on migrant labour. For example, the US Bureau of Labor Statistics reports that migrants constitute 23 per cent of manufacturing workers and 20 per cent of workers in the service sector (reported in IDB 2006). In the UK, the Labour Force Survey shows that 13 per cent of the total UK workforce is foreign born (Coleman 2010). Foreign labour, and therefore remittances, tend to be geographically concentrated where demand is highest, often for menial work or sectors where the domestic labour force is undersupplied. In the UK around a third (37.4 per cent) of migrant workers are based in London where the need for labour in the service sector, hotels, restaurants, etc. remains high (ONS 2010). Construction is often a key employer of migrant labourers. This was astonishingly highlighted when remittances from Louisiana leapt to US$208 million in 2006 compared with $61 million in 2004 (IDB 2006). This was because the vast number of construction workers required for rebuilding Louisiana in the wake of Hurricane Katrina included many migrant workers who then sent money home.

Sometimes concerns are expressed by the public and politicians about migrant workers taking 'our' jobs and sending 'our' money home. In terms of how much migrants send home, the IDB (2006) survey reports that about 10 per cent of a migrant's income is remitted, with the other 90 per cent being spent where they have settled. Moreover, contrary to media and popular opinion, migration is not, in fact, bad for the economy and nor are migrants spongers or welfare queens. The available economic evidence tends to support the view that migration is good for the economy overall. In the UK, for example, a recent study shows that for every £1-worth of public services used by migrant workers from Eastern Europe they paid £1.37 in taxes. This compares to the 80p in taxes paid per £1 of government services used by people born in Britain (Dustmann *et al.* 2010). Nevertheless, such evidence may well miss the point if public attitudes toward immigration policy are driven more by cultural concerns about the composition of the local population than economic concerns about wages or taxes, as more recent research has shown (Card *et al.* 2012).

Migration cannot, however, be reduced to just the laws of supply and demand. Other important factors include geography, culture and history (see de Haas 2010 for a review). Locational proximity is clearly important and so is cultural proximity such as a common language. So whereas most Mexican migrants go to the US, most South American migrants choose Spain, and Caribbean migrants tend to travel to the UK (Fajnzylber and López 2007). This underlines the importance of historical colonial ties.

Figure 8.1 *Top 20 immigration countries.*

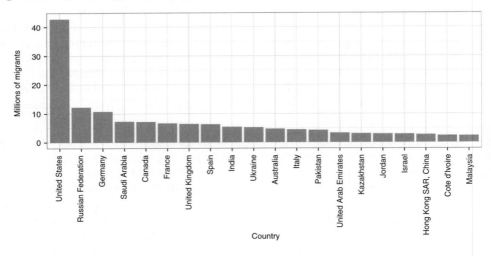

Figure 8.2 *Top 20 immigration countries by percent of population.*

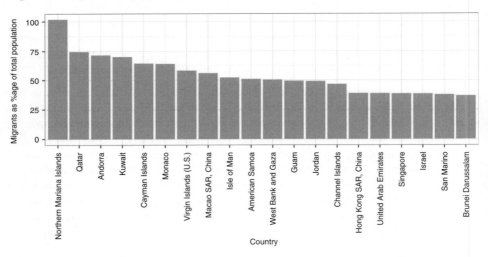

In terms of migration patterns, i.e. who migrates, whether they tend to be the more or less educated, tends to be quite varied (Fajnzylber and López 2007). For example, migrants to the US from Mexico and Central America tend to be those with fewer years of formal education, whereas in Latin America and the Caribbean they tend to be those who have been in formal education for longer. In part these differences reflect the relative ease and cost of travel from Mexico to the US compared with travelling from further south in the continent; migrating over larger distances tends to be the preserve of professional and educated classes. Among Mexican migrants, only about 4 per cent have tertiary education, whereas it is around 30 per cent for South America and about 70 per cent for India, China, Philippines, Egypt, Iran, Indonesia, Pakistan and Malaysia.

Why and how do migrants remit?

There are a number of different theories and explanations of why migrants remit. The four most common explanations are altruism, self-interest, a social contract and insurance. First, altruism. In conditions of household poverty family members choose to migrate in order to earn extra income. This is because they feel a responsibility towards their family's welfare. As such remittances are driven by a moral sensibility and indeed love. Second, others point to a narrower more self-interested logic. The migrant sends their remittances home to invest into assets such as land, or financial assets in order to save the accumulated wealth, maybe to start a business or build a house. From a purely rational perspective, then, the family actually provides a trustworthy and cheap intermediary. A third alternative is that the decision and ability to migrate and the subsequent flow of remittances represent an implicit family contract. As opposed to the altruistic version it is argued that remittances represent an implicit or explicit quid pro quo. That is, the future migrant is invested in by the rest of the family, for example through having their education prioritised and paid for and/or covering the costs of the migration from household savings. In return, once migrated, the migrant sends back remittances as a way of 'repaying the loan'. Another version of this theory is that the migrant acts a form of insurance for the family. Having a source of income which is detached from economic conditions in the developing country is a way of diversifying the household's income and reducing the risk of negative income shocks.

Box 8.3

A typical remittance transfer

A typical remittance transaction takes place in three steps:

Step 1: The migrant sender pays the remittance to the sending agent using cash, check, money order, credit card, debit card, or a debit instruction sent by e-mail, phone, or through the Internet.

Step 2: The sending agency instructs its agent in the recipient's country to deliver the remittance.

Step 3: The paying agent makes the payment to the beneficiary. For settlement between agents, in most cases, there is no real-time fund transfer; instead, the balance owed by the sending agent to the paying agent is settled periodically according to an agreed schedule, through a commercial bank. Informal remittances are sometimes settled through goods trade. The costs of a remittance transaction include a fee charged by the sending agent, typically paid by the sender, and a currency-conversion fee for delivery of local currency to the beneficiary in another country. Some smaller money transfer operators require the beneficiary to pay a fee to collect remittances, presumably to account for unexpected exchange-rate movements. In addition, remittance agents (especially banks) may earn an indirect fee in the form of interest (or 'float') by investing funds before delivering them to the beneficiary. The float can be significant in countries where overnight interest rates are high.

Source: Ratha (2012)

Migrants can use a number of different channels – formal and informal – to get money home. These include: hand delivery while travelling home either by themselves or through friends, using informal networks such as the *hawala* or *hundi* system, or through banks or specialist remittance service providers (RSPs) such as Western Union. Charges can be expensive. Commercial banks are often the most expensive; the average total cost for sending US$200 is 12 per cent, as of 2010. Post Office and money transfer organisations (MTOs) were 6.72 and 7.09 per cent respectively (World Bank 2010c). Addressing the high costs of sending remittances has been a key target for the World Bank and policy community more generally. This is discussed at greater length in the policy section below.

Volume of remittances

Remittance flows are estimated to have reached $351 billion in 2011, which after a slight downturn in 2009 (the first recorded decline since the 1980s) represents an 8 per cent increase over $325 billion in 2010 (Ratha *et al.* 2010; Mohapatra *et al.* 2011). The World Bank also notes that remittance flows were relatively resilient to the crisis. Remittance flows, unlike private capital flows, have continued to act as a vital source of external finance in the developing world through the global downturn. Their stability, and even counter-cyclical nature, is a noted and celebrated characteristic of remittance flows.

Looking at the overall flows of remittances it is clear just how much they have increased over time, putting the development community's recent interest into context (see Figure 2.4, p. 40). For example, just two decades ago the value of remittances in Latin America and the Caribbean, which is the region with the largest receipt of remittances in the world, was only a tenth of what it is now (Fajnzylber and López 2007). As shown by Figure 8.3, the Latin America and Caribbean, East Asia and Pacific regions are the largest remittance receiving regions. And the LAC is the largest receiving region per capita, with an average of US$102 per year. However, some technical and political initiatives have also helped increase the recorded amounts of remittances, for example better accounting methods and pushes to get remittances through formal channels, a lot driven by anti-money laundering and terrorism concerns (see Box 8.5).

The top recipients of remittances in 2011, as with previous years, were India ($58 billion), China ($57 billion), Mexico ($24 billion) and the Philippines ($58 billion) (see Figure 8.5). Other large recipients in absolute terms include Pakistan, Bangladesh, Nigeria, Vietnam, Egypt and Lebanon. However, when remittances are measured as a share of GDP, then it is smaller countries such as Tajikistan, Lesotho, Samoa, Moldova, Kyrgyz Republic, Nepal, Tonga and Lebanon that rise to the top of the list of remittance recipients (Mohapatra *et al.* 2011). In all of these cases remittances were 20 per cent of the country's GDP or higher. For small countries remittances can account for a large proportion of national income. For example, in Haiti remittances represented 52.7 per cent of GDP, Jamaica 17 per cent, Honduras 16 per cent and El Salvador 15 per cent of GDP. In terms of remittances per capita, Jamaica received roughly $550 per person. And for many countries remittances are the single most important source of

Figure 8.3 *Migrant remittance inflows by region, 1990–2011.*

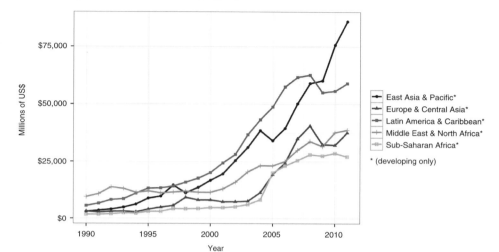

Figure 8.4 *Migrant remittance inflows by income group, 1990–2011.*

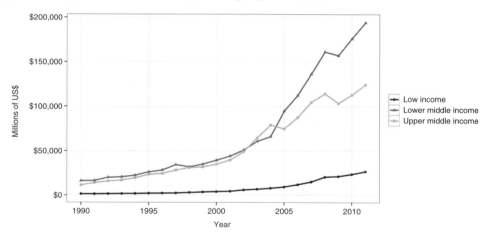

external finance, by some margin. For example, remittance flows are four times as large as FDI flows in Guatemala and Honduras, three times as large in El Salvador and twice as large as FDI flows into the Dominican Republic. Even for a large country, such as Colombia, in 2003, remittances were roughly twice the volume of FDI flows.

Drivers of remittance flows

A number of factors influence the aggregate flows of remittances. While the total inflow of remittances received increases with the number of migrants abroad, as one

Figure 8.5 *Top 10 remittance receiving countries in total US$ and by percent of GDP.*

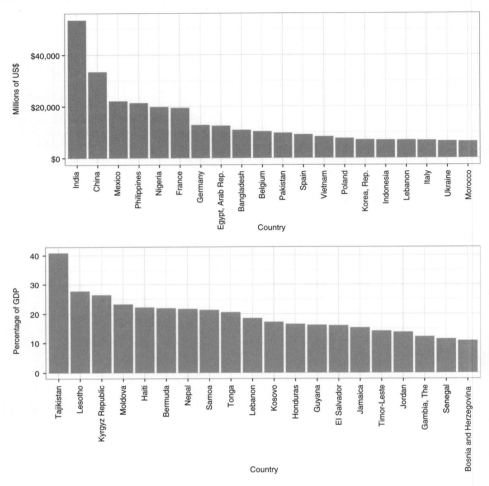

would expect, the amount of remittances per migrant tends to fall as the total number of migrants increases. In addition, the complexion of the migrant population matters. Notably, better educated migrants, on average, tend to send back less, while there is no significant difference in the amounts sent back by men and women (Fajnzylber and López 2007).

On the supply side, the business cycle in the sending country matters. Economic growth in the country where the migrants have settled and send remittances from tends to increase the volume of remittances sent. And this relationship also works in the opposite direction. For instance, since the onset of the economic crisis in September 2008 and the recession in the US, remittances to the Latin America and Caribbean region have declined strongly (Ratha *et al.* 2010). Plus, sectoral issues matter too, i.e.

Figure 8.6 *Remittances, ODA and FDI into selected countries.*

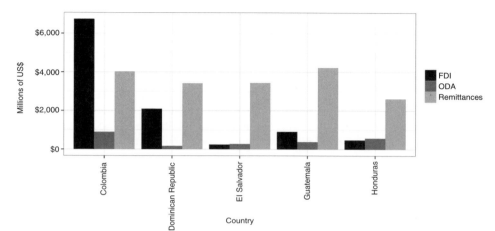

Figure 8.7 *Top 20 remittance sending countries.*

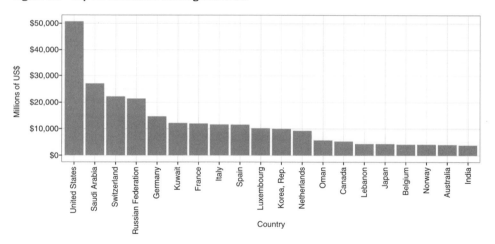

because many Mexican migrants in the US work in the construction sector, the housing slump has hit remittances especially hard.

Countries which have a more diversified range of migrant destinations enjoy more resilient remittance flows (Ratha *et al.* 2010). For example, India and the Philippines have migrants in both the US and the Gulf region as well as across Europe, East Asia and Australia. The diversification in migrant destinations, with different business cycles, means that overall remittance flows are more stable. In contrast the Latin American/Caribbean region's migration is almost exclusively to the US, and as such remittances to Latin American/Caribbean countries are subject to greater fluctuation as they are closely tied to economic conditions in the US.

Higher barriers to labour mobility – whether at the border or in terms of labour market regulation – tend to result in more stable remittance flows. In countries where it is easy to move to and from – for example within the EU and between Russia and the Commonwealth of Independent States – migrants tend to return home during economic downturns, but where it is more difficult to re-enter, migrants tend to stay put and make sacrifices in terms of their own living standards (Ratha *et al.* 2010).

There is a demand aspect to the drivers of remittances too. Remittances have long been noted for their stability, but many have identified natural counter-cyclical tendencies to remittance flows (Maimbo and Ratha 2005b). This is because remittances quickly respond to perceived need back home. For example, after a natural disaster such as the Haiti earthquake or Hurricane Mitch, or an economic or financial crisis such as in East Asia in 1997–78 or Mexico in 1995, migrants send more money back to their families in order to help cope with the economic dislocation (Savage and Harvey 2007). After the tsunami in Sri Lanka which destroyed more than 99,000 homes, killed over 31,000 people and caused an estimated $1 billion worth of damage (about 4.5 per cent of Sri Lanka's GDP) the Central Bank of Sri Lanka reported a sharp increase in remittances (Deshingkar and Aheeyar 2006). So much so that Sri Lanka's balance of payments went from a $205 million deficit in 2004 to a surplus of $179 million in 2005. Similarly, the importance of remittances in reducing households' vulnerability in Northern Pakistan after the 2005 earthquake has been documented (Suleri and Savage 2006). However, it is also possible that remittances respond to business and investment opportunities which – if this is the reason to remit – makes them pro-cyclical like other private capital flows (Fajnzylber and López 2007). The evidence suggests that, on average, flows tend to be counter-cyclical, but not exclusively so.

The World Bank (2006c) shows that remittances have a negative relationship with the output level of the recipient country and a positive relationship with the output level of the sending country. Plus, counter-cyclicality is larger for UMICs. Though, again, there is an important degree of heterogeneity across countries, so while 16 of the 26 countries in the Latin American/Caribbean region had counter-cyclical results, El Salvador, Paraguay and Venezuela were pro-cyclical. The World Bank (2006c) also examined the behaviour of remittances after severe negative economic shocks associated with macroeconomic crises: sudden stops and currency crises. They show that in the four years after a sudden stop, remittances increased by 0.75 per cent of GDP, and 2.74 per cent after a currency crisis, providing much needed emergency income and an effective insurance policy for migrant-sending households.

Sayan (2006) examines remittance flows into 12 countries over the period 1976–2003 and finds that remittances are counter-cyclical over the whole group, and the increase in remittances lags the fall in GDP by one year. However, looking at individual countries a more complicated pattern emerges with some countries being counter-cyclical, some a-cyclical and others pro-cyclical, i.e. where workers were increasing their remittance during periods of economic growth back home. It is likely that the different individual drivers of remittances explain these differences, with self-interested investment motives leading to pro-cyclical behaviour, and altruistic motives leading to counter-cyclical behaviour.

The other 'demand side' factor that appears to matter is the level of financial development in the migrants' home country. The fact that more remittances are recorded in countries with a more developed financial system may simply be because they are recorded rather than reflecting actual differences in inflows. Interestingly, in light of the above discussion, remittances co-move directly with output fluctuations in recipient countries with a higher degree of pro-cyclicality for countries with shallower financial systems (Giuliano and Ruiz-Arranz 2005).

Foreign exchange movements also strongly affect recorded remittance flows (Ratha *et al.* 2010). First, there is the valuation effect where the dollar denominated valuation of remittances is affected by currency changes. For example, part of the decline in remittance flows to Poland in the past few years can be explained by the depreciation of the British sterling against the US dollar. The same amount of pounds sterling are being sent back to Poland, it is just that once converted into US dollar amounts this records a lower amount; however the crucial issue, for the recipients in Poland, is how the złoty has fared against the British pound. Second, foreign exchange movements can also have real effects on remittance flows. If a currency depreciates then migrants who are interested in investment back home tend to remit more, taking advantage of the cheaper land and other asset prices, and vice versa.

Who receives remittance flows?

The information and research based on balance of payments data, as presented above, only gives an aggregate figure and says nothing about who receives remittances within the different countries. For that it is necessary to look at household surveys. National surveys from 11 Latin American countries (which together account for more than two-thirds of all remittances to the region) provide some important insights. First, the survey data show that remittances are a common income source for households in these countries, with 10–25 per cent of households reporting having received remittances (Fajnzylber and López 2007). The type of household which receives remittances varies across countries. Survey data which tracks the variation across households according to their non-remittance income shows a range of different patterns. In Mexico, remittances predominantly go to poor households, with 61 per cent of recipient households occupying the bottom quintile and only 4 per cent in the top one. Whereas in Peru only 6 per cent of recipients are in the bottom quintile and 40 per cent goes to the richest quintile.

From Figure 8.8, it is possible to identify three broad patterns: (1) countries with a pro-poor distribution of remittances (sloping from left to right, like Mexico), (2) countries with a pro-rich distribution (sloping from right to left, like Peru), and (3) countries where the richest and poorest households receive remittances, but not the middle quintiles (a U shape, like Bolivia). However, when we take it all together the overall picture is that remittances are regressive with respect to distribution of income in this

Figure 8.8 *Households receiving remittances by quintile of the non-remittances income distribution.*

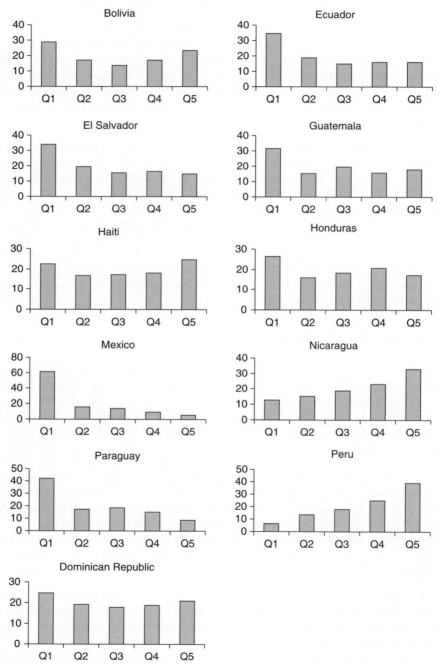

Source: Fajnzylber and López 2007: 7

region (Fajnzylber and López 2007). Across the 11 surveyed countries, the poorest 60 per cent of the population receive only 25 per cent of total remittances.

Development impact

Orthodox proponents claim that 'Remittances constitute one of the broadest and most effective poverty alleviation programs in the world' (IDB 2006: 1). At the macroeconomic level remittances provide a valuable source of foreign exchange for developing countries, increasing economic growth and reducing poverty. The microeconomic channels through which remittances work facilitate increased household savings and expenditure, investment in human capital, labour supply and entrepreneurship. As shown above, although remittances are regressive and although the distribution varies across countries, in absolute terms remittances *do* go to poor households and *do* increase poor households' incomes. One could expect this to mean that remittances reduce poverty (though not inequality) and increase consumption and investment. But critics point out that remittances incur costs as well and that these may well outweigh the benefits, such as the costs associated with brain or brawn drain, a loss of competitiveness through a reduction of labour supply and real exchange rate appreciation, and at the household level a loss of household income from absence of the remitting migrant (see de Haas 2010 for a review). Not to mention the private and social costs to families and communities with missing family members, and the dangers of often risky and expensive migration for individuals.

External balances

As documented above, remittances have emerged as a crucial source of external financing for some developing countries. Remittances have allowed countries to reduce their current account deficits, even allowing countries like the Philippines, Bangladesh and Nepal to maintain current account surpluses. Other developing countries have been able to use remittances to build up their international reserves so as to act as an insurance against external shocks.

Moreover, the extra (and stable) income has allowed developing countries to access international capital markets. For example, Bangladesh received its first ratings from Standard & Poor's and Moody's, which were comparable with emerging markets; the rating agencies cited remittance flows as one of the key reasons for the rating given. And in April 2010, despite challenging global economic conditions, Bangladesh was able to successfully issue a $750 million bond (Ratha *et al.* 2010). In addition to the judgements of the rating agencies, remittances help contribute to IFI evaluations of sovereign creditworthiness such as the joint World Bank-IMF LIC Debt Sustainability Framework which takes remittances into account. Plus, many of the IMF's Article IV assessments of countries' economic performance now include remittances. All of which serves to reduce sovereign spreads and the cost of borrowing for developing countries.

On the other hand, the same Dutch disease effects as outlined in Chapter 4 are held to apply to remittances, i.e. extra consumption in the non-tradables sector fuelled by the extra income from remittances will cause inflation and a real appreciation in the exchange rate. The evidence suggests that the effect varies hugely; one study found that a doubling of remittances will, on average, result in a real exchange rate appreciation of somewhere in between 3 and 24 per cent (World Bank 2006c).

Economic growth

The evidence for the impact of remittances on economic growth has been largely inconclusive. Some find that there is a positive impact on growth (Faini 2002), others negative (Chami *et al.* 2005), while others suggest that there is no significant relationship between remittances and growth (IMF 2005). The lack of a generally positive finding may be because of the time it takes for remittances to be converted into increases in human and physical capital. But it may also be because of their counter-cyclical quality. A simple study of the relationship between remittance flows and development would suggest that increased flows are related to negative outcomes. But this would be to miss the point, since – as with development aid – causality also runs in the opposite direction, with a lack of development attracting greater inflows.

This means that reverse causality and other sources of endogeneity can confound empirical analysis. Fajnzylber and López (2007) used a wider range of controls as well as time-varying instrumental variables (which are used to estimate causal effects where experiments are not feasible) and found that remittances have a statistically significant positive impact on growth. The study suggests that the average Latin American country witnessed an increase in remittances from 0.7 per cent of GDP (in 1991–95) to 2.3 per cent of GDP (in 2001–05) which resulted in an increase in GDP/capita growth of 0.27 per cent. This is relatively small in economic terms. The authors note that remittances probably increase growth through increases in domestic investment, estimating an increase of investment/GDP of 2 per cent.

Despite their modest impact on growth rates, as noted above, the additional advantage of remittances is their counter-cyclical character, helping to reduce shocks and growth volatility. The evidence suggests that output volatility is a significant brake on growth, so remittances could play an important role here. A study looking at the macro-smoothing role of remittance inflows found that the larger the remittance flow the less volatile economic output is (World Bank 2006c). A one standard deviation increase in remittance inflows reduces the standard deviation of growth by more than 10 per cent, with the effect being stronger for wealthier countries.

Poverty and inequality

How do remittances affect households' poverty levels and domestic inequality? Adams and Page (2005) performed a cross-country analysis to test for the effect of remittances on poverty reduction. They found a positive and significant relationship between remittances and poverty reduction. A 10 per cent per capita increase in official remittances is associated with a 3.5 per cent decline in the share of poor people in the population. Other studies looking at sub-Saharan Africa also confirm a positive relationship (Gupta *et al.* 2007). Likewise, in Latin America remittances have a positive though modest impact on reducing both poverty and inequality. Cross-country estimates suggest that a 10 per cent increase in share of remittances to GDP/capita is associated, on average, with a reduction of 3.5 per cent of the population living in poverty. Extreme poverty is reported to fall by more than 35 per cent in Mexico, El Salvador and the Dominican Republic (Fajnzylber and López 2007).

In terms of domestic inequality, according to the World Bank (2006c), if remittances did not exist, inequality would be higher. However, the changes in the Gini coefficient are small. Crucially, the reduction in inequality is more significant, especially in those countries, such as Mexico, where migrants tend to come from poorer households. The World Bank (2006c) calculates what inequality levels and poverty levels would have been in the absence of migration, thus making sure that the results don't overstate the impact on poverty (by making the common mistake of assuming that migrants wouldn't have been productive if they stayed at home). Reductions in inequality, as measured by Gini coefficients, were observed in Haiti (7.7 per cent), Gutatemala (2.9 per cent), El Salvador (2.1 per cent), Nicaragua (1.8 per cent) and Honduras (1.1 per cent). There were increases in Mexico and the Dominican Republic. Likewise, Rodriguez (1998) finds that remittances increase inequality in the Philippines. Surprisingly, countries with the largest poverty- and inequality-reducing effects from remittances are not the ones with more poor migrants and where we would expect remittances to go more to poor households. Furthermore, where remittances are regressively distributed, such as in Haiti and Honduras, remittances do appear to reduce inequality and poverty. Instead, what seems to be important is that the four countries where inequality and poverty are reduced are all countries with high remittance to GDP ratios (Fajnzylber and López 2007). Finally, with respect to international inequalities, while remittances tend to decrease domestic poverty, and sometimes inequality, they tend to increase income inequality at the global level (World Bank 2006c).

Switching from macroeconomic data to household survey data in Latin America suggests that poverty headcounts are only reduced in 6 out of 11 Latin American countries and reduce the poverty gap in only 3 out of 11 (Fajnzylber and López 2007). Two cases – the Dominican Republic and Nicaragua – showed an increase in poverty for households receiving remittances. One interpretation of this is that very poor households lose more than they gain from migration as the remittances sent back are less because the migrant is underskilled for the destination country.

Microeconomic channels

The debate at the micro-level tends to revolve around what households do with their remittances, whether they consume, save or invest them. The IDB (2006) survey found that senders and recipients tended to agree on the major uses of remittances. Specifically these were food, health, utilities, education, clothing, housing and business.

Expenditures on education, health and consumer durables should be seen as investment or savings (see Attanasio and Szekely 2000).

There is also evidence that remittances encourage savings and are not 'just' consumed. However, intriguingly, this tends to be truer for lower-income households, with richer ones reducing their savings rates. Receiving remittances also increases access to the financial system, with recipient households more likely to use deposit accounts, but causes no change in credit practices. Remittances increase children's educational outcomes, however, this tends to be true for households where the parents have low levels of schooling. Health outcomes, for children, are increased in low-income households. Entrepreneurship is increased for individuals within low income groups.

Saving, investment, and consumption

Adams (2005) reports that, in Guatemala, families that receive remittances spend less of their total income on food and non-durables, and more on housing, education, health and non-durables. Fajnzylber and López (2007) found that this held for other Latin American countries they studied, though again the exact balance in where these increases fall in terms of durables, education and health expenditure varies from country to country.

Household expenditure behaviour also varies across income groups (World Bank 2006c; Fajnzylber and López 2007). Poorer remittance-receiving households tend to increase expenditure on non-durables rather than invest in human capital expenditures such as education and health. Hence, on average, wealthier households tend to save less than non-receiving households, but spend it on more investment- or savings-like expenditures. The World Bank (2006c) suggests that savings rates are higher in poorer households as they use remittances as a form of self-insurance for anticipated shocks. However, wealthier households tend to consume their remittance income, although Mexico proves to be a significant exception. Poorer households tend to use the extra income to invest in health, durables, education, etc., however richer households tend to increase their expenditure on food and non-durable goods. The lessons which can be drawn from this are twofold. One, cross-country variations are significant. And two, these relationships are not iron economic laws, they are strongly conditioned by socioeconomic circumstances, cultural norms and practices, as well as being amenable to policy interventions.

Health and education

Evidence from Mexico, El Salvador and Sri Lanka has found that remittance-receiving households tend to spend more on education, such as tuition, and have lower drop-out rates (Cox-Edwards and Ureta 2003; Hanson and Woodruff 2003). Plus, the children born in these households tend to have better life chances (higher birth weight) (Woodruff and Zenteno 2001). Household surveys in Guatemala and Nicaragua find that children from households which receive remittances tend to be healthier than they would be if they didn't receive remittances, especially so within the poorest households (Fajnzylber and López 2007). Plus, in 6 out of the 11 countries in the region surveyed, remittances are positively and significantly related to higher enrolment rates. Plus, the positive effect is larger for children whose parents have a low level of schooling.

Financial development

What of the effect of remittances on financial development? Something which, as we saw in Chapter 2, is considered especially important by orthodox authors. Some have argued that remittances have actually acted as a substitute for financial system development, directly alleviating credit constraints but bypassing the formal system (Giuliano and Ruiz-Arranz 2005). However, the orthodox hope is that remittances – by providing a flow of money that wouldn't otherwise exist – will generate new demand for the use of financial services (Orozco 2005). Indeed, studies tend to find that remittances are positively related to financial development (Aggarwal *et al.* 2009). A 1 percentage point increase in remittances as a share of GDP results in roughly a 5 percentage point increase in bank deposits and credit as a proportion of GDP (World Bank 2006c; Fajnzylber and López 2007). Interestingly, there is a smaller impact on financial development in the Latin American region than in other developing countries. Fajnzylber and López (2007) suggest that the history of financial crises in the region has undermined trust in the financial system. They also suggest that remittances tend not to go through the formal banking system. And that this may in part be due to the lower number of bank branches per square area in Latin America (Beck *et al.* 2005), plus higher fees (Beck *et al.* 2006).

Survey data from El Salvador and Mexico shows that remittance-receiving households are more likely to have a deposit account (Fajnzylber and López 2007). The data from El Salvador shows that if remittances are disbursed through the formal banking system, recipients are twice as likely to have a bank account. However, the same does not hold true for households' demand for credit, which is not related to remittances. Likewise, municipal evidence from Mexico shows that where more of the population receive remittances there is a higher number of deposit accounts, which are larger, and a higher number of branches per capita (controlling for other characteristics) (World Bank 2006c).

International public policy

In this section we consider what the policymaking community has sought to do to harness the developmental impact of remittances flows, the logic for doing so, and some critiques. Broadly speaking, the liberal institutionalist agenda – on the assumption and increasing evidence that remittances have a positive, though modest and heterogeneous, impact on poverty – has sought to do two things: (1) to facilitate remittance flows, decreasing the costs and thus increase the volume, and (2) to increase the developmental impact of each dollar remitted.

In terms of increasing remittances impact, the global development community faces a dilemma of how much to guide how individuals decide to spend their own money. Direct interventions are thorny and neoliberal and liberal institutionalists are squeamish about trying to guide household decisions. Meanwhile, critical reformists and especially radicals are concerned about the direct exploitation of the poor for reasons of mopping up public expenditure or extending neoliberal financialisation. The idea that public authorities should be getting individuals to invest or save private money is anathema to many. Dliip Ratha (2005) captures the orthodox view perfectly:

> Fundamentally, remittances are private funds that should be treated like other sources of household income. Efforts to increase savings and improve the allocation of expenditures should be accomplished through improvements in the overall investment climate, rather than targeting remittances.

Plus, the idea of trying to get poor households to save more doesn't make much sense either. As Maimbo and Ratha (2005b) have argued, saving rather than spending by remittance recipients actually reduces, not improves, the welfare of the poor.

The logic of the orthodox agenda dictates that senders should make greater use of the formal financial system to reduce the costs of sending remittances *and* to increase their effectiveness. This has been the main thrust of the international public policy response and the evidence suggests some success too. For example, a 2006 survey (IDB 2006) shows that the number of migrants using the formal financial system increased from 8 per cent in 2004 to 19 per cent in 2006. The following sections unpack this further.

International public policy responses

In line with the boom in interest in international remittances there have been a number of national and multilateral initiatives from the various organs of global governance and regional development banks. Plus new inititatives and organs have been set up. For example, the Global Migration Group and the Global Forum were established following a proposal in the UN's 2006 Report from the High-level Dialogue on International Migration and Development (Annan 2006). The Global Migration Group is an inter-agency group of 14 UN and related organisations which work on matters relating to migration. The Global Forum on Migration and Development meets annually for a voluntary and non-binding discussion amongst governments, IOs and civil society investigating the ways in which migration can contribute to the development goals. It has been held in Brussels in 2007, Manila in 2008, Athens in 2009, Puerto Vallarta in 2010, Geneva in 2011 and Mauritius in 2012.

Various organisations have been focused on increasing remittance volumes, leveraging the impact of inflows, banking the unbanked, and financial development. Reflecting the size and importance of remittances to Latin America, the Multilateral Investment Fund (MIF) of the Inter-American Development Bank has done a great deal of work in trying to harness and leverage remittances since 1999 in terms of research, advocacy, funding and policy interventions (Orozco and Fedewa 2006). Its work has been targeted at reducing the transaction costs of remittances and expanding the links between remittances and credit unions in order to encourage financial development. The MIF has worked with the International Fund for Agricultural Development (IFAD) which has an $18 million multi-donor Financing Facility for Remittances which funds projects and research activities which help to reduce remittance costs, and bank the rural unbanked, as well as promote productive activities for rural recipients such as job creation. USAID has been interested in remittances since 2000, and in 2002, through its Latin American regional office, sought to expand the financial services available to remitters, such as microfinance, credit unions, and cooperatives (Orozco and Fedewa 2006). The United States Agency for International Development (USAID) has also been involved in providing funding for projects to use new technologies to lower transaction costs.

But it is the World Bank which has secured itself at the centre of the international public policy agenda on migrant remittances. It has a dedicated Migration and Remittances Unit led by Dilip Ratha, who is credited with making sure that migration and its potential rewards have risen to 'a top-of-the-agenda concern in the world's development ministries' (De Parle 2008, see Box 8.1). The Unit produces regular data, policy briefs and academic publications (Maimbo and Ratha 2005a; Özden and Schiff 2006; World Bank 2008). Plus, in November 2004, the World Bank, alongside the Bank for International Settlements, set up a taskforce whose job was to look at the international coordination of remittance payment systems. The Task Force has produced a set of 'General Principles for International Remittances Services' (CPSS/World Bank 2007) and a Guidance Note which clarifies the details of how to implement the General Principles (see Box 8.4). The Global Remittances Working Group (GRWG) was set up by the World Bank at the behest of the G8 at its Tokyo Summit in 2008. Its role is to facilitate the flow of remittances and to advise the international community on policy best practice. The GRWG has four working groups on: (1) data, (2) interconnections with migration, development and policy, (3) payment and market infrastructure and (4) remittance-linked financial products and access to finance.

Box 8.4

CPSS/World Bank General Principles

The General Principles and related roles

The General Principles are aimed at the public policy objectives of achieving safe and efficient international remittance services. To this end, the markets for the services should be contestable, transparent, accessible and sound.

Transparency and consumer protection

General Principle 1. The market for remittance services should be transparent and have adequate consumer protection.

Payment system infrastructure

General Principle 2. Improvements to payment system infrastructure that have the potential to increase the efficiency of remittance services should be encouraged.

Legal and regulatory environment

General Principle 3. Remittance services should be supported by a sound, predictable, nondiscriminatory and proportionate legal and regulatory framework in relevant jurisdictions.

Market structure and competition
General Principle 4. Competitive market conditions, including appropriate access to domestic payment infrastructures, should be fostered in the remittance industry.

Governance and risk management
General Principle 5. Remittance services should be supported by appropriate governance and risk management practices.

Roles of remittance service providers and public authorities
A. Role of remittance service providers. Remittance service providers should participate actively in the implementation of the General Principles.
B. Role of public authorities. Public authorities should evaluate what action to take to achieve the public policy objectives through implementation of the General Principles.

Source: CPSS/World Bank (2007: 4)

Increasing the volume of remittances

Remittance services are expensive. Once the global development community identified remittances as a flow to be maximised, the high costs of remitting emerged as one of its primary targets. Policymakers chose not to use direct forms of regulation such as price controls and instead sought to bring down costs through (1) increasing transparency, (2) increasing competition, and (3) reducing barriers.

The nature of remittance costs penalises small transfers. They are also opaque with lots of hidden charges and uncompetitive exchange rates. Plus the cost savings from new technologies and electronic forms of transfer have not been passed on to the customers of remittance service providers (World Bank 2006b). The World Bank (2010c) estimates that if the cost of sending remittances could be reduced by 5 percentage points relative to the value sent, remittance recipients in developing countries would receive in excess of an extra US$16 billion per year.

In 2009 at the L'Aquila summit, the G8 endorsed what has become known as the '5×5 objective': 'a reduction of the global average costs of transferring remittances from the present 10% to 5% in 5 years through enhanced information, transparency, competition and cooperation with partners, generating a significant net increase in income for migrants and their families in the developing world' (G8 2008: para. 134). Significantly, this is the first time a quantifiable target has been set. The methods recommended are those identified in the CPSS/World Bank (2007) General Principles, i.e. fostering market transparency, improving the payments system infrastructure, enhancing market competition, legal and regulatory reform.

Enhancing market competition

The costs of sending remittances home have been estimated at around 4–10 per cent (Orozco 2003). World Bank research reveals that the global average cost of sending $200 home is 8.72 per cent of the sent amount, i.e. just over $17 (World Bank 2010c).

However, some corridors have average costs at 22 per cent, and some individual providers charge over 40 per cent. It is noticeable that costs of sending remittances between countries and the volume of remittances along those corridors are highly correlated (World Bank 2010c), suggesting that competition does affect prices.

The main way in which authorities have tried to enhance competition is to increase the number of remittance service providers by reducing the barriers to entry, i.e. regulation, licensing and registration requirements. The orthodox view is that overly burdensome regulation serves to increase costs and fuel the growth of informal money transfer services (Fajnzylber and López 2007). Yet, there is a tension in this strategy as it runs up against trends towards tighter regulation of money transfer services as set out in Box 8.5.

Box 8.5

Terrorism, remittance services, and the Financial Action Task Force

The September 2001 attacks on the World Trade Center in New York, and the Pentagon in Washington, DC, precipitated a global 'war on terror'. One part of the response by Western governments was to target the ways in which terrorists were financing their operations.

Moves to stop money flows and confiscate terrorist groups' assets led to an increasing interest in money laundering in general and the use of informal remittances and money transfer operations in particular. For example, only weeks after the attacks, in November 2001, the largest Somali remittance transfer company, Al-Barakaat, was summarily shut down by the US Treasury because there were allegations that it had links with al-Qaeda (Deans *et al.* 2008). In the event, no links were found, Al-Barakaat was cleared, and the US Treasury was found to have acted without proper grounds. Plus, remittances to Somalia didn't stop; alternative transfer companies and routes were found. The more general strategy has been to increase the monitoring and policing of money laundering and the funding of terrorist activities in an attempt to move informal remittances 'above ground' and choke off money flows to terrorist organisations. The Financial Action Task Force (FATF) has been the fulcrum of such attempts.

The FATF is an inter-governmental forum that was created by the G7 in 1989 to tackle money laundering with respect to drug trafficking. It is housed in the OECD's headquarters in Paris and facilitates policy diffusion of best practices through processes of self-assessment, mutual evaluation, cross-country reviews, and naming and shaming (Roberge 2011). The FATF responded immediately to the attacks by meeting in an extraordinary session on 29–30 October 2001 in Washington, DC to consider what steps it needed to take to prevent and combat terrorist financing. The outcome of the meeting was a decision to expand the FATF's mandate to include terrorist financing (FATF 2001). The immediate outcome was to issue eight Special Recommendations on Terrorism Financing. Remittance transfer companies have been forced to purchase new technology to comply with the new regulations which include: all organisations that use informal money transfer systems or networks should be subject to all the same rules, administrative and civil and criminal sanctions as banks are; remittance providers should include accurate and meaningful originator information (name, address and account number) on transfers and related messages that are sent, and the information should remain with the transfer or related message through the payment chain; and non-profit organisations should be subject to extra scrutiny from authorities as they are 'particularly vulnerable' to being abused for the purpose of concealing or obscuring 'the clandestine diversion of funds intended for legitimate purposes to terrorist organisations' (FATF 2001, Annex A: 2). In sum, remittances, both formal and informal, have been subject to a great deal more scrutiny and regulation than prior to September 2001.

Increasing market transparency

Even with enhanced competition within the industry, in many cases prices remain opaque and customers do not know which is the best deal. The World Bank argues that the lack of transparency on the remittance market is 'the single most important factor leading to high remittance prices' (World Bank 2010d). Remitters face the complexity of prices which are comprised of a sending fee, an exchange rate margin and sometimes a collection fee. As such, prices often contain different direct and indirect costs and fees and, importantly, opaque exchange rate differentials. These charges can also vary depending upon the type of service and speed of transfer chosen (World Bank 2010d). This complexity means that consumers cannot be sure of getting the best price, hindering competition as direct comparisons cannot easily be made. This is not innocent, of course, there is money to be made by remittance service providers who can and deliberately do mislead customers.

As such, public authorities have worked to publicise comparative prices and quality of service to make them available to remitters. The World Bank publishes a Remittance Prices Worldwide Database which was launched in September 2008 (see http://remittanceprices.worldbank.org/). The Bank argues that publication of the pricing data serves four key purposes: 'benchmarking improvements, allowing comparisons among countries, supporting consumers' choices, and putting pressure on service providers to improve their services' (World Bank 2010d). The data covers 178 remittance corridors between countries and provides information on the transaction fees and the exchange rate margin. While this is useful for policymakers, professionals and researchers, there are further more direct ways of 'supporting consumers' choices'. In the UK, the Department for International Development (DFID) set up and funded a scheme called 'send money home' which provides user-friendly information for remitters (see Box 8.6).

Box 8.6

'Send money home' website

The send money home initiative was launched in March 2005 by the UK DFID. It was designed to reduce transfer costs for remittance transfers by creating competition between remittance providers and arming migrants with information about their rights and ability to compare prices across providers. The initiative was driven by a concern about the stranglehold that a small number of big providers had on the sector and that they were charging exorbitant rates that badly informed migrants were paying, unnecessarily and unfairly. The DFID study behind the initiative compared the costs of sending £100 and £500 to Bangladesh, China, Ghana, India, Kenya and Nigeria, using 18 different specialist money transfer services and banks. The cost of sending £100 varied from £2.50 to £40, 2.5–40 per cent of the initial amount, while fees for transferring £500 ranged from £4 to £40, 0.8–8 per cent. It's clear that costs vary hugely and that people were penalised for sending smaller amounts. Finally, the transfer times across the different service providers varied from 10 minutes to 10 days (Cater 2005).

The send money home service was launched with 900,000 information leaflets to 20 different diaspora communities. The website went live – essentially a price comparison website – and users could sign up for email updates. The website also answers a set of frequently asked questions relating to transferring money 'back home' from the UK, such as 'How can I send my money?', 'If I send £X, how much will be received by my friends or family?', 'What documents/information do I need to send money?', 'What protection is in place to make sure the money is received?' and others. Since then the service has become a privately run, free web service which caters to a wider audience of people wishing to know the cheapest and best way to send money overseas.

A World Bank report which compares a range of national price comparison websites concluded that between 2005 and 2009, the sendmoneyhome.org website helped in reducing the costs of sending £100 by an average of about 48 per cent (World Bank 2010e). The World Bank also has its own, well-maintained, global remittance price comparison website, found at http://remittanceprices. worldbank.org/.

Improving the payments system infrastructure

Improving the quality and access to cross-border payment and settlement systems and the interoperability of the systems is also crucial for keeping down costs and improving the flow of remittances. This is partly a technological issue, but also requires cross-border bi- or multilateral legal cooperation. On which point, Ratha has proposed an International Remittances Institute, which would be responsible for monitoring flows of people and remittances and making sure they flow cheaply, safely and maximise their productivity (Banerjee 2010).

Remittance sending used to be a labour intensive physical transmission process, often relying on courier services. Now it's based around wire services, offering cash-to-cash transfers which MTOs have dominated. The emergence of formal financial providers has increased the number of account-to-cash and account-to-account transfers. Many banks and formal financial service providers have begun to partner up with remittance transfer service providers (Orozco and Fedewa 2006). Such partnerships seek to maximise the benefits of each partner's comparative advantage in terms of greater numbers of branches, distribution network, marketing clout on the one hand, and trust and word of mouth within the migrant community on the other. Plus, with these partnerships, customers do not need to have an account at the bank, and the exchange rate is superior and the service faster.

Technological advances, especially mobile telephone banking, have massively reduced the costs of remitting and the ease and reach of remittance services. For example M-Pesa, the Kenyan mobile-phone based money transfer service, recently branched out from microfinance to remittances (Hughes and Lonie 2007). There are other examples of this including Bangalink in Bangaldesh, EasyPaisa in Pakistan and Zain (a Kuwaiti mobile operator) which has launched Zain Zap, a mobile remittance service which covers 15 African countries and 42 million subscribers. These services have also sped up transfer times from four to five days to one day. But perhaps more importantly from the perspective of encouraging financial development, these branchless banking providers also offer a bundle of other services such as savings and payment services (Ratha *et al.* 2010).

Policy, legal and regulatory reform

In line with liberal institutionalist insights, it has been argued that despite an inconclusive relationship between remittance inflows and economic growth, if governments were to make a series of liberalising and stabilising macroeconomic policy reforms it would help leverage the impact of remittances. That is to say, the impact of remittances is conditional on the quality of government policy in education, institutional quality and the macroeconomic environment (Fajnzylber and López 2007).

Developing countries have taken a number of more focused initiatives to help boost the flow of remittances. For example, the Pakistan Remittance Initiative (PRI) was established to encourage the flow of remittances by subsidising the cost of remittance transfer services (http://www.pri.gov.pk). Going a step further, the Philippines has established a series of agencies such as the Philippines Overseas Employment Administration which is a government agency which promotes and monitors the employment of Filipino migrant workers, for example, through providing information on job opportunities, securing visas, monitoring illegal recruitment and providing assistance to migrants who need help when overseas.

One thing almost all commentators are agreed upon is that remittances shouldn't be taxed by recipient governments, no matter how tempting it is to take a slice of private income streams to invest in much-needed public goods. This is partly an ethical argument insofar as it would be double taxation, with the worker's income already having been taxed by the host government. But secondly, it is also partly a policy argument. Having worked so hard to get remittances into the formal system and 'above ground', the fear is that a remittance tax would reduce remittances through raising the cost and/or drive remittances underground and out of the formal financial system (Ratha *et al.* 2010). Nevertheless, there are examples of this happening. And, in 2009, a sending government (Oklahoma in the US) introduced a tax on money transfers of $5 (or 1 per cent for transactions > $500). The proceeds are to go into a fund to fight money laundering and drug trafficking.

Host governments can also play a positive role through introducing pro-remittance regulation. For example, three days after the earthquake in Haiti, the US authorities granted temporary protected status (TPS) to all Haitians in the US. Thus, for 18 months the estimated 200,000 irregular Haitian migrants in the US could work and send money home through formal remittance channels without fear of being deported. It has been argued that this measure benefitted Haitains 'more than any other aid and assistance' (Ratha 2010).

A further interesting proposal – which hasn't been taken up – is RemitAid™, a scheme for remittance tax relief for international development (Faal 2006). The proposal combines the characteristics of existing government tax relief schemes with a common fund framework which would pool together private remittances committed through the scheme. The tax relief would incentivise remitters and the pooled money would be invested into productive and public projects in a way that private household-household remittances can't be.

Increasing the development impact of remittances

In addition to attempts to increase the overall volume of remittances by making it easier and cheaper to transfer money back home, the second plank of the policy community's strategy was to increase the development impact of remittances. The challenge, as constantly articulated by the World Bank, is to leverage the impact of remittance flows without directly interfering with remittances, migrants or receiving households (e.g. see Ratha *et al.* 2010).

Banking remittances

Remittances on their own provide an important stream of income for many households in developing countries, but some worry that without developing broader and deeper access to other financial services, the remittances will be little more than a palliative (Orozco and Fedewa 2006). Many remitters and many recipients do not have access to a bank account. Yet many households have an unfulfilled demand for savings, credit, insurance and pensions. As such, remittances also offer a way of banking the unbanked and getting poor households into the formal financial system.

A common assumption used to be that remittances were cash-to-cash transfers outside the financial system. The IDB (2006) suggests that this is wrong. Instead most remittances do go through a financial services provider, but they do not go into a bank account. The challenge is integrating remittance services with other banking services. Instead remittance services have been treated as a parallel business (with parallel queues inside banks!), as they were not seen as profitable customers. However, mainstream banks have now realised that while there might not be many profits to be made in the remittance business, the longer-term strategy of getting the unbanked using other financial services such as deposit accounts, loans and credit cards will generate returns (Orozco and Fedewa 2006).

Some banks, which offer remittance services, now offer customers access to other financial products and services in order to turn them into full bank clients (Orozco and Fedewa 2006). It is a gradual process, usually beginning with a savings account, sometimes promoted through prize draws for household appliances. These efforts are usually accompanied by demonstrations on how to use the services, such as ATMs, and some financial literacy information. The Consulate General of Mexico has the New Alliance Task Force which is a coalition of banks, community organisations and government agencies which seek to improve financial access and literacy, such as holding classes on money skills and using financial products and services such as ATMs (Orozco and Fedewa 2006).

Migrants are interested in a number of different financial products (IDB 2006). For instance, direct payment methods to be able to pay their family's household bills or school fees; products that could use the remittances as collateral for a mortgage or housebuilding loan; insurance products such as life, health, or buildings insurance for their family; and a channel into microfinance institutions for savings accounts or loans.

In *A Scorecard of the Industry*, migrantremittances.com found that the main barriers to migrants not accessing the formal financial system were regulatory and legal barriers, cultural factors and a lack of appropriate products and services. The US is an interesting example of a policy intervention designed to address problems in getting migrant remittances into the formal banking system. It was recognised that a proper and accepted form of identification was the primary barrier for immigrants to using the formal financial system (Orozco and Fedewa 2006). In many cases it is impossible to open a bank account and access the formal financial system without identification. For the many irregular migrant workers this is a problem as they do not have a legal right to be in the country. In the US, in 2002, the Department of the Treasury advised that the Mexican consular identity card, as issued by the Mexican government, could be used to open a bank account in the US. The Federal Deposit Insurance Coporation promoted the use of the *Matricula* with banks in the US in a two-year education drive to encourage US banks to accept it. The Mexican consulate reported that in the year following the initiative a survey of over 30 banks in the US Midwest revealed that over 50,000 new bank accounts had been opened with an average balance of US$2,000, i.e. more than US$100 million in deposits had been made. Plus, over 35,000 migrants had participated in financial education classes.

Leveraging remittances

Most discussions of remittances are implicitly referring to individual or family remittances, but some remittance flows are collective or pooled remittances. Some migrants have formed philanthropic organisations to help funnel funds into small-scale local development projects back in their hometowns (Orozco with Lapointe 2004). It is estimated that there are some 2,000 such hometown associations in the US which are collectively investing $30 million. They often fund projects to improve health, education or infrastructure. This can be a significant amount, for example one survey looked at 62 communities in Mexico and found that the average amount raised was $23,000 per year per town, which equated to around a fifth of the municipal budget for public works. The hometown association will coordinate with a community leader, perhaps the municipal government, to manage the disbursement of funds and monitor progress. Hometown associations create jobs, improve access to health and education services, improve social capital and increase participation within the community. They also serve a cultural function of keeping migrants connected to their communities back home, maintaining a level of belonging there, as well as with fellow migrants from home. But they can also create political and cultural struggles (for example see Page 2007).

The Mexican government has sought to increase the flow and impact of these collective remittances through its Citizen Initiative programme, also known as Iniciativa Ciudadana 3x1 or *tres por uno*. It is a matching-grant programme which was brought in in 2002 where, for each dollar of remittances, the federal government adds one dollar, the state government adds one dollar and the municipal government adds another dollar. The scheme was originally a state scheme in Zacatecas called *Dos-por-uno* which had its origins in the late 1980s before being emulated and scaled up by the national government (Iskander 2005).

Another method of leveraging remittance inflows has been through remittance securitisation. If a developing country can get remittances into the formal financial system and ensure a large and steady volume of remittances it can use these to raise significant development finance (Kektar and Ratha 2001, 2004/05). Instead of physical assets being used as collateral, the future income streams from remittances can be pledged to an investor in return for a loan. By packaging the income stream as a financial asset (a bond) and selling it, the borrower (issuer of the bond) can unlock future income immediately. Countries such as Brazil, Mexico and Turkey have been the leaders in remittance securitisation.

In a slightly different way, migrants and remittances represent an investment opportunity for many developing country governments. Many migrants state that they are interested in investing in their home country (IDB 2006). Mostly this is through buying a family home or investing in a small business. Additionally, diaspora bonds represent a way of governments tapping into this desire (Ketkar and Ratha 2007). Israel and India have issued diaspora bonds and raised over $35 billion. The bonds tap into the wealth of the diaspora, but also tap into the diaspora's patriotism and desire to invest in the future of their country. The money raised by Israel and India was used to support balance of payments needs and finance expenditure on infrastructure, housing, health and education projects (Ratha *et al.* 2010). The potential for this sort of borrowing is huge, especially for developing countries in times of need. Giving the example of Haiti, Ratha *et al.* (2010) note that if the 200,000 Haitians in the US, Canada and France were each to invest $1,000 in diaspora bonds, they would raise around $200 million, which could then be increased by expanding the bond sale to supporters of Haiti such as private charitable organisations. Moreover, institutional investors could be attracted if the bond issue were to be guaranteed by multilateral or bilateral donors.

Political economy of remittances

As should be clear, the remittance and development policy agenda is very much informed by orthodox thinking, at least at the international level and within development circles. At the national level, political and cultural concerns trump any welfare gains from the freer movement of people, no matter how large (Card *et al.* 2012). For example, Michael Clemens has documented the 'trillion dollar bills on the sidewalk' in referring to the global gains from liberalising labour flows, which dwarf the benefits from free trade in goods and services or capital movements.

Unlike the development debates on aid, debt, FDI and so forth there has been much less heterodox work done on migration, remittances and development. What insights we can derive tend to make four points: (1) focusing on remittances is addressing the symptoms of underdevelopment, not the causes, (2) migration and remittances are not pro-poor in the sense that they produce regressive income redistribution, (3) the orthodox position ignores the human costs of remittances, and (4) the global governance of migration and remittances is extending neoliberal individualisation and financialisation through a new global but private regime.

First, migration isn't the solution to underdevelopment so much as migration is caused by underdevelopment (De Haas 2007). Sceptics argue that 'if migration brought development, Mexico would be Switzerland' (De Parle 2008). Migration and remittances can act as a palliative, helping individuals, households and perhaps even communities out of poverty. This is not to be sniffed at. Indeed, the right to mobility is fundamentally important in its own right and not just to bring about development (UNDP 2009). However, as critical reformists highlight, migration and remittances are not substitutes for development, in the sense of structural change. Focusing on remittances as a solution to the problem of development is focusing on the symptoms rather than the underlying causes (Kapur 2005).

Second, and related, as the World Bank (2006c) notes, remittances actually tend to increase global inequality. As was highlighted in the section on the geography of remittance flows, the vast majority of remittances go to MICs, not the poorest countries. That said, this is largely a consequence of the barriers to mobility that face the poorest households. Radical and progressive policies of enabling migration among poor households could address this and would receive support from liberal economists. But their political feasibility is the bigger question. The closest we have is the Mode 4 of the WTO's General Agreement on Trade in Services (GATS), which regulates the movement of people who travel to another country to deliver services. The Mode 4 provisions could be used to build a regime to encourage temporary migrant labour benefiting both developing and developed countries, but it hasn't received sufficient support among the WTO membership.

Third, there are considerable-and-all-too-often-overlooked costs from migration which precedes and enables remittance flows: the very possibility of remittances. At the national level there is the longstanding debate around brain drain or brawn drain. If one considers the number of migrants relative to the total population, small countries tend to come 'top' of the rankings. For example, over 80 per cent of college degree holders from Haiti, Jamaica, Grenada and Guyana live abroad, compared to 10 per cent for South America more generally (Fajnzylber and López 2007). Around 30 per cent of the labour force of the Caribbean islands has migrated, whereas for Grenada almost 50 per cent of the population has migrated. Such brain drain has serious implications for the development of small developing countries. Radical critics point to the role of migration in the development of underdevelopment.

At the household level, the costs can be even more acute. Economically, the costs are felt through a loss in income as the migrant is often the entrepreneurial or economically active member of the household. And the welfare impact on children growing up in a house with one or maybe two parents absent is potentially the longer-term problem. When a mother from the Philippines is in New York or Hong Kong looking after a wealthy family's children, what happens to her own?

Fourth and finally, other critics have argued that remittances can become an excuse for Western governments to do less (Hudson 2008). Despite the fact that the likes of Dilip Ratha have made it clear that remittances should not be seen as development aid, it is clear that conservatives and popular critics of aid point to remittances in justification for regressive policy change. For example, work done by the likes of the Hudson Institute in highlighting and promoting philanthropy and remittances as 'private aid' can muddy the waters, suggesting that they should count as official development assistance (*Washington Post* 2006). Worse still, remittances can be referred to as 'cost free', when in fact this ignores the considerable costs borne by migrant workers who are typically less well-off, involved in working long hours in dangerous, unpleasant, unwanted and largely invisible-but-crucial jobs, often more than one at a time, to be able to send money home (Shelley 2007). These conditions, barriers and costs will get worse, not better, without a massive overhaul of the international migration regime. Looked at through this lens, development is a process of exploitation, not assistance. Indeed, it is argued that remittances represent the extension of neoliberal financialisation into developing countries with the accompanying embedding of norms of individual responsibility for individual welfare, now on a global scale (Hudson 2008; Kunz 2011).

Conclusion

The chapter has examined the phenomenon of migrant remittances, and it is clear why the desire to harness this flow of financial resources has leapt up the agenda for the likes of the World Bank. The volume of money involved is huge, outstripping other official and private sources of foreign exchange, both in the aggregate for developing countries, but very much so for individual countries. Despite some traditional concerns about the extent to which remittances result in development and investment as opposed to consumption and welfare spending, the international community – guided by a strong liberal sensibility – has been careful not to encroach and commandeer these resources, recognising them as private flows. Policymakers have proceeded on the assumption and increasing evidence that remittances have a significant and positive impact on poverty, though the evidence tends to show that the effects are modest and highly heterogeneous.

The existence of remittance flows is celebrated as a function of the entrepreneurial spirit and drive of those who wish to increase their and their families' livelihoods and opportunities. But it is also crucial to recognise that migrant labour, which remittances are premised on, is fundamental to many rich country economies working. The migration mechanisms behind remittances which allow them to flow are often cruel and produce their own costs.

In sum, remittances are not necessarily pro-poor in terms of going to the poorest countries. They do however reduce poverty and occasionally domestic inequalities, but crucially do *not* close international inequalities. They have improved individuals' and individual households' consumption and welfare, reducing the impact of economic shocks, but have been less effective in boosting overall or national development. Remittances are often hailed as an engine for development, but it is doubtful that they are a substitute for sound development policies of liberal or interventionist flavours.

Discussion questions

1   What is driving the international policy community's remittance euphoria?

2   Given remittances' positive, but modest and heterogeneous, effect on poverty, and the ambiguous effects on inequality and growth, as well as the negative effect on international inequalities, are they pro-development flow? Can they be?

3   To what extent do you agree with Dilip Ratha when he argues that what households do with their remittances shouldn't be subject to any direct intervention?

4   Should remittances be channelled through the formal financial system as opposed to migrants being able to use informal channels?

5   What else should be done to assist migrants from developing countries?

Further reading

● There are few book-length treatments of remittances and development. Two edited collections are worth mentioning. Although the data will always be out of date, the key issues are well covered. Maimbo and Ratha's (2005a) *Remittances: Development Impact and Future Prospects* covers the key issues of interest to the World Bank with impacts, migration, transfer systems and financial development all covered.

● Meanwhile the OECD's (2005a) *Migration, Remittances and Development* is a useful collection of short essays which came out of an OECD conference in Marrakech and covers a remarkable range of issues to do with migration and remittances.

● Finally, Toby Shelley's (2007) *Exploited: Migrant Labour in the New Global Economy* is a blistering and angry polemic revealing the exploitation of migrants, but also the dependence of key sectors in the OECD economies on cheap migrant labour. Shelley, a journalist with the *Financial Times*, argues that this is a logical consequence of economic liberalisation, managed by big business and aided and abetted by rich world governments. A fine corrective to too much abstract liberal optimism.

Useful websites

● The World Bank's *Topics in Development* web portal on migration and remittances is superb. There is a wealth of data and information here on remittance statistics, different policy initiatives, plus research papers on migration, development and remittances. You can also access all the data contained in the World Bank's *Migration and Remittances Factbook* via these pages; the *Factbook* is also freely available via issuu.com. Very useful on the latest data and research. Available at: http://go.worldbank.org/0IK1E5K7U0.

● *Migrant Voice* is an innovative charity which works to transform how migrants are viewed in media and policy circles, to correct the negative bias towards migration in mainstream media reporting and challenge the stereotypes of migrants as 'passive, disempowered and marginalised "victims" '. It seeks to empower migrants so that they can interact with and shape public debate over the issues surrounding immigration and integration. The website provides lots of resources and information relating to migrants' lives through their voices. Excellent and insightful. Available at: http://www.migrantvoice. org/.

- The website for the annual Global Forum for Migration and Development provides all the latest high-level policy discussions on remittance-related issues. The Global Forum is a 'voluntary, inter-governmental, non-binding and informal consultative process' which is open to all UN members and observers. The Forum allows governments to share their experiences in migration and development. Its aims are twofold: (1) enhance dialogue and international cooperation, and (2) foster practical and action-oriented outcomes. The website includes lots of background documents from each year's annual meeting. A good window into the current discussion in the world of international public policy. Available at: http://www.gfmd.org/.

9 Microfinance

Learning outcomes

At the end of this chapter you should:

- Understand the emergence of modern microfinance institutions and the demand for them, in particular the financial needs and services required by poor households.
- Appreciate the problems associated with financial exclusion, why the poor tend to be excluded from formal financial services and why they are undersupplied.
- Understand the mechanisms by which microfinance has traditionally worked, such as small loans, frequent repayments, group lending, rotating savings, as well as more recent formats such as individual lending and the use of dynamic incentives.
- See the differences between microcredit, savings and insurance, and the evidence base and debates around each.
- Appreciate the ongoing dilemmas and decisions in the design and mission of microfinance in terms of outreach and sustainability, the double-bottom line, and microfinance plus.
- Be aware of the political debates around scaling-up microfinance, its regulation, and the critical reformist and radical critiques of microfinance's role in shoring up the neoliberal global development architecture.

Key concepts

Microfinance, access to finance, development, microcredit, microsavings, microinsurance, impact evaluation, poverty, public policy

Introduction

Microfinance is the provision of financial services – such as deposit accounts, loans, payment services, money transfers and insurance – to poor and low-income households and enterprises which do not have access to the formal financial sector. The most common form of microfinance is microcredit, where small loans of anywhere between $50–$500 are made, usually (but not exclusively) for self-employment. The orthodox logic behind making small loans is that they help release the productivity of the entrepreneurial poor who lack sufficient access to financial services (Demirgüç-Kunt *et al.* 2008). Notably, microfinance loans are often made to individuals without collateral and so 'group lending' contracts have characteristically been used. Microfinance schemes also tend to use group meetings and a scheme of regular small repayments to ensure the well publicised high repayment rates, of 95–98 per cent, remain high. Box 9.1 tells the story behind the emergence of the modern microfinance movement, of the famous Grameen Bank and its charismatic founder Muhammad Yunus, and highlights many of these features. However, the idea of microfinance is not new – for instance, the German credit cooperatives of the nineteenth century had a similar joint liability structure based on the ideas of Victor Aimé Huber, and similar ideas and institutions existed in Britain before that. These initiatives were also often driven by religious or charitable and philanthropic motives to assist the poor (see Roodman 2012, Chapters 3 and 4 on the history of microfinance).

As the sketch in Box 9.1 suggests, what with a Nobel Peace Prize and so on, the international community seized upon microfinance as something of a magic bullet. However, in the past few years the pendulum has swung the other way. The year 2010 was something of an *annus horribilis* for the microfinance industry. A spate of suicides in India; concerns over overheating, overlending, bubbles and indebtedness, bad collection practices, commercialisation and profiting from the poor; the release of an extremely negative film; and Yunus's unseemly departure from Grameen all took their toll as more and more people questioned the value and purpose of microfinance. Ultimately this has probably been a healthy recalibration of expectations. Ironically, those working in microfinance have always been much more cautious about the potential for and role of microfinance in development. To navigate between the old and new extremes we need to ask and answer a series of questions.

The first section of this chapter looks at the current supply of financial services for the poor. What is the extent and what are the costs of financial exclusion? How big is the microfinance industry? The second section looks at the demand for microfinance services, as well as some of the key forms of microfinance, specifically microcredit, savings and insurance. To what extent have each of them reduced poverty? The third section outlines three key debates within the microfinance community about how to design and deliver microfinance: Is group liability the only or indeed best way of lending to the poor? Should microfinance institutions (MFIs) concentrate on social goals such as poverty outreach or on financial sustainability? Can microfinance be more than just delivering financial services? The fourth and final section considers the political debates surrounding the microfinance industry: How should it scale-up, especially in terms of financing microfinance? What are the regulatory issues? What is microfinance's place within the politics of neoliberalism?

Box 9.1

$27 in Jobra

In 1976 Muhammad Yunus, an economics Professor at Chittagong University, Bangladesh, was visiting the village of Jobra. Emotionally moved by the famine in Bangladesh and frustrated by the experience of teaching elegant but abstract economic theory in the classroom, he had taken to spending time trying to understand the realities of poverty in the real world. As the story goes, he got chatting with a woman who was making a bamboo stool. Shocked that she was only earning a couple of pennies a day from selling what were beautiful pieces of furniture he asked her why. She explained that she had no money to buy the bamboo, so a local moneylender lent her the money (about 25 cents) on the condition that, at the end of the day, she sold him all the furniture at a price set by him. Hence, at the end of the day, the woman had only two pennies. Essentially she was in bondage, and all for just 25 cents a day.

Yunus returned the following day to see how many more women were in this situation. He found 42 people, which summed up to $27. For $27 he realised he could end their poverty trap and so he set about lending them the money from his own pocket. The women repaid Yunus, with interest, something he hadn't asked for or expected. But it made Yunus realised the extent of the unrealised demand for local development financing for poor households and – crucially – a sustainable and valuable opportunity to provide it.

After failing to persuade a local bank to make small loans to other poor households – the bank considered them too risky – Yunus decided to act as the guarantor and in December 1976 raised a loan from the government Janata Bank in order to make more loans. Again, they were repaid. Eventually, in October 1983, he turned his action research project into an independent bank, called the Grameen Bank, meaning village bank. The model of lending was to make loans to individuals in groups, called solidarity groups. The members of these groups acted as co-guarantors of each other's loans and provided support for each other's enterprises. From that initial loan in Jobra the Grameen Bank has grown.

As of October 2011, the Grameen Bank has 8.35 million borrowers, 97 per cent of which are women. They are served by 2,565 branches, providing services to 81,379 villages, which is just over 97 per cent of all the villages in Bangladesh (Grameen Bank 2011). Cumulatively, the bank has made loans totalling about US$6 billion and is now completely self-reliant (Yunus 2006). Its repayment rate is 97 per cent (Grameen Bank 2011).

The past 30 years have witnessed what many refer to as a 'microfinance revolution' (Robinson 2001). The Grameen model has been replicated and adapted in many countries across the world. The UN declared 2005 the year of microcredit. In 2006, the Nobel Peace Prize was awarded jointly to Muhammad Yunus and the Grameen Bank 'for their efforts to create economic and social development from below'. And in 2007, the Microcredit Summit Campaign reported that the world had met the goal agreed in 1997 to reach 100 million of the world's poorest with microloans (Daley-Harris 2009). Including family members of the borrowers, it is estimated that, as of December 2010, these loans have impacted upon the lives 687.7 million people (Maes and Reed 2012). Just under 1 in every 10 people on the planet. One of the goals of the Microcredit Summit Campaign is 'Working to ensure that 175 million of the world's poorest families, especially the women of those families, are receiving credit for self-employment and other financial and business services by the end of 2015' (Maes and Reed 2012: 3). The Campaign is on track to meet its goal of expanding microcredit to more and more people; progress is displayed in Figure 9.1.

Based on surveys it has carried out, the Grameen Bank has stated that 68 per cent of its borrowers have crossed the poverty line (Grameen Bank 2011). And stories from around the world back this,

Figure 9.1 *Growth trajectory of poorest clients reached, 2005–2010.*

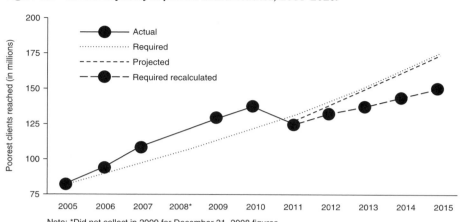

Note: *Did not collect in 2009 for December 31, 2008 figures.

Source: Adapted from Maes and Reed 2012: 35, Figure 5.

of individuals taking out a small loan to enable them to buy small machine tools, transport, pay off moneylenders, which has allowed them to pull themselves and their family out of poverty. Buoyed by stories of success, wealthy philanthropists such as George Soros, Pierre Omidyar and Natalie Portman have pledged hundreds of millions of dollars to microcredit institutions around the world. And now private capital is flowing into the sector through investment funds and bond issues. Microfinance has become a global industry.

The current demand and supply of finance

The portfolios of the poor

As outlined in Chapters 1 and 2, from a finance for development perspective, the poor are held to be 'financially constrained'. Access to finance offers poor households important opportunities for increasing their welfare and is a prerequisite for poverty reduction. Try to imagine not being able to transfer money and make payments, or not being able to deposit money into a bank account in order to save it, or to borrow money either to take advantage of a commercial opportunity which suddenly presented itself, or to get a short-term bridging loan to cover household expenses until the next market, or not being able to borrow a small amount of money to cover the costs of a sudden medical procedure. All of these severely limit an individual's freedoms.

Traditionally, poor individuals and households have depended on informal credit markets to fulfil their demand for credit. This is often through borrowing from relatives. Or, for example, farmers take advance payments for delivering food in order to finance the harvest and be able to bring it to market. Or farmers have pooled savings and created rotating credit through rural credit cooperatives. Moneylenders have also filled the gap between the demand for credit and the formal financial system; either through accumulated personal wealth or borrowing from the formal banking system and on-lending at higher rates to those who cannot access such finance. Agriculture credit unions or banks have also been a traditional form of credit provision. Nevertheless, there has remained an undersupply of formal financial access for the poor. Microcredit providers, along the lines of the Grameen Bank model, have stepped into this gap.

The ability to access a loan and use that money can create significant opportunities for the poor. Poor households can make good use of loans to invest and increase their productivity, such as buying fertiliser for crops or a trailer to take goods to market. However, many loans are not used for 'productive' investments and are instead required to pay medical bills or for a family wedding. There are longstanding debates on what capital should be spent on: whether it should only be invested in profitable business opportunities, or whether it is okay for borrowers to use it for consumption purposes (and indeed where the line lies). Evidence from Indonesia suggests that, on average, about half the microfinance loans issued by BRI (The People's Bank of Indonesia, the second largest bank in Indonesia) for business purposes were used for consumption (non-entrepreneurial) purposes (Johnston and Morduch 2008: Table 6). This again underlines a fundamental characteristic of money: it is fungible.

Some recent survey data (shown in Table 9.1) from Andhra Pradesh shows both the range of different reasons why people take out loans, plus how different sources are used for different purposes (Meka and Johnson 2010). Famers buying agricultural inputs tend to access bank loans, individuals borrowing to cover their consumption costs tend to use self-help groups (SHGs), but when it comes to health costs or to pay for marriage-related expenses people tend to turn to informal sources. These tendencies are clearly driven by the access borrowers enjoy to less or more formal types of finance as well as the flexibility and speed that the different sources provide. It is also noted that a lot of loans are used to repay old debt. What the survey data doesn't tell us is whether this borrowing-to-repay is because borrowers are taking advantage of cheaper interest rates or whether they are falling into debt traps and having to take out new loans to pay off old loans and getting further into debt.

The work of Collins, Morduch, Rutherford and Ruthven on the *Portfolios of the Poor* uncovers and underlines the reality of poor people's financial lives. In sum, their financial lives are sophisticated. Based on work in Bangaldesh, India and South Africa with over 250 households, the authors got the families to use financial diaries to gather information on their financial transactions (see Box 1.1, p. 6). The empirical details are fascinating and show how poor households navigate their financial needs and resources, borrowing, saving, exchanging cash in formal and informal contexts and networks in order to deal with sudden emergencies such as health problems or natural disasters, or raise lump sums for school fees, weddings or to buy land. Poor households are shown to be very astute and sophisticated, but lack the tools and resources needed to easily – in Stuart Rutherford's (2000) memorable and insightful phrase – turn small sums into usefully large sums.

Table 9.1 *Usage of loan money by lender type*

	Bank	Microfinance institution	Self-help group	Informal
Start new business	2%	3%	2%	1%
Buy agricultural inputs	58%	13%	19%	20%
Purchase stock	3%	10%	4%	3%
Repay old debt	15%	25%	20%	7%
Health	11%	11%	19%	25%
Marriage	4%	5%	2%	12%
Funeral	0.1%	0.2%	0.5%	2%
Other festival	1%	4%	4%	5%
Other improvement	10%	22%	13%	14%
Unemployment	0.0%	0.0%	0.1%	0.8%
Purchase land	1%	1%	1%	1%
Education	4%	4%	6%	5%
Purchase jewellery	1%	1%	2%	0.4%
Consumption	27%	32%	50%	25%
Buy livestock	3%	6%	6%	2%

Source: Meka and Johnson 2010: 24.

Financial exclusion

The World Bank's *Finance for All* reports that roughly 40–80 per cent of people in the developing world lack access to the formal banking sector (Demirgüç-Kunt *et al.* 2008). More than 80 per cent of households in Western Europe and North America have a financial account. Compare this to 60 per cent in Chile (the best in Latin

Figure 9.2 *Fraction of households with an account in a financial institution.*

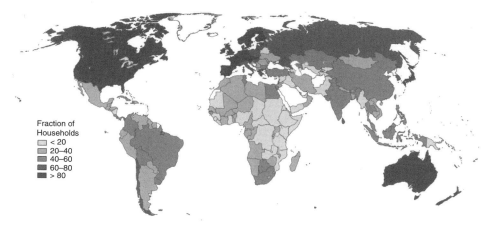

Fraction of
Households
☐ < 20
■ 20–40
■ 40–60
■ 60–80
■ > 80

Source: Demirgüç-Kunt *et al.* 2008: 35.

America) and 20 per cent in Nicaragua (the worst in Latin America). In Asia the range goes from 40–60 per cent, and access is below 20 per cent in many sub-Saharan African countries.

Karlan and Morduch (2010) use these figures to estimate that there are some 2–3 billion unbanked and underbanked adults in the world (from a world population of roughly 7 billion). The poor have been systematically excluded from and ignored by banks and other financial institutions. Spain, for example, has 96 branches per 100,000 population and 790 branches per 10,000 square kilometres, while Ethiopia has less than one branch per 100,000 people and Botswana has only one branch per 10,000 square kilometres (Demirgüç-Kunt *et al.* 2008). This clearly matters for individuals, but how does it affect economic development? Using household survey and aggregate data on loan and deposit accounts, Demirgüç-Kunt *et al.* (2008) estimate an indicator of access to finance (or rather use of financial services) and show that access to finance is positively correlated with economic development. Their results suggest that once a country nears a GDP/ capita of $20,000, the use of financial accounts closes in on 100 per cent.

Without access to finance, poor households or entrepreneurs need to rely on accumulated wealth or income for consumption, to protect them from negative shocks such as health, to invest in their education or to take advantage of a commercial

Figure 9.3 *Economic development and use of financial services.*

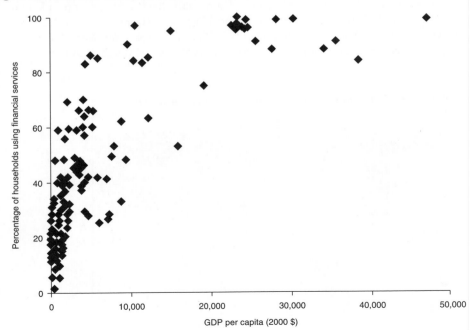

Figure 9.4 *Distinguishing between access to finance and use.*

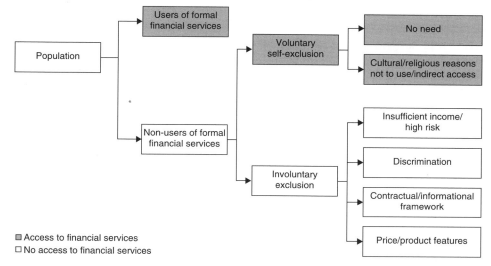

☑ Access to financial services
☐ No access to financial services

Source: Demirgüç-Kunt *et al.* 2008: 29.

opportunity. As with Demirgüç-Kunt *et al.* above, orthodox scholars have found that countries with more developed and deep financial systems have lower levels of absolute poverty (Honohan 2004). As such, better access to finance tends to be pro-poor. Beck *et al.* (2007) have found that, controlling for other variables, roughly 30 per cent of cross-country variation in poverty rates is attributable to the variation in countries' level of financial development. And, in line with diminishing marginal returns, the growth potential is large for firms, especially small and poor firms, which have previously lacked access to financial services.

Exclusion can be a function of geography, such as not having a local bank branch. Documentation is another barrier, either because of excessive demands or poor households lacking the necessary documents. Cost is another barrier. For instance, it has been estimated that the running cost of a Ugandan bank account is 30 per cent of its average per capita GDP, compared with 0 per cent in Bangladesh. Bangladesh has 229 deposit accounts per 1,000 population, Uganda has only 47 per 1,000 population. The final barrier is inappropriate products. Many poor households need financial services provided in small increments (hence the micro- in microfinance). However, in Nepal it is reported that the minimum amount a person can borrow is 12 times the average income (Demirgüç-Kunt *et al.* 2008)!

At this point it is important to distinguish between voluntary and involuntary exclusion. This is illustrated in Figure 9.4. Many people, non-poor and poor, do not wish to take up the opportunity of particular financial services. As such, non-use must not be conflated with non-access. But if there is demand and the ability to repay, but no services because of discrimination or the costs of providing services to that

Table 9.2 *Stated reasons for not availing of savings account among the financially excluded*

Reason	Percentage of households citing reason
No or not enough savings for bank account	37
Don't want/need	24
Save through other means	1
Bank/procedure related	49
Have no idea about banks or bank products	28
Don't have proper documentation	16
Fees/expenses	5
Applied but rejected	3
Procedures/application too difficult to understand	2
Takes too much time	1
Banks not trustworthy	1
Branch officials not friendly/courteous	0.5
Branch too far	0.2
Other reasons	2

Source: Meka and Johnson 2010: 18.

community, then we can speak of financial exclusion. So, to illustrate, the survey evidence reported above from Andhra Pradesh suggests that 24 per cent of households don't want or need a savings account, while 37 per cent of financially excluded households were excluded because they don't have sufficient funds to open a savings account. Whereas nearly half of the identified problems related to the banks and their procedures, such as not having required documentation, the expense, lack of awareness and so forth (Meka and Johnson 2010). Geography and lack of trust did not feature in this particular Indian state, but may well matter more elsewhere.

Why don't banks lend to the poor?

The unbanked – those without access to financial services, or insufficient access (the underbanked) – present a huge challenge for established financial service providers. Poor households represent a large untapped market, but the perceived risks and costs and limited profitable opportunities to be earned from banking the unbanked have traditionally dissuaded the formal and commercial sector from engaging with poorer households. Standard economic theory illustrates why (Armendáriz and Morduch 2010). First, lenders suffer information asymmetries, which initially means that they do not know who are the risky and who are the good customers (the adverse selection problem). This results in higher interest rates across the board, driving many otherwise creditworthy customers out of the market.

Furthermore, lenders cannot monitor and enforce contracts easily if and when customers default on their repayments (the enforcement problem). These are standard problems for lenders, but they become larger constraints when dealing with poor households for two reasons.

First, the transactions of the poor are too small to generate enough profit to cover these costs (Johnston and Morduch 2008; Cull *et al.* 2009). This is because the running costs of administrating many small accounts for the financial service provider – i.e. those identified above, such as screening, monitoring and enforcement – are far larger than for a smaller number of larger accounts; which is the situation for a financial service provider with wealthier customers.

Second, by definition, many of the poor do not have any economic assets to act as collateral to cover the risks of possible defaults. Without collateral, lenders rely on the interest rate to screen clients for creditworthiness. This is important because, as Stiglitz and Weiss (1981) famously showed, information problems mean the price mechanism (i.e. interest rates) fails to work. Their argument proceeds as follows. Theoretically, in a world of imperfect and costly information, banks can just increase the interest rates they charge to cover this riskiness and the lack of collateral. However, this seemingly logical response can be counterproductive and actually increase the bank's exposure to the risk of defaults. First, higher rates attract riskier projects and entrepreneurs who think they can repay the higher rates (adverse selection), and second, those who take on a loan at a higher rate will take on more risky projects to pay off the loan. Banks recognise this and respond by restricting access to finance rather than letting market demand lead to more loans but at higher rates. Ultimately, finance is rationed. Thus financial repression is not exclusively associated with excessive government intervention, as argued by neoliberals (McKinnon 1973; Shaw 1973, see Chapter 2). There is a case for a different type of intervention.

The supply of microfinance

Despite these commonly accepted barriers to banking the poor, the Grameen Bank, with its celebrated high repayment rates of 96-plus per cent, has demonstrated that it *is* possible to successfully provide financial services to the poor and it *is* possible for the poor to successfully manage debt. As evidenced by the fêted success of Yunus and the Grameen Bank, microfinance had become an important element of development architecture and strategies from the 1980s onwards. And by the turn of the century it was a 'vast global industry' (Arun and Hulme 2008: 3). The microfinance industry now incorporates millions of clients, hundreds of thousands of employees, and involvement from development agencies, governments, banks, consultancies, NGOs and cooperatives.

The Microfinance Summit Campaign has produced an annual report for the past 14 years which purports to be the largest collection of data from MFIs available (Maes and Reed 2012). Recent figures suggest that, as of 31 December 2010, the 3,652 MFIs that reported to the Campaign had reached 205.3 million clients (Maes and Reed 2012: 3), of which 56.5 per cent of these people were among the poorest. The poorest are defined by the Campaign as those 'whose income is in the bottom 50 percent of all those living below their country's poverty line, or any of the 1.2 billion who live on less than US$1 a day adjusted for purchasing power parity (PPP), when they started with the program' (Daley-Harris 2009: 44). And 82 per cent of these were women (see Box 9.2).

Moreover, assuming that each household is made up of five people, this means 687.7 million people in the poorest households have been directly affected by microfinance since 1997. These figures, collected annually since 1997, undoubtedly include much double counting. But the scale of microfinance is clear – 687.7 million is more than the total population of the EU and Russia combined (Maes and Reed 2012).

Box 9.2

Gender and microfinance

As has been noted elsewhere, microfinance tends to have a woman's face (Roodman 2012). Indeed, some institutions, such as India's SKS, lend exclusively to women. The explanations for the gender targeting of microfinance are several. First, women tend to suffer from less access to the formal financial system, so there is greater demand among women for microfinance services than men (Agarwal 1994). Second, women tend to have higher repayment rates and so make better clients than men (D'Espallier et al. 2011). Third, women tend to utilise loans more effectively in terms of improving household welfare than men do (Pitt and Khandker 1998). Fourth, women are said to work better in groups than men do, and are generally more compliant than men, which holds an attraction for lenders (Rahman 1999; Roodman 2012). However, as Roodman (2012) notes, the gender targeting argument also depends on what type of microfinance you're talking about. For solidarity group members it is 99.6 per cent women, for village banks it is 86.6 per cent, but for individual loans women only represent 46 per cent of all clients. The explanation is relatively straightforward: men tend to be the ones who engage in larger commercial endeavours. Given that these endeavours also create more employment, it has been suggested that gender targeting is at the heart of the economic growth versus poverty reduction schism that characterises microfinance (Kevane and Wydick 2001).

Evidence suggests that microfinance can be empowering (Hashemi et al. 1996). However, famously, Goetz and Gupta (1996) found that gendered power relations within the household mean that access to finance and control of the credit can often diverge. Despite women being able to get loans it was the men who were still behind the household spending decisions (hence the need for savings commitment devices – Anderson and Baland 2002 – see the discussion in the 'Microsaving' section). Access to finance, on its own, doesn't necessarily translate to empowerment – entrenched social and cultural norms can work against this and they too have to be addressed. Indeed, there is evidence that microfinance can result in gender conflict as men feel their traditional roles as head of the household are challenged or undermined (Mallick 2002). For example, see the studies about microfinance and gender training in South Africa discussed in the 'Microfinance plus' section (Pronyk et al. 2006; Kim et al. 2007).

Despite the huge growth in microfinance since 1972, it is only in the likes of Bangladesh, Indonesia, Uganda, Kenya and Bolivia where supply is anywhere near enough to create a competitive market (Arun and Hulme 2008). The geographical spread of microfinance clients is heavily skewed towards Asia. This is both in absolute terms, but also in the proportion of households living in absolute poverty covered. In Asia there is 78.5 per cent coverage, whereas Africa only has a coverage of 12.7 per cent (Daley-Harris 2009: 30). Plus, within countries there is a high degree of variation in the quantity and quality of MFIs available to the population (Arun and Hulme 2008). For example, in India there are relatively few MFIs in the poorer north and east of the country; likewise in Indonesia the poorer eastern provinces are less well supplied than Java and the Western Isles. And more generally across Latin America and Africa, there is a sustained urban-rural gap.

The logic of microfinance

Precursors to microcredit

The evidence presented above suggests that access to formal finance still trails demand, and badly. But where the poor do not have formal access to financial services they don't necessarily go without: they borrow informally. Banerjee and Duflo's (2007) work, based on surveys of poor households' financial activities, shows that reported borrowing rates vary enormously, from only 11 per cent of households in rural East Timor to 93 per cent in Pakistan. Crucially they show that much of the poorest's borrowing is informal, mostly from relatives, shopkeepers and other villagers, and only 6.4 per cent of borrowing is from a bank or cooperative. The problem with informal credit is that it is expensive: around 4 per cent a month (Banerjee and Duflo 2009). Nevertheless, demand for credit through informal channels remains high for a number of reasons, not just the orthodox supply-side issues highlighted thus far.

Moneylenders are much more flexible than the formal banking system, and especially more so than microfinance schemes, and so can be highly attractive for individuals seeking fast credit or fewer strings attached. The price borrowers pay for this flexibility is a very high interest rate, often at rates that exceed 100 per cent a year. These profits mean that moneylenders can be profitable where formal banks cannot. The other advantage moneylenders enjoy over banks is their greater knowledge of the market and lower administrative costs. But the likes of Yunus argue that the risks of falling into a debt trap are high. This is why proponents of formal microfinance argue that microfinance has a very specific role to play in helping the poor and removing their reliance on informal and potentially usurious moneylenders (Robinson 2001; Yunus 2006).

In the 1970s, many developing country governments recognised the need for poor households to have access to credit. They also recognised the high transaction costs that banks faced. The solution was to provide government-subsidised loans to the poor. The loans were designed to assist the poor and break the monopoly of moneylenders in poor and rural communities. However, the schemes are now widely considered, by and large, to have been a failure. The most famous critique of rural subsidies credit comes from the Ohio School (Adams *et al.* 1994). The problem they identified was that the collateral issue wasn't resolved so defaults were common. And because the government subsidised the banks they had no incentive to resolve the problem. And because interest rates were kept low on loans, they were also low on savings with the result that no deposits were built up. This much is in line with the neoliberal analysis of financial repression discussed in Chapter 3. Furthermore, it is well-documented that many of the subsidised loans ended up going to politically well connected individuals rather than the poor households they were initially intended for. Or the loans went to low-risk borrowers, i.e. the non-poor, in order to ensure repayment, resulting in a regressive redistribution of incomes (Adams and Von Pischke 1991). The result was that most schemes ended up running out of money without alleviating poverty.

Group-based lending as the solution

Microfinance was designed to maintain consistently high repayment rates, bypass moneylenders, avoid the problems faced by banks and avoid repeating the other failures of the rural credit programmes. Karlan and Morduch (2010: 4706, emphasis in original) highlight how 'Yunus's fundamental insight was to show that the poor are bankable *if the right lending mechanism is used*'. Recall the problems of adverse selection, enforcement and lack of collateral – the big innovation (or rediscovery) to address these problems was group-based lending.

Group-based lending is where borrowers are jointly liable for repaying the loan. The approach can solve the problems identified above by transferring the transaction costs involved in adjudicating individuals' credit-worthiness and monitoring and enforcing contracts from the lender to the clients. This group- or joint-liability means that the group is collectively responsible for the repayment of a loan – if an individual cannot meet a payment then the other group members become responsible for making that payment. There are a number of mechanisms through which this works. First, clients screen other potential group members to make sure that the group is only made up of trustworthy and creditworthy individuals (Ghatak 1999). This underlines how group-based lending can make use of information which is available at the community level, but is difficult for outside lenders to access. Second, the group effectively creates an inbuilt insurance scheme – assuming cooperative relationships – whereby if one borrower cannot pay, the outstanding amount is covered by others or the group collectively. Third, many have highlighted the effectiveness of peer pressure in reducing the risk of defaulting. Peer pressure can operate through positive encouragement, but it is often the more subtle social sanctions which 'encourage' borrowers to make payments: the threat of unwanted stigma, such as the loss of reputation, loss of access to contacts and resources, and social isolation.

This last point about peer pressure highlights that while there is a clear economic logic to the group-based nature of microcredit it is also a distinctly social one. From this we can surmise that trust, social capital and social networks play an important role in the success of microfinance. For example, Karlan (2007) shows that groups that are ethnically and geographically more proximate enjoy lower default and higher savings rates.

The group nature of microfinance schemes – both credit and savings – is that it provides the discipline and support structure to allow individuals to borrow, repay and/or save. The most notable manifestation of this is the group meetings, usually held weekly, which coincide with repayments. The common practice is making a year-long loan, and having 50 weekly repayments commencing a couple of weeks after the original disbursement. Regular smaller repayments mean that the borrower is repaying from their income stream, not the final returns on the investment project. This is especially desirable where clients do not have access to a savings account.

The meetings serve a number of purposes: (1) small and frequent repayments are easier to manage for many clients, (2) the frequency cements a greater discipline of repayments, especially for those without a history of borrowing, and (3) regular repayment makes it harder to accidentally overlook when a payment is due (Karlan and Morduch 2010). The meeting structure is very important, but such frequent and regular meetings are time and resource consuming for both the MFI and the clients. For example, Armendáriz de Aghion and Morduch (2000) highlight a study of a Chinese microfinance programme where 8 per cent of clients had to walk over an hour to get to each meeting, making the average meeting and travel time over 100 minutes.

Impact assessment

Microcredit

The $72 billion question: has microcredit worked? Researchers look for improvements in productivity, income and poverty measures. Proponents of microfinance underline the powerful potential of providing the poor with access to finance. There is little doubt that microfinance has given people access to capital they wouldn't otherwise have had. It has been assumed that this will – if used wisely – have a positive impact on the economic and social conditions of borrowers. An assumption founded on the logic of decreasing marginal returns, as above. These sorts of assumptions appear to be borne out by the before and after stories which characterise microcredit and decorate the websites and brochures of lenders. On the other hand, similar to arguments against aid, there are neoliberal critics from the political right as well as from the heterodox approaches who doubt the effectiveness of micro-loans versus larger-scale industry and salaried employment. Critics of microfinance point to how it creates new poverty traps and microdebt (Dichter 2007; Fernando 2006; Bateman 2011).

Box 9.3

Microfinance interest rates

So how much does it cost to borrow from an MFI? A survey of MFIs showed that interest rates tend to lie between 9 and 40 per cent plus per year (Cull *et al.* 2009). Many critical reformists feel that this is way too high for a poverty-focused intervention. Instead, the poor should benefit from subsidies. However, the standard neoliberal counter argument is that interest rate caps or ceilings (1) distort incentives for individual borrowers, meaning that people take out loans for activities which have poor returns, and therefore (2) the increased demand results in collective credit rationing (Goldsmith 1969; McKinnon 1971; Shaw 1973).

Indeed, the weight of opinion suggests subsidising interest rates is bad, which goes back to the arguments from the critique of Adams *et al.* (1984). High interest rates allow MFIs to cover their considerable costs. Most MFIs charge between 20 and 40 per cent a year. However, on average, NGOs charge about double that of microfinance banks (Cull *et al.* 2009). This is because it is more expensive for NGOs making smaller loans (Conning 1999). Yet, Morduch (1999a) argues that completely avoiding subsidised credit – because it leads to mistargeting and subsidy traps – is mistaken. It is perfectly possible to partially subside interest rates to make them cheaper (to say 20 per cent) without loans ending up in the wrong hands. The conventional wisdom is that microfinance clients are relatively insensitive to interest rates. However, recent research has suggested that small increases in interest rates strongly affect both demand for loans and sustainability or profitability: clients are strongly sensitive to changes in interest rates (Karlan *et al.* 2009). In Mexico, the increase in the number of new clients from a small 0.5 per cent decrease in the monthly interest rate resulted in larger profits with larger revenues outweighing the increase in costs of administering the new loans.

Defenders of high interest rates point out that these rates are still much lower than borrowers would otherwise be paying to moneylenders, so it still represents a significant improvement. Plus, making many small loans is, by its very nature, an expensive business. Survey data confirms that the more an MFI lends to poor clients the higher the interest rates need to be to cover their costs (Cull *et al.* 2009). Moreover, defenders further argue that the demand for loans and the proven ability of poor households to repay loans at these rates shows that they are not too high; the so-called 'market test'. This is not just luck, though, there is an economic logic backing higher rates. This is because the poor will enjoy greater returns to capital, and as such they are capable of repaying at higher interest rates because of the principle of declining marginal returns to capital (Rosenberg 2002). That is to say, lending to the poor offers greater productivity gains per unit of capital than those already with access to finance.

However, the principle of declining marginal returns assumes that all other factors are held constant, which in the real world is generally not the case (Morduch 1999a). The poor face severe constraints across the board; they lack education, skills, or contacts to set up a profitable business and therefore make repayments. In such circumstances the only outcome may well be increased indebtedness (Adams and Pischke 1992). As such, there is a recognition that microfinance may well be more beneficial for non-poor, low-income households rather than those living in poverty (Hulme and Mosley 1996). The non-poor are more capable of accessing other resources and taking entrepreneurial risks. Instead, grants and other welfare interventions are better suited to the poorest, not loans.

It's worth making a point, here, about repayment rates and financial sustainability. Although the repayment rates of Grameen *et al.* sound very impressive, they're also actually very necessary (Rosenberg 1999). A repayment rate of 95 per cent can mean an annual loss for the MFI of 40 per cent a year. This is because the MFI will be making further lending partially on the back of repayments of existing loans, and with loans over a short period, perhaps several times a year, this 5 per cent can be multiplied several times over and a seemingly impressive repayment rate can actually signal a portfolio in trouble.

One crude – but common – measure of success is the demand for loans and the ability of borrowers to repay. So, according to this logic, as long as there is demand for, uptake of and, crucially, continuing high repayment rates of microcredit loans then microfinance is reaching its market and clients are utilising the loans effectively. The so-called 'market test' is also a defence of microcredit interest rates and a rejection of micro-debt trap arguments. The longstanding justification of high interest rates is that because of the higher returns to capital available to poor households they are both willing and able to pay higher rates. Karlan and Morduch (2010) criticise the 'market test' as a simple proxy for success, i.e. that if there is uptake and repayment then the poor can be banked (at current interest rates). This ignores the possibility that many borrowers may be making their interest repayments but falling further into debt either by not paying off the principal or by cross-borrowing from moneylenders or other microfinance loans. Clearly a more detailed understanding of how borrowers manage their loans and the impact of taking out loans is necessary.

One study by McKenzie and Woodruff (2008) asks the most foundational of questions: To what extent are the poor financial constrained? And what difference does providing access to finance make? To answer these questions they examined microenterprises in León, Guanajuato, in Mexico. They randomly selected enterprises from a sample to receive a 'treatment' of cash or equipment worth 1,500 peso ($140). Their results showed that the enterprises that received the extra finance increased their monthly profits by about 20–33 per cent. Moreover, when they examined the difference between those enterprises that already had access to finance elsewhere with those that didn't, they found that the return for the constrained firms was even higher at 70–79 per cent. Importantly, for those firms without constraints the effect of the cash injection was negligible. This would suggest that some microenterprises are financially constrained and where this is the case it acts as a significant drag on their potential profits.

Another study conducted in Hyderabad, India, looked at the effect of making small loans to women (Banerjee *et al.* 2009). The women put the money to use in different ways. Those women who already owned a business invested the loan in expanding their working capital. Those women who didn't have a business (one in eight of the borrowers) started a business that they wouldn't have done without the loan. The rest of the women tended to spend the money on durables such as sewing machines. However, despite all this, there was no effect on health, education or women's empowerment. But, given that the study was only 18 months from the disbursement of the loan it may well be that it was too soon to see any such benefits.

One concern running through the microfinance debates, but through development finance more generally, is that capital should be invested, not consumed. Karlan and Zinman (2010) address the question of whether consumer credit improves social and economic welfare. The assumption is often that if loans are used for, say, food consumption or repaying other debt rather than investment, they are likely to be damaging. Hence they looked at previously rejected loan applicants in South Africa. They chose their sample by examining the effect of a loan by reconsidering the group of people who had been marginal rejects on their original loan application, which gave them a sample of 787 applicants who were randomly allocated into a treatment and a control group. The applicants were subsequently surveyed to see whether the access to credit improved their economic and social well-being. Karlan and Zinman found that after 6–12 months the borrowers benefited, with increases in employment, income, food consumption, as well as boosts to subjective indicators such as optimism. Moreover, over the period of 15–27 months after the initial loan, when some critics suggest that clients fall into debt traps as the costs of borrowing sink in, there were still positive effects for the borrowing group.

To summarise, after decades of optimism, if not downright hype, there is a healthy recalibration of expectations for the developmental potential of microcredit. The impacts are positive – if correctly and carefully measured – but tend to be modest at best (and sometimes neutral). Success is hugely contingent on the economic environment and design of the scheme, but also the products. Indeed, a focus exclusively on credit and providing loans for business purposes is not necessarily in the interests of the poor who have many other financial needs (Collins *et al.* 2009). As such, there has been a broadening of interest away from purely microcredit to microfinance more generally, i.e. from the provision of small loans to a wider range of services, including savings products, transfer services and insurance. Again, it's worth highlighting that savings and insurance services for the poor have a long history, both in the West and in developing countries, through traditional group savings mechanisms such as chit funds in India.

Microsaving

Stuart Rutherford's (2000a, 2000b) foundational argument about microfinance is that poor households need mechanisms to convert small deposits into large sums, because this is both the nature of most households' income (small) but also the nature of outgoings (unpredictable and sometimes large). For example, large sums are needed for one-off expenditures such as weddings, school fees, house improvements and so forth. Rutherford's argument is that borrowing and saving are essentially the same thing. Borrowing means the expenditure is upfront whereas with savings the expenditure comes at the end. Borrowing is generally more expensive, but borrowers benefit from accessing the cash more immediately. But essentially both are about helping the poor to collect a 'usefully large sum'.

The importance and potential of savings is illustrated well by Karlan and Morduch (2010) through the example of vegetable vendors in Chennai, India. The traders borrow money each morning to pay for their stock, repaying the lenders in the afternoon having sold the vegetables, paying up to 10 per cent interest a day. And this was routine behaviour with half of the vendors surveyed confirming that they had been doing this for 10 years. If the vendor were to save just one rupee each day, and borrow that much less the following day, within 50 days the vendor would be able to finance their stock without borrowing each day and see their daily profit margin boosted.

It is a commonly held view that the poor do not and cannot save. Contrary to this 'common sense', the poor both can and do save. For example, in one study from Indonesia it was found that approximately 90 per cent of clients save but do not borrow (Johnston and Morduch 2007). Meanwhile Banerjee and Duflo's (2007) survey data underline the fact that poor households are not just spending on survival, but also non-survival goods. Households living on under a dollar a day spend between

56 and 78 per cent of their income on food. Saving *is* possible; individuals may be poor, but through choices (often difficult ones, not always ideal, but they are choices that are made), households do have disposable income to spend on alcohol, tobacco and festivals. But poor households lack opportunities and mechanisms to save.

Banerjee and Duflo (2007) and Collins *et al.* (2009) show that savings rates are low, below 14 per cent. Again, similar to the evidence on borrowing, it is not that the poor do not save it's just that it is often through informal mechanisms such as lending money to family members, investing in livestock or jewellery, putting small savings under the bed, with deposit collectors or savings clubs. All of these methods tend to represent a bad deal for savers with low returns and risks of high losses. One alternative is money collectors. For example, the *susu* (savings) collectors in Ghana or the *tontines* in Cameroon are mobile bankers and their services are very popular (Aryeetey and Steel 1995). Money collectors go around the market at the end of the day collecting deposits from market traders who wish to save money. The accumulated savings are then returned at the end of the month. The money collectors do not pay out interest on the deposits and charge a handling fee equal to one day's savings, around 3.3 per cent a month. This is equivalent to an annual return for the saver of *minus* 54 per cent. However, this is the value users attach to the savings (and commitment) service provided by the *susu* collectors. Rotating savings and credit organisations (ROSCAs) are another popular form of saving. Here a community group with a common goal of saving and withdrawing a large lump sum join together to pool their resources. Each week they all pay in an agreed sum of savings which one person, in rotation, gets to withdraw each time.

Both of these mechanisms – money collectors and ROSCAs – fulfil the goal of turning small deposits into usefully large lump sums, which was Rutherford's insight. However, Wright and Mutesasira (2001) detail the large risks of losses that people face in the informal sector. The reason why savers put up with these low returns and high risks of losses is because they value the commitment characteristic of such savings mechanisms and these are often the best options available. Gugarty (2007) highlights the importance of such savings mechanisms as a form of self-control. For example, if it is known that an individual has money there can be an awful lot of pressure from others – friends, family, fellow villagers – to part with that money (Rutherford 2000b). Interestingly it has been documented how savings mechanisms can fulfil the purpose of 'other-control' as well as self-control. For example, research has shown how ROSCAs are used as a form of controlling husbands (Anderson and Baland 2002). Wives who wish to save up money often join a ROSCA to protect the household savings – effectively putting the accumulated capital outside the home and outside their spouse's control. In supporting their claim Anderson and Baland (2002) show that married women are far more likely to join a ROSCA.

To see whether access to a savings accounts is undersupplied and produces positive impacts Dupas and Robinson (2008) conducted an RCT in rural Kenya where they offered free savings accounts in a local bank to a randomised sample of 122 active 'micro-entrepreneurs'. The savings accounts did not pay any interest plus savers had to pay a fee to withdraw money, meaning the de facto interest rate was negative. Nevertheless, take-up was high, especially among women, compared with a randomly selected control group of 81 microentrepreneurs. The findings suggest that access to formal savings accounts has positive effects for women but not men. Specifically, for women, it increased average daily productive investment by around 40 per cent, increased daily expenditure levels on foodstuffs by 13–38 per cent and reduced vulnerability to health shocks. The findings underline two things: one that formal savings opportunities are undersupplied, and two, that women face greater savings constraints than men (because of demands on their money and a desire to spend money on family rather than save).

Meanwhile, research from a commitment savings product in the Philippines found that there is a positive relationship between savings and women's empowerment. Women were more likely to influence decisions over their children's education, goods purchases and family planning (Ashraf *et al.* 2010). Collins *et al.* (2009: Chapter 6) highlight the example of a successful commitment savings device called a 'pensions' saving device. Depositors are required to make monthly fixed deposits into the scheme and they receive their savings (plus interest) after 5–10 years, depending on the product. Savers can use the returned lump sum to pay for anything they wish.

The argument for encouraging poor households to use commitment savings accounts is not because poor individuals are any less rational than wealthy ones, nor more likely to give in to temptation, but that the costs of mistakes – made by all, wealthy and poor – have a much greater impact on poor people because of their lack of insurance or alternatives. In short, mistakes are more costly (Karlan and Morduch 2010). But because the circumstances many poor households face – the challenges, the nature of their income and outgoings – are so varied, the key is to provide tailored savings products, not generic savings accounts (Collins *et al.* 2009).

Microinsurance

Exposure to risk follows the social gradient: the poor are more exposed and more vulnerable to risks such as poor crop harvests, ill health and the insecurity of informal employment (Dercon 2004). As Karlan and Morduch (2010: 4760) observe, for the poor, 'temporary shocks translate into long-term losses'. The impact of risks to the poor is especially acute for children and is often translated via health outcomes (Dercon 2004). For example, losses in height and education in schooling years translate into worse employment outcomes in the future (Dercon and Hoddinott 2004). The evidence on this tells us that temporary shocks can result in long-term, permanent effects on poverty. This is a situation that calls out for insurance.

In a context where governments are not willing or able to protect their populations from such risks through welfare safety nets, households could go through microinsurance programmes. Morduch (2006) notes that crop insurance is especially important given the majority of the developing world depends upon agriculture, but such insurance is desperately undersupplied. This is for the same reasons of moral hazard and adverse selection that bedevil microfinance more generally – that scale and transaction costs hit providers' profitability. To make matters worse, Morduch goes on to note that while insurance is critical, we know precious little about its supply or demand or effectiveness. As such, savings and credit products remain the core risk management instruments (Karlan and Morduch 2010).

Many poor households insure themselves against risks through informal risk-coping mechanisms, such as networks of family and friends, or through holding savings in the form of grain, livestock or land, or borrowing from moneylenders (Townsend 1994). While this may work for small-impact or local risks such as health shocks, it doesn't necessarily cover for large and collective losses, for example a bad harvest where a household's neighbours and community are also likely to be affected (Morduch 1999b). Under such circumstances, informal risk-sharing mechanisms tend to fall down. Microinsurance is a particular area of microfinance where lots more work and policy is needed.

Designing and delivering microfinance

This section considers three key debates within the microfinance industry. There are a series of choices that MFIs need to make; and they are all connected to one another. These are: (1) the model used, whether it should be a group-based contract or individual-based approach? (2) What is the purpose of the MFI? Should MFIs concentrate on maximising the depth of their poverty outreach, or focus on becoming commercially viable and financially self-sustaining? (3) What range of services should MFIs provide? Should MFIs concentrate on their core business and just be about making loans and offering financial services, or is there a case for providing other developmental interventions such as nutrition or education programmes through integrating these other welfare services? There are other issues for MFIs, but these are among the key ones.

Box 9.4

Microfinance best practices

Commitment to applying good practice in microfinance comes from the highest levels of donor countries and agencies. In June 2004, the G8 endorsed the 'Key Principles of Microfinance' at a meeting of heads of state in Sea Island, Georgia, USA. Developed (and endorsed) by CGAP's 28 public and private member donors, the Key Principles are translated into concrete operational guidance for staff of donors and investors in these Good Practice guidelines.

1   Poor people need a variety of financial services, not just loans. In addition to credit, they want savings, insurance and money transfer services.

2   Microfinance is a powerful tool to fight poverty. Poor households use financial services to raise income, build their assets and cushion themselves against external shocks.

3   Microfinance means building financial systems that serve the poor. Microfinance will reach its full potential only if it is integrated into a country's mainstream financial system.

4   Microfinance can pay for itself, and must do so if it is to reach very large numbers of poor people. Unless microfinance providers charge enough to cover their costs, they will always be limited by the scarce and uncertain supply of subsidies from donors and governments.

5   Microfinance is about building permanent local financial institutions that can attract domestic deposits, recycle them into loans and provide other financial services.

6   Microcredit is not always the answer. Other kinds of support may work better for people who are so destitute that they are without income or means of repayment.

7   Interest rate ceilings hurt poor people by making it harder for them to get credit. Making many small loans costs more than making a few large ones. Interest rate ceilings prevent microfinance institutions from covering their costs, and thereby choke off the supply of credit for poor people.

8   The job of government is to enable financial services, not to provide them directly. Governments can almost never do a good job of lending, but they can set a supporting policy environment.

9   Donor funds should complement private capital, not compete with it. Donors should use appropriate grant, loan and equity instruments on a temporary basis to build the institutional capacity of financial providers, develop support infrastructure, and support experimental services and products.

10   The key bottleneck is the shortage of strong institutions and managers. Donors should focus their support on building capacity.

11   Microfinance works best when it measures – and discloses – its performance. Reporting not only helps stakeholders judge costs and benefits, but it also improves performance. MFIs need to produce accurate and comparable reporting on financial performance (e.g. loan repayment and cost recovery) as well as social performance (e.g. number and poverty level of clients being served).

Source: CGAP (2006).

Group versus individual approaches

Microfinance is often considered synonymous with the group lending model and microcredit popularised by the original Grameen Bank model and the likes of BancoSol in Bolivia. And understandably so, since joint liability constituted such an important breakthrough with respect to the problems of lack of collateral, adverse selection and enforcement. However, the joint liability model – where loans are made to individuals but the group is responsible for repayment – is far from the only game in town. First, there are some other group-lending models, and second, individual lending is increasingly popular.

One common alternative to the Grameen Model is the SHG (self-help group). These groups tend to be larger, often between 10–15 and sometimes up to 20 members, and tend to be more autonomous than the Grameen groups. They are formed around a common social identity or goal, sometimes operating similar to ROSCAs in that they mobilise members' savings, but often borrowing from a bank and then lending internally to their members. The SHG has proved particularly popular in India. India's development bank – the National Bank for Agriculture and Rural Development – was formed in the early 1980s to provide credit to rural areas and provides incentives for banks to support SHG formation and lending. Another group-based model is the Village Banking Model pioneered by FINCA International. Its methodology is based on a system of cross-guarantees where individuals guarantee loans of another member. In addition, like the SHGs, the emphasis is on the group administering their internal financial affairs such as creating bylaws, and being responsible for the bookkeeping, fund management and loan supervision.

The major alternative to group models – whether joint liability, cross guarantees or self-help – is the individual lender-borrower contract where MFIs lend to individuals outside of groups. Interestingly and importantly, Grameen dropped joint liability and by 2002 had redesigned itself as Grameen Bank II. Grameen Bank II switched from group loans to personal loans alongside savings and pension products, but retains the centrality and importance of groups for support and organisation. Similarly, by 2005, only 1 per cent of BancoSol's loan portfolio was group contracts (Giné et al. 2010).

Why the change? A couple of common limitations of group lending are the costs of regular meetings for both clients and MFIs. Plus problems arise when some individuals want or need to borrow and risk more than others in their entrepreneurial schemes (Armendáriz de Aghion and Morduch 2000). Lenders have found that group lending is less well suited to wealthier, low-income borrowers; such clients prefer individual loans (and contracts) rather than being tied to smaller loans and being jointly liable with others borrowers. Furthermore, individual lending helps avoid situations of strategic defaults by individual members of a group in the knowledge that others will be forced to cover them (moral hazard).

A common concern with the switch to individual lending is that the advantages of the group monitoring and enforcement mechanisms are lost. For example, this would affect outreach insofar as individual lending is traditionally to the relatively less poor, to those with assets, skills or economic opportunities. This relates to the issues, below, of poverty reduction versus financial self-sufficiency. Second, does the break-up of joint liability lead to higher default rates as the group discipline and peer pressure is disbanded?

So how have lenders addressed these issues? Assuming that collateral is still not available, a number of other ways of addressing the common problems of lending to the poor have emerged (Armendáriz de Aghion and Morduch 2000). First, lenders tend to spend more time gathering information, screening clients and monitoring repayment. Second, clients are incentivised to repay their loans through the threat of losing out on future, larger loans. The non-refinance threat – where future loans are conditional on repayment performance – is referred to as 'dynamic incentives'. The evidence on the positive impact of dynamic incentives is good. A lender who offered random clients a discount on future loans, on the condition that they repaid their existing loans, saw a 10 per cent reduction in defaults overall, and the repayment success and size of the discount were positively related (Karlan and Zinman 2009).

Third, the importance of regular payments is crucial. Although the exact details vary – sometimes weekly, sometimes monthly – regular smaller repayments mean that the borrower is repaying from their income stream, not the final returns on the investment project.

Fourth, group meetings are still important. For example, like Grameen, ASA has abandoned group lending, but retains public meetings and still employs a group structure. This allows ASA and Grameen to retain the benefits of bringing clients into one place; it reduces transaction costs for the MFI, facilitates monitoring, enables peer pressure and allows for training activities such as financial literacy. They have concluded that groups are still important, but not for joint liability.

Giné *et al.* (2010) used a field experiment in Lima, Peru, to test this conclusion. Over seven months they played nearly 300 games with microentrepreneurs in the marketplace to see how they responded to different types of contracts and incentives. A number of interesting and relevant findings emerged. They confirmed that dynamic incentives reduced defaults (and risky project choices). And joint liability contracts actually increased the rates of risky project choices. This is because borrowers perceived that if their project failed they would be bailed out by their group members. However, repayment rates were higher in the group contracts, suggesting that the collective insurance mechanisms work.

A double bottom line? Poverty reduction and financial sustainability

There is a huge debate in the microfinance community about the commercialisation of microfinance, the extent to which it is happening and whether it should be embraced. The answer to the first is broadly yes, and the answer to the second is very, very contested. Microfinance is often said to have a 'double bottom line'. That is, in addition to making profit (the traditional bottom line of a company's accounts), microfinance also has a social bottom line: reducing poverty. Orthodox proponents identify the 'fortune at the bottom of the pyramid', arguing that the best way to eradicate poverty is through the win-win proposition of releasing the entrepreneurial capacity of the poor and the creation of profits and new markets (Prahalad 2004). This view suggests that, contrary to standard models of aid using subsidised grants, microfinance can be a financially self-sufficient model of poverty alleviation. The basis of the proposition is that the poor do not simply demand access to cheap credit, instead they just want access to credit full stop, and are willing and capable of paying high borrowing costs (high enough to make lending to the poor profitable) (Morduch 1999a). Moreover, the steady commercialisation of the sector and the successful initial public offerings by the likes of Banco Compartamos in Mexico in 2007 and SKS in 2010 suggest that private financing can help grow the microfinance industry. Commercial success depends upon the profitability and the financial sustainability of MFIs. It also has important implications for the outreach and social purpose of MFIs.

Morduch (1999a: 617) famously took on this 'enticing "win-win" proposition'. He argues that a better approach is to recognise a schism within the microfinance community between socially-minded and financially-minded programmes. On the one hand there are the not-for-profit or NGO deliverers of microfinance such as the Grameen Bank, Opportunity, Finca International and Accion International. On the other hand there are the commercial banks and financial institutions which provide microfinance services such as Unibanco, Indonesia Bank Rakyat and ICICI, which on-lends to smaller MFIs. The microfinance industry remains divided around these two alternative approaches. They differ with respect to the purpose of microfinance: whether it is primarily to provide affordable credit and other financial services to the poor in order to reduce poverty, versus a primary concern with becoming financially sustainable. The guiding philosophy clearly has important ramifications for an MFI's views on the role of subsidies and who the target clients of MFIs are. The debate between outreach and sustainability is often seen as one of the classic microfinance trade-offs (Cull *et al.* 2007, 2009).

Research by Cull *et al.* (2009) shows that the socially-minded and financially-minded approaches are not equivalent and fulfil different roles (contrary to the simpler win-win view). They use the Microfinance Information Exchange Dataset which, at the time, covered 346 institutions with nearly 18 million active clients with a combined total of $25.3 billion in assets to see how the industry is divided. Table 9.3 shows the distribution of lenders by type.

As Table 9.3 indicates, three-quarters of the institutions in Cull *et al.'s* (2009) sample are NGOs, with banks representing only 10 per cent of all lenders. However, banks controlled over half of the total portfolio, with the NGOs accounting for only around a fifth. Meanwhile, in terms of outreach, in numbers of clients, the NGOs account for 51 per cent of total clients with the banks reaching only 25 per cent. These figures are broadly in line with a bigger self-reported dataset in Gonzalez and Rosenberg (2006). They show that NGOs served about a quarter of microfinance clients in 2004, with banks and other financial institutions reaching 17 per cent of clients.

Both studies show that banks lend more but serve fewer customers. Given that loan size is a rough proxy for wealth, with smaller loans implying poorer clients, then NGOs serve more poor people. For example, if a lender increased the number of small loans

Table 9.3 *Distribution of microfinance institutions by type*

	% of institutions	% of assets	% of borrowers	% of female borrowers	% of subsidised funds
Bank	10	55	25	6	18
NGO	45	21	51	73	61
Non-bank financial institution	30	19	17	16	18
Credit union	10	4	6	4	3
Rural bank	5	1	1	1	0
	100	100	100	100	100
Numbers total	346 institutions	$25.3 billion	18 million	12 million	$2.6 million

Source: Cull *et al.* 2009: 174, Table 1.

Table 9.4 *Profitability of microfinance institutions*

	Institutions		Active borrowers	
	Number in sample	Percent profitable	Number (millions)	Percentage served by profitable institutions
Institution type				
Bank	30	73	4.1	92
Credit union	30	53	0.5	57
Non-bank financial institution	94	60	2.6	75
NGO	148	54	8.9	91
Lending method				
Individual	105	68	7.2	95
Solidarity group	157	55	7.4	85
Village bank	53	43	1.6	67
Total	**315**	**57**	**16.1**	**87**

Source: Cull *et al.* 2009: 178, Table 2.

(below $300) from 50 to 60 per cent then this is associated with a rise in the percentage of poor clients reached from 52 to 59 per cent (Gonzalez and Rosenberg 2006). Thus it is apparent that at the moment NGOs retain a significant role in terms of outreach, especially if one considers column 4 in the table showing that NGOs serve three-quarters of women in the sample.

The results on the profitability of the different types of organisations are very interesting. Cull *et al.* (2009) show that most NGOs are in fact profitable (in the sense of having total revenues exceeding total costs). Note that this does not however imply a move towards commercialisation, which is where the profits accrue to shareholders rather than being ploughed back into the MFI. Looking at the figures (in Table 9.4) it is apparent that banks are more likely to be profitable than NGOs. But because of the larger number of NGOs in the sample, more people are served by a profitable NGO, specifically 8 million versus just under 4 million (91 per cent of 8.9 million versus 92 per cent of 4.1 million).

Table 9.4 also shows how institutions lend and the levels of repayment. Solidarity group-lending is based on the Grameen model and the 'village bank' approach is also a joint liability group model. Cull *et al.* (2009) report that two-thirds of banks lend individually and three-quarters of NGOs use group-based methods. The figures in Table 9.4 also show that individual methods are more profitable. This squares with village banks and solidarity group methods being used to target poorer customers. But, notably, portfolio performance was similar across banks, financial institutions and NGOs; all the different institutions do quite well in terms of loan repayments.

All of this tells us that the question about which model is better – financially self-sustainable or subsidised – is probably the wrong one to ask. The answer is that a mix of programmes with different goals, interest rates and designs is not only desirable but also feasible, sustainable and complementary.

It is also worth considering the view of Elizabeth Rhyne (1999). She suggests that there is actually only one objective: outreach. She argues that sustainability is not an end in itself, but is a valuable means of ensuring future outreach. This also matters in terms of the minimalism versus microfinance-plus debate.

'Microfinance plus' versus microfinance minimalism

'Microfinance plus' is a term used to describe MFIs which carry out non-financial welfare-focused services in addition to, and often integrated with, their financial services. The welfare services can be directly related to the aims of microfinance, such as providing financial literacy education or livelihood counselling, in an effort to maximise productivity. Or, they can be health or educational interventions such as nutrition, health and sanitation or domestic violence prevention training, or delivering insecticide-treated bed nets, all of which aim to maximise social and developmental impact and operate parallel to the narrower financial aims of the MFI.

BRAC is a good example of an MFI that integrates health and education services with microfinance services through its village organisations. It trains health promoters who travel around local households to provide basic health information and products such as soap and medicines or distribute reading glasses. BRAC also trains model farmers from its microfinance groups (BRAC 2010). These model farmers pass on the technical support they received from BRAC and use their own farms for demonstrations to help share best practices with other famers within their community. Another example is employment generating activities. For example, GrameenPhone works with local entrepreneurs to acquire mobile phones and sell phone services within their communities. GrameenPhone is the largest cellular operator in Bangladesh, with about 63 per cent of the market (Sullivan 1997).

The reason why MFIs are identified as a good way of delivering 'plus' services in the first place is because they have regular and institutionalised interaction with poor households. Proponents argue that microfinance distribution networks should be leveraged for providing multiple services. It is more efficient, reduces transactions costs and benefits from existing loyalty, trust and social capital. More neoliberal critics argue that this is detrimental to the successful scaling up of microfinance. MFIs are engaging in delivering services which they do not have expertise in. There is little doubt that microfinance plus adds to an MFI's costs, but the question is the extent to which the developmental benefits exceed these, and more so than delivering separate welfare projects.

What is the evidence? On the more narrow definition of 'plus', Karlan and Valdivia (2011) have examined the marginal impact of adding entrepreneurship training to microloans to women in Peru. The treatment group received 30–60 minutes of training during their usual weekly or monthly banking meeting over one to two years. The findings were positive. Compared to the control group, the clients who received the training enjoyed improved business outcomes – higher profits in the good months, and fewer losses in the bad months – plus the repayment and client retention rates were higher for the MFI.

On the more holistic definition of 'plus', an increasing number of studies have looked at the effects of integrating health interventions and microfinance (Pronyk *et al.* 2007; Leatherman and Dunford 2010). This is particularly important because the WHO and World Bank report that illness is routinely cited by the poor as a key cause of poverty (Dodd and Munck 2002). Dohn *et al.* (2004) reveal some remarkable results of using microfinance to deliver health interventions. Their results show that where mothers received microfinance only, there was no change in diarrhoea incidence; where mothers received just education the diarrhoea incidence decreased by 29 per cent; but for those who received both education and microfinance, diarrhoea incidence fell by a huge 43 per cent. Given that diarrhoea is the second leading cause of child deaths in the world, this is a hugely important finding. Meanwhile, research conducted in South Africa looked at a microfinance scheme that was combined with a gender training curriculum which covered issues such as gender roles, violence and knowledge about HIV. The reduction in reported violence was over 50 per cent (Pronyk *et al.* 2006). The researchers attribute the reduction of violence to increased female empowerment, communication, autonomy and challenging gender norms, along with the value of a woman's contribution to the household (Kim *et al.* 2007).

In sum, the evidence tends to suggest that microfinance can work, but it works even better when aligned with a health or educational intervention. However, this comes with costs, costs which need to be absorbed by the MFIs and/or their donors. But it underlines a perspective on finance outlined by Alex Counts, the Grameen Foundation president and CEO, that microfinance should serve as a *platform* rather than a *product*. He argues that the real assets of MFIs are not their 'loan portfolios, but their high-quality relationships with the world's poor' (Counts 2008: 48).

Policy and politics

Scaling up microfinance

Understandably, microfinance has been touted as part of the solution for reaching the MDGs (Morduch and Haley 2002; CGAP 2003; Littlefield *et al.* 2003; UNCDF 2005). Financial access can directly help to reduce poverty and hunger and increase employment (MDG1), increase education (MDG2), empower women (MDG3), improve health outcomes (MDGs 4, 5 and 6) and to a lesser extent improve sanitation and slum conditions (MDG7) and channel development assistance to the poorest (MDG8). For example, the Monterrey Declaration underlined the importance of microfinance, especially for women and in rural areas, in enhancing the impact of the financial sector on poverty reduction (paragraph 18, page 4). Meanwhile, the UN Millennium Project is cited as saying that 'Microfinance is one of the practical development strategies and approaches that should be implemented and supported to attain the bold ambition of reducing world poverty by half' (UNCDF 2005).

Moreover, as outlined at the beginning of this chapter, up to 3 billion people lack access to formal financial services, which suggests that microfinance needs to serve more than just the poor. Microfinance needs to provide the whole range of financial services to low-income clients, not just the poor or very poor (M. Robinson 2001).

Yet, as the World Bank reports, there are only eight countries in the world where microfinance borrowers represent more than 2 per cent of the total population (Honohan 2004; Demirgüç-Kunt *et al.* 2008). The same is true for savings. The UN Millennium Project (2005: 37–38, 148) underlines the importance of scaling-up microsaving when it notes that saving rates of the LDCs are only 6.7 per cent of GNP, which compares unfavourably with sub-Saharan Africa's average of 11 per cent. All of this implies a massive expansion in the number of households covered by microfinance.

As the above section highlighted, there are ongoing debates about the future of microfinance and how it should be delivered. Regardless of which direction individual MFIs choose to proceed, the larger context is of 'scaling up' microfinance. This, then, raises a series of further questions: Where will the financing come from and does this imply further commercialisation of the industry? And within such an agenda, what scope is there for MFIs to improve their operations, including new technologies? What is the future of microfinance regulation? Meanwhile, radical critics highlight the negative role of microfinance in expanding the global development architecture.

Financing microfinance

As Greeley (2006) suggests, if scaling-up – to meet the MDGs – is the name of the game, then commercialisation and the accessing of private capital becomes an almost irresistible logic. Currently, for NGOs, 39 per cent of funding comes from donations and another 16 per cent from soft (non-commercial) loans. For microfinance banks, these two categories accounted for just 2 per cent of funding, whereas commercial borrowing and deposits contributed 84 per cent of total funding (Cull *et al.* 2009). Either donors and individuals will have to continue subsidising and acting as social investors, or private investors will need to be tapped up. While NGO providers still outnumber commercial microfinance providers – and serve over half of all clients versus the quarter served by commercial outfits – most observers argue that the direction of change is towards commercially-based MFIs (Arun and Hulme 2008; Karlan and Morduch 2010).

An increasing number of global investment banks have already moved into the microfinance sector, such as ABN-AMRO, Barclays, Citigroup, Morgan Stanley, Deutsche Bank and Société Générale (Arun and Hulme 2008; Reille and Forster 2008). There has been aggressive expansion into microfinance with the recognition that there are profits to be made in supplying financial services to the poor: the market at the 'bottom of the pyramid' in one commentator's terms (Prahalad 2004). CGAP reports that between 2004 and 2006 the stock of foreign investment more than tripled to US$4 billion (Reille and Forster 2008).

There are three major categories of external investors: donors, retail and/or social investors and institutional investors. And there are various ways in which financing has been mobilised, from equity financing by investors to MFIs raising money through bond issues or securitising their loan portfolio. As Figure 9.5 shows, just over half of all funding comes from development finance institutions (DFIs) such as the IADB, EBRD and the IFC. Then there are retail investors, who comprise high net worth individuals such as the founder of eBay, Pierre Omidyar, who gave US$100 million to create a fund, but also individuals who invest through mutual funds or more directly through websites such as Kiva, which ostensibly allow for peer-to-peer lending (cf. Roodman 2012). Finally, institutional investors are now a serious force in microfinance with nearly US$550 million invested in 2006, representing a $100 million increase since 2005 (Reille and Forster 2008). And it is not just external finance either. As Arun and Hulme (2008) report, ICICI bank in India increased its microfinance clients to 1.5 million in 2005, from just 10,000 in 2001. This equals a portfolio worth US$265 million – with ICICI lending to MFIs at 9.5–11 per cent and the MFIs lending to their clients in SHGs at 16–30 per cent (*The Economist* 2005).

Roughly half of all total microfinance investment is done so via microfinance investment vehicles (MIVs). MIVs are essentially managed microfinance funds that offer investors a more efficient way to invest in microfinance rather than direct investments. As of 2006 collectively, MIVs had about US$3 billion under management (Reille and Forster 2008). The DFIs were the early MIV investors, but they have become relatively less important over time as private investors have increased their share.

Figure 9.5 *Foreign investment in microfinance by source.*

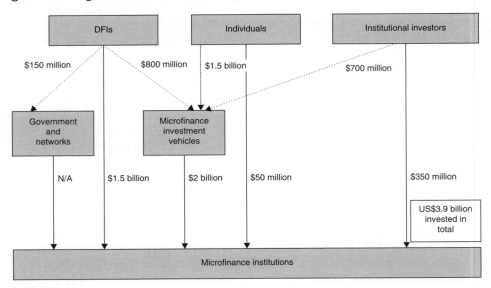

Source: Adapted from Reille and Foster 2008: 2, Figure 1.

The key financial innovation for financing microfinance is securitisation. For example, the Indian bank ICICI has sought to create a secondary market in microfinance. The future promises to pay are rebundled together to provide a new income stream to sell to an investor. The pricing of the asset is based on the past performance of the portfolio and management quality of the MFI. In 2006, BRAC received US$180 million over six years from Citigroup and three other investors following the creation of a special-purpose trust to purchase BRAC's receivables from its microcredit portfolio and issue certificates to investors.

One huge problem with this – as with all private finance solutions – is that funds do not go to where poverty is highest and microfinance is most needed. The majority of foreign investment goes to the top 150 MFIs that are based in roughly 30 countries. Africa and Asia receive only 6 and 7 per cent respectively of this foreign investment (Reille and Forster 2008). The second problem with relying on wooing international investors is the need to look profitable. Currently the returns on equity, from the survey conducted by Cull *et al.* (2009), are just 3 per cent for NGOs and 10 per cent for microfinance banks – meanwhile Citigroup's return on equity in 2004 was 16 per cent. So there is a profitability gap. However, change is clearly afoot with the top 50 microrate MFIs increasing their median return on equity from 14 to 23 per cent between 2003–08 (Reille and Forster 2008: 13). Critical reformist and radical critics note that turning to private investors will result in 'mission drift' as MFIs focus more on hitting their financial targets rather than on their social impact (Greeley 2006). Mohammed Yunus has been a particularly vocal critic of this – see Box 9.5 on the Compartamos Affair. The real question is, will private investors tolerate profits being reinvested into clients as, say, subsidised interest rates or funding greater outreach to hard to reach clients (Reille and Forster 2008)?

Box 9.5

The Compartamos affair

On 20 April 2007 Compartamos Banco, a Mexican microfinance institution, listed one-third of its shares: 20 per cent of the shares were offered on Mexico's stock exchange with the other 80 per cent in New York (Randewich 2007; *The Economist* 2008b). The IPO – which is when a company floats on the stock exchange – raised $467 million. But Compartamos was certainly not the first for-profit MFI, BancoSol in Bolivia had been run along commercial lines for some years already, but Compartamos was the first IPO. Compartamos – meaning 'let's share' in Spanish – was established in 1990 by Jose Ignacio Avalos Hernandez, a wealthy Catholic businessman who had earned his fortune from his family pharmaceutical and cosmetics company. The original lending model was a group-based one similar to the Grameen Bank, but in 2000, seeking greater scale, the non-profit organisation decided to become commercial. The next step was to tap into global financial capital to fuel further growth. It was a defining moment for the microfinance movement/industry. Compartamos has since been followed by SKS in India, another for-profit MFI.

Critics, such as Muhammad Yunus, argued that this was the opposite of what microfinance was meant to do: 'Microcredit was created to fight the money lender, not to become the money lender' (Bloomberg 2007). Compartamos not only secured enough profit to be self-sufficient, but the bank's shareholders had been making annual profits of 100 per cent on their investment year after year. Radicals underline how this is the imperialism of capitalist finance further commodifying the economy and society. Radicals and critical reformists note how the dynamics of debt and profit, when added to financialisation, mean that borrowers are increasingly squeezed as interest rates rise and bottom lines are inflated to attract more global capital. Compartamos had profits of $57 million in 2006 and its aggressive growth and profits meant that the IPO was 13 times oversubscribed (Reuters 2007).

Liberal defenders of the commercialisation of finance argue that this is proof that the 'win-win' proposition is viable (*The Economist* 2008a). The way in which Compartamos defended itself was that it began as a socially concerned organisation in the same way the Grameen Bank did, including the same group lending model, but that it was now convinced that the best way of fighting poverty was with profits (*The Economist* 2008b). More specifically, Compartamos defended its high interest rates (somewhere between 70–100 per cent) by noting that they allowed the bank to expand quicker thereby reaching more potential borrowers (from 60,000 to 900,000 in the eight years before going public). Compartamos also added that it loaned to those without access to formal credit, but not the poorest of the poor; the poorest, it agreed, were far better served by public income support. The final argument was that the 'big win' of the initial public offering served a bigger purpose of attracting lots of new interest and capital into the microfinance industry, making it possible to reach more borrowers, increase competition and lower interest rates over the long term.

A number of debates about microfinance were crystallised in the Compartamos Affair. First, should banks be profiting from the poor or do profits help secure the mission? Second, are high interest rates really necessary (see Box 9.3)? Does aggressive growth and profit-taking result in indebtedness and undermine development, such as in Andhra Pradesh? And once an MFI is publicly listed, can it still balance its social and financial missions? The IPO undoubtedly makes it harder to do so, since more shares are now in the hands of private commercial interests whose only consideration is to maximise investor profits (Rosenberg 2007).

Regulating microfinance

Liberal institutionalists, such as all of the work coming out of the World Bank on financial access, emphasise the importance of a stable macroeconomic environment as a prerequisite for an effective financial system. With large or variable inflation rates, people do not have confidence in the future purchasing power of the currency, which impedes both saving and borrowing (Demirgüç-Kunt *et al.* 2008). Key here are property and creditor rights, the independence of the judiciary and regulators, but most especially the stability and security provided by having a statute supplying explicit clarification of the law covering different financial services. Hence, if scaling-up microfinance is to proceed successfully then more thought is going to have to be given to the regulation of microfinance.

Regulation of MFIs is an area that many governments are trying to deal with (Arun 2005). For liberal institutionalists, in terms of prudential regulation, the aim of the game is to stop banks making loss-making credit decisions (Demirgüç-Kunt *et al.* 2008). But, from such a position, there is a tightrope to walk between having protective regulations and stymieing innovation. M-finance (mobile finance) is a good example.

The low population density in Africa raises problems for standard methods of microfinance delivery, though new technologies may help to resolve this (Arun and Hulme 2008). Advances in mobile technology are providing better access to financial services for poor households (Sullivan 2007). The potential benefits of M-banking (mobile banking) are lowering transaction costs and increasing outreach. The potential downside is that it undermines the social nature of microfinance, for example the group meetings. In terms of regulation, liberals point to the fact that South Africa has insisted that M-finance services are provided by a bank whereas in Kenya phone companies are able to provide the services. The result: take-up has been far greater in Kenya (Demirgüç-Kunt *et al.* 2008).

Meanwhile, some developing countries have introduced affirmative regulatory policies which require financial service providers to provide appropriate services to poor and excluded households. For example, South Africa has consumer protection against extortionate or predatory behaviour and measures to work-out overborrowing, plus a financial sector charter which includes a basic bank account scheme (Demirgüç-Kunt *et al.* 2008). A famous example of such a charter was in India between 1977–90. All commercial banks could only open a new branch in an area that already had one on the condition that they opened four more in areas without a bank. The aim was to expand access to finance in rural areas. It has been calculated that branch expansion accounted for 60 per cent of rural poverty reduction during the period of the policy (Burgess and Pande 2005; Demirgüç-Kunt *et al.* 2008). Since then NABARD has been providing incentives for SHGs in order to extend rural credit.

Microfinance and the global development architecture

In a well reported comment, Vijay Mahajan, the chief executive of Basix, has suggested that microcredit 'seems to do more harm than good to the poorest' (cited in Tripathi 2006). As Karnani (2007: 38) puts it: 'We should not romanticize the idea of the "poor as entrepreneurs" '. Problematically, for visions of the poor pulling themselves up by their bootstraps, only a quarter of households living on less than $2 a day (per person) include an entrepreneur (Banerjee and Duflo 2007). There is the 'fallacy of composition' at work here: just because the model of making a small loan works for one or two people, this does not mean that it will succeed economy-wide (Bateman 2010). In fact, it will undermine itself as the market will become saturated by microenterprises and prices (and incomes) will fall.

Bateman (2010) has argued that the oversupply of microfinance credit is now creating problems of over-indebtedness; what Adams and Pischke (1992) memorably called 'microdebt'. Moreover, as well as the strain placed on individuals, microfinance can also strain households and communities. Rahman's (1999) fieldwork shows how the logic of joint liability and the demand for repayment produce a toxic cocktail of peer pressure and social coercion that results in escalating verbal aggression and physical violence and the break-up of families as they are forced to migrate to earn money to remit and repay the loans. And contrary to the results reported from South Africa in the microfinance-plus discussion, other research has found that the shift in resources and challenging of gender norms has increased domestic violence (Schuler *et al.* 1998).

If this is indeed the case, then why has microfinance become so central? Radical critics argue that microfinance is, at best, a passing of the responsibility buck by Western donors and, at worst, a conscious strategy to bring the developing world into the circuits of global capitalism (Soederberg 2004; Fernando 2006). For example, Weber (2002, 2004) identifies the focus on poverty reduction as a reaction by the World Bank and IMF to criticisms of structural adjustment. That is, the turn to poverty reduction and the promotion of microfinance is a reaction to the acknowledged failures of austerity measures that structural adjustment entailed. She argues that the promotion of microcredit should be understood as a crisis management tool. The logic being that the provision of credit at the local level effectively 'mops up' the economic slack and lack of effective demand caused by structural adjustment. As Weber (2004: 380) argues: ' "banking on the poor" seems to have become integral to managing the crises of global capitalism'.

A different set of critics, such as Karnani (2007: 36), argue that creating jobs and increasing worker productivity are far better investments in reducing poverty:

> To understand why creating jobs, not offering microcredit, is the better solution to alleviating poverty, consider these two alternative scenarios: (1) A microfinancier lends $200 to each of 500 women so that each can buy a sewing machine and set up her own sewing microenterprise, or (2) a traditional financier lends $100,000 to one savvy entrepreneur and helps her set up a garment manufacturing business that employs 500 people. In the first case, the women must make enough money to pay off their usually high-interest loans while competing with each other in exactly the same market niche. Meanwhile the garment manufacturing business can exploit economies of scale and use modern manufacturing processes and organizational techniques to enrich not only its owners, but also its workers.

There have been two responses to the concerns about microdebt and the negative effects of microfinance: one from the microfinance industry (graduation) and one from the Global South (cash transfers). Recognising that grants are often more appropriate for the very poor, and not loans, 'graduation' is an important way in which MFIs have retained a poverty focus (Arun and Hulme 2008). Here MFIs provide financial support through aid to the poorest of the poor, i.e. those who would not necessarily benefit from microloans anyway. During a period of one to two years the beneficiaries of this aid are provided with support, skills and access to financial services such that they can transition out of poverty to making productive use of the full range of MFI services. For example, BRAC launched its 'Challenging the Frontiers of Poverty Reduction: Targeting Ultra Poor, Targeting Social Constraints' (CFPR) initiative in 2002. The scheme targets the most vulnerable and food-insecure women, providing them with free assets, training, health services, allowances and heavily subsidised loans. Early evaluations are positive with sustainable improvements in income, employment and food security identified, but less so in education (Matin et al. 2008; Das and Misha 2010). This is a concrete recognition that lending to the poorest is less suitable than providing aid.

Cash transfers have been described as the 'southern challenge' and a rejection of the paternalism and ineffectiveness of the global 'aid and development industry' (Hanlon et al. 2010: 4). Cash grants can take a number of forms, such as hard cash, child benefits or family grants. They are not given as an emergency palliative, but as a guarantee of providing an adequate standard of living. It is an approach which has emerged and been pushed by the likes of Brazil, South Africa, India, Mexico and China, and now schemes operate in at least 45 countries. The big debate is between those who argue for conditional cash transfers (CCTs) – meaning that the poor get the money only if they fulfil certain obligations such as keeping their children in school, getting vaccinations and so forth. CCTs have been proved to work and not induce idleness, a favourite concern of sceptics (Fiszbien and Schady 2009). The alternative, championed by Hanlon *et al.* (2010) is just to give the poor money, i.e. without strings. They survey the evidence and argue that such programmes are affordable and recipients use the money well, they reduce unemployment and facilitate economic growth and welfare improvements. The evidence is that the cash transfers produce a virtuous spiral with poor households transforming the original grant into greater income.

Conclusion

The financial exclusion of poor households is an important limit on their freedoms. Lack of access to the formal financial system is both a cause and an outcome of poverty. Suppliers of formal financial services don't tend to cater to the needs of the poor. This is a problem for individuals, but also for national development as cross-country econometric studies have suggested a large, causal and statistically significant relationship between financial development and poverty reduction (Beck *et al.* 2007). Poor households have a huge unmet demand for financial services, which is currently met through informal channels. Microfinance has begun to fill this hole.

Microfinance has been an extraordinary success in terms of identifying a solution to a problem, globally replicating its model and scaling-up, and becoming a central plank of the global development community's FfD agenda. Early optimism has given way to a more measured evaluation of its success in poverty reduction. The initial results from more rigorous trials and impact evaluations do, however, suggest potential benefits from microfinance if well designed. In many ways we are only just learning what does and doesn't work at the household level.

As Rosenberg (2009) has pointed out, the *Portfolios of the Poor* project highlights that what poor people value is the ability to *cope* with poverty regardless of whether it lifts them out of poverty or not. The problem with poverty is not just one of low income, but the volatility of that income and the vulnerability of poor households, which means that they constantly have to borrow and save in order to make ends meet (Collins *et al.* 2009), more so than richer households. Plus, informal financial services tend to be unreliable, which is the value-added of microfinance:

> When we hear that microcredit may not lift people out of poverty, we tend to be disappointed, and regard consumption-smoothing as a 'mere palliative.' But we react this way only because our own basic consumption needs are seldom if ever threatened. As Portfolios demonstrates, poor people see it very differently.

(Rosenberg 2009)

The likes of Rosenberg have argued strongly that the value of microfinance lies less in its direct translation into higher income, or education or health outcomes, but more in that it is a far cheaper form of development financing than aid and grants, which go into providing social service provision. Following the microfinance model – if done correctly – initial subsidies can be recycled and leveraged many times over. He cites the example of BancoSol in Bolivia. It started with a few million dollars of subsidies in the 1990s, but now represents a loan portfolio of $200 million which serves 300,000 active savers and borrowers. As Greeley (2006: 1) notes, 'The challenge for the industry is to manage scaling-up without losing sight of its social purposes'.

But, based on critical reformist and radical critiques, I would add a second, third and fourth challenge. The second is scale-up without doing harm – something which has received much more attention of late because of the problems in India, and which implies prudential regulation, customer protection and alternatives to microfinance such as cash transfers and graduation schemes. The third, at the same time, is for governments and donors not to abdicate responsibility. The fourth is to integrate a sustainable microfinance industry into a bigger development strategy, not assuming that a millions small loans will release a successful Schumpeterian gale of free enterprise. Roodman (2012) makes an excellent point when he identifies that the historical emergence of microfinance is during periods of modernisation. That is to say, microfinance is a temporary transitional phenomenon – as countries grow wealthier the need for small-scale financial services for the poor disappears. And as such, the microfinance revolution in developing countries was, in some sense, inevitable (Dichter 2007; Roodman 2012). No country got rich *because* of microfinance. Economic development is a result of industrialisation, and does not happen on the back of small loans. This does not mean that microfinance is pointless, it provides essential services to poor people who need them. But like other sources of finance, on its own, it is no panacea.

Discussion questions

1 Where would you start in providing financial services to an underserved community: credit, savings or insurance?

2 How should microfinance be evaluated? How do we know whether it works or not? What should the measure of success be?

3 What are the major choices to be made in rolling out a microfinance programme? What would your choices be and why?

4 Is microfinance a neoliberal Trojan horse or a progressive developmental intervention?

5 Assuming that microfinance continues to be scaled-up what, in your view, is its fate?

Further reading

- David Roodman's (2012) *Due Diligence: An Impertinent Inquiry into Microfinance* is by wide agreement probably the single best book written on microfinance. It is organised around the simple (but oh-so-important) question 'Does microfinance work?': Does it reduce poverty? Improve freedoms? Has it produced a sustainable industry which promotes economic development? As well as answer these questions, Roodman provides an excellent historical review and background on microfinance. Definitely worth reading.

- For the inside story into modern microfinance see Muhammad Yunus's book *Banker to the Poor: The Story of the Grameen Bank*. Although some of the content is now somewhat dated, it is still a revealing read. It provides an insight in Yunus the man, why he established the Grameen Bank and the barriers he faced. It is also a very hopeful and inspiring book.

- For a thoroughly critical and radical take look at Milford Bateman's (2010) *Why Doesn't Microfinance Work?: The Destructive Rise of Local Neoliberalism*. He argues that microfinance creates poverty traps and *is* a Trojan horse for forwarding neoliberal practices. It is a controversial book and many (including Roodman) disagree with much of the analysis. But Bateman was well ahead of the curve in terms of questioning the unalloyed benefits of microfinance and he asks a lot of the right questions. It is well worth engaging with – whether one agrees with him or not.

Useful websites

- The best publicly available data on microfinance comes from the MIX database, available at: www. themix.org. MIX is a non-profit organisation headquartered in Washington, DC. Although it only contains information on the institutions that report to it, the MIX Market does contain the profiles of more than 2,000 microfinance institutions in over 100 countries and covers over 90 million borrowers. The data covers over 10 years now and is free to search. The data includes institutions' quarterly financial information, social performance information, funding structure data and so forth.

- The Consultative Group to Assist the Poor (CGAP) is an independent policy and research centre that works to advance financial access for the world's poor. It describes itself as 'an evidence-based advocacy platform to advance poor people's access to finance'. CGAP was established in 1995 and is housed at the World Bank and is supported by over 30 development agencies and private foundations. Its main roles are to provide market intelligence, promote standards, innovation and advisory services to governments, microfinance providers, donors, and investors. CGAP is available at: http://www.cgap.org.

10 Conclusions: the political economy of global finance and development

Learning outcomes

At the end of this chapter you should:

- Appreciate the capital fundamentalism of the orthodox mainstream, and the underplaying of systemic and political factors in the analysis of finance and development.
- Be able to put global finance in its place, in terms of its significance in improving the wealth, welfare and freedoms of poor countries and poor people in developing countries.
- Develop your own view on the most relevant theoretical approach to adopt and/or insights from a combination of approaches.
- Understand the range of alternative policy strategies available to developing countries to manage financial crises.

Key concepts

Politics of reform, political economy, Tobin Tax, Robin Hood Tax, bancor, global reserve currency, International Clearing Union

Introduction

Over the preceding nine chapters the book has sought to introduce key concepts and the major debates related to global finance and development; to place them in their intellectual and political context in terms of a spectrum of positions from neoliberal, liberal institutionalist, critical reformist to radical; and to present the empirical evidence which proponents of the different positions draw upon to back their arguments. There is an old quip, attributed to the Irish playwright George Bernard Shaw, who was also a co-founder of the London School of Economics and Political Science, that 'If all economists were laid end to end, they would not reach a conclusion'. Lest the reader feels the same having read the arguments for and against liberalising or regulating the different flows of global capital, the following discussion is designed to summarise and draw out a few key conclusions from the preceding chapters.

The second section then returns to the more practical question of what is to be done. Political economy, as the interaction of power and wealth, or the production and distribution of resources (economics) and the contestation and deliberation over collective values and goals (politics), implies both the empirical and positive analysis of what is, as well as the normative analysis of what ought. But there is a third purpose of political economy – one which stretches back to classical political economy, and has been very much kept alive by the critical and radical traditions – which is to help bring about positive change. The second section of the chapter reflects on some of the insights from the practical dimension of political economy.

What is the evidence?

This section covers two things. First it uses the theoretical lenses to reflect on the FfD orthodoxy. And second, it reflects on the empirical evidence, in the aggregate, to see to what extent, or not, the different approaches capture reality and are relevant for students of finance and development. The original premise of the book was that headline development debates – such as the MDGs – are dominated by 'capital fundamentalism'. That is to say, poverty is framed as a problem of a lack of capital and the solution and route to development is through the transfer of capital from developed to developing countries. In its most traditional sense this is the entire logic of development aid. But this capital infusion paradigm also includes the promotion of private capital flows and microfinance, even though they work within a for-profit model. The structure of the book was to present evidence about the workings of global finance in terms of the drivers of capital flows, the consequences of these flows and attendant financial and political structures for development outcomes, in a chapter-by-chapter comparative treatment of the main sources of development finance. It also sought to frame the questions asked and the answers provided through different theoretical lenses introduced in Chapter 3.

What do the lenses tell us about the financing for development vision?

Robert Heilbroner cautions against a naïve comparison and evaluation of the different theoretical lenses. This is because all of the lenses are inherently political in the sense that the very questions they pose vary, and the way in which finance and development are framed betrays different philosophical commitments or views of the world. As such,

> There is very little use in arguing which of these general perspectives is correct, or even more correct. What is important is to see all visions as expressions of the inescapable need to infuse 'meaning' – to discover a comprehensible framework – in the world. Visions thereby structure the social reality to which economics, like other forms of social inquiry, addresses its attention.
>
> (Heilbroner 1990: 1112)

Nevertheless, the following section does risk doing exactly that, i.e. trying to argue which of these general perspectives *is* more 'correct'; or perhaps more useful. However, before we get there, this section asks, 'What do the theoretical lenses – taken together – reveal about the "capital fundamentalism" of mainstream MDG-style international development policy?' In short, what do the lenses tell us about the FfD vision?

The FfD vision is a popular one, a largely intuitive one, but is also largely apolitical and theoretically simplistic. The poverty trap logic for development finance assumes away a lot of important factors; for example, the intervening role that good and bad political institutions play in radically altering the impact of geography or history on a country's development, and assuming that financial resources come without political consequences or are free from the operations of power or exploitation. As such, the different approaches reveal important limitations of the mainstream agenda. And for those who think this is anachronistic given that we will soon be in a post-MDG world, it is salutatory to reflect on just how far the agenda will shift. Policymaking is typically characterised by its incremental nature and is a science of muddling through (Lindblom 1959). This was graphically revealed in the way in which the MDGs were created out of a bricolage of existing international development targets (Hulme 2009). The fact that the UK prime minister David Cameron was chosen as one of the co-chairs of the UN's panel to develop the post-MDG agenda fits well with the mainstream. His credentials for the job are that the UK has displayed an impressive commitment to increasing aid towards the 0.7 per cent goal (i.e. the aid paradigm at work), plus he has been behind attempts to develop a broader understanding of poverty and wellbeing in the UK (i.e. progressive on measures), but has also championed the role of business and the private sector in delivering public services at home (in the NHS) and abroad (through the position of the UK's DFID), plus he is hostile to the incorporation of issues of inequality into the debate. The FfD mainstream is being challenged but it remains the mainstream.

There is a hierarchy of capital flows: in terms of development impacts, but also access. As other studies have suggested, comparing the evidence across different capital flows suggests that not all finance is created equal, at least in terms of its relationship with development (UN DESA 2005; Spratt 2009). Furthermore, there also appears to be a dynamic relationship between capital flows and a country's level of development, whereby the optimal mix of development financing changes as a country gets wealthier (in the broadest sense). Abramovitz (1989) famously argued that the advantages of economic backwardness are only accessible if the developing country possesses social capabilities which are sufficiently developed to adopt the new technologies – there is a threshold and a notion of 'appropriate technology'. A similar argument can be advanced for finance – there are more and less appropriate mixes of finance depending on the social capabilities of a country. Development is only secured when the appropriate capital inflows combine with sufficient institutional and other contextual factors, including smart and activist policies.

It is easy to overstate the benefits of capital mobility. This statement is clearest when expressed from a critical refomist or radical perspective. However, in a sense, it is even more powerful when approached from an orthodox perspective. For example, despite the capital account liberalisation cheerleading undertaken from an orthodox perspective, good liberal scholars have time and again demonstrated that the benefits are much less than usually claimed and are relatively less important than other policy measures or types of liberalisation that could be promoted. For example, Dani Rodrik's work with his colleagues shows that the institutional environment is a much better predictor of growth rates than economic openness is (Rodrik *et al.* 2004). But a more startling and compelling piece of evidence comes from Michael Clemens (2011). His argument is that even minor changes in labour mobility (in the region of 5 per cent) have the potential to increase global welfare by more than by eliminating all remaining barriers to trade and capital put together. Table 10.1 reproduces Clemens's summary of the best recent economic work calculating the gains from eliminating barriers to trade, capital and labour (Clemens 2011: 85). The relative gains are clear, but it is also clear that the global gains from eliminating remaining capital barriers are small. Furthermore, these figures say nothing about the distribution of gains and indeed costs between rich and poor countries and/or between rich and poor people, and the costs associated with moving from the current situation to the hypothetical world of free capital mobility – i.e. a series of financial crises in line with the historical record of countries going through liberalisation – and finally who by and how this compensation and protection is provided. Finally, it's also worth noting that the estimated gains (such as they are) rely on orthodox neoclassical modelling and the assumption of equilibrating markets, an assumption strongly contested by critical refomists and radical scholars.

Finance goes hand-in-hand with development, rather than being a prerequisite. The proverbial chicken and egg question is an old (and often tedious) one. Nevertheless, the issue of endogeneity is crucial – i.e. not only do financial flows lead to development but development causes financial inflows. We can think of various renditions of this argument. For example, it is not just that capital inflows fund growth, but that domestic

Table 10.1 *Efficiency gain from elimination of international barriers (percentage of world GDP).*

All policy barriers to merchandise trade	
1.8%	Goldin *et al.* (1993)
4.1%	Dessus *et al.* (1999)
0.9%	Anderson *et al.* (2000)
1.2%	World Bank (2001a, b, c)
2.8%	World Bank (2001 a, b, c)
0.7%	Anderson and Martin (2005)
0.3%	Hertel and Keeney (2006)
All barriers to capital flows	
1.7%	Gourinchas and Jeanne (2006)
0.1%	Caselli and Feyrer (2007)
All barriers to labour mobility	
147.3%	Hamilton and Whalley (1984)
96.5%	Moses and Letnes (2004)
67%	Iregui (2005)
122%	Klein and Ventura (2007)

Source: Clemens 2011: 85

growth and industrial investment allow domestic actors to save more, which leads to capital deepening and financial development. Or, in addition to external capital flows boosting the resources available for investment, it is really growth (or more probably the perceived *potential* for growth) that drives and attracts capital inflows. A very good reason to be suspicious of the stronger claim that external financial resources are necessary (even if insufficient) for development is provided by Eswar Prasad, Raghuram Rajan and Arvind Subramanian. Based on their research, they conclude that a lack of domestic financial resources does not present a major constraint on poor countries' development (Prasad *et al.* 2007). Indeed, they find that those countries that have relied less on external capital have actually enjoyed faster long-term growth. This is a hugely significant finding. A simplistic reading would be that countries are better off de-globalising. And there is something to that. But, as suggested in the following section, the more interesting and relevant question is *how* governments can manage the domestic-global boundary, rather than whether they should close or open it more. Liberal institutionalists emphasise the importance of correct sequencing of reforms and the role of good governance as a buffer, while critical reformists suggest the centrality of state-led financing. But again, both of these answers beg the (political) question of how.

The FfD perspective reduces finance to a flow or resource and fails to see finance as a structure. As one moves from a neoliberal to liberal institutionalist to critical reformist to radical perspectives, the role of political and social relations becomes more and more important. They also change. Liberal institutionalists identify the need for complementary and market-supporting institutions to facilitate the effective functioning of markets. Critical reformists go further, and instead of adding politics alongside markets they open up the box of the market to get inside and offer a critique and provide an alternative, non-equilibrating view of how markets really work. Finally, radicals step back to show how the system of markets, *in toto*, serves to produce and reproduce political patterns of domination and exploitation. In all three visions studying finance is less about a focus on just capital as a flow. So, respectively, this is on the way finance interacts with the political environment, or how capital emerges from and responds to the social structure of uncertainty, or that capital is the embodiment and institutionalisation of inequality. Understanding global finance, its drivers and development outcomes, means reconceptualising it from 'a thing' to a relation or structure.

The mainstream approach has neglected politics even more than institutions and policies. This is true for both international power politics and the domestic politics of reform. As Goran Hyden (2008) has convincingly shown, international donors and IFIs have ignored the way in which donor influence controls and shapes the aid regime and how ignoring political realities is a recipe for failure. Meanwhile, Andy Sumner's survey of FDI and development consistently showed that the countries that were able to extract the greatest benefits from FDI were those that engaged in interventionist and highly selective policies – essentially, those governments who were willing and able to exercise control over the domestic economy and polity (Amsden 2001). On which point, an important caveat in Prasad *et al.*'s (2007) story – about the negative relationship between external capital and long-term growth – is that they note the negative relationship between capital inflows and economic growth only holds for non-industrial, non-transition countries and furthermore that there are episodes of fast growth and large current account deficits, for example East Asia. So, the more interesting question is not just what explains this (which we have already documented is interventionist policies designed to finance and coordinate industrial activity and regulating capital inflows), but *how* to successfully emulate it. This takes us into questions of politics – agency and coalitions, cooperation, persuasion and contestation.

What does FfD tell us about the theoretical lenses?

Heilbroner's (1990) point aside, how can we summarise the case for rejecting or adopting the different approaches? I think the safest and strongest claim can be made is about neoliberal approaches to finance and development. Empirically they have proved a failure. The attempts at radical privatisation and liberalisation associated with structural adjustment were disastrous for many countries, and in particular they hit the poorest and most vulnerable the hardest. There is a straightforward explanation for failure: the neoclassical assumptions about the way in which markets work and individuals behave are plain wrong. Markets are not spontaneous and individuals are not fully-informed utility maximisers. But a further, and more sociological, explanation is that this mistake is compounded by the fact that the whole neoliberal agenda is a social experiment in projecting textbook abstractions about how the world works into the world itself – for example by forcing people to become calculative 'entrepreneurs' through changes in land tenure systems designed to increase productivity and efficiency. The attempts to create 'rational economic man' often cut against established social norms, conventions, practices and patterns of conduct (Williams 1999). Such virtualism is likely to fail and prove socially corrosive (Carrier and Miller 1998; McMichael 1998; Watson 2005).

It would be fair to say that the mainstream agenda has acknowledged and understood this, at least the failure of unreconstructed neoliberalism, if not the reasons. The current orthodoxy is still informed by neoclassical foundations, but it now incorporates an important role for institutions, equity and social safety nets, as well as a series of auxiliary issues such as social capital, gender and, most recently, even a return to the possibility of (pro-market) industrial policy (Lin 2012). This position is, of course, what we have characterised here as liberal institutionalism. I would agree that many of the 'on average' findings of liberal institutionalists are probably correct and all very well, but they also suggest a number of problems. One, there is a great deal of variation around the line of 'best fit'; that is to say, institutional or legal structures only explain so much of development outcomes. Understanding positive and negative outliers is important for future research and policy. For example, Peru, Pananma, Tanzania, Algeria and India are all doing much better than one would predict given the common measures of good institutions and governance (Grindle 2010). Two, how do countries acquire the 'right' institutions? It's all very well saying that Zambia needs to become more like Australia in terms of its governance arrangements, but how can this happen? This question, a question of politics, is turned to in the following section. Third, the types of institutions which are studied are strongly biased towards liberal institutions – i.e. prioritising the protection of private property rights. This is based on the assumption that, suitably institutionalised, political stability and market forces will work to channel savings into investment.

However, while protection or private property may help, they are certainly not the whole story. As Pranab Bardhan (2005) has persuasively argued, this is to get your institutions wrong. It is important to unbundle institutions and pay much more attention to the role of institutions that secure democratic participation rights and ones that help address coordination failures in early industrialisation. These two critiques – the failure of savings to transform into investment and development, and the central role of coordinating institutions – are of course the major critiques of liberal institutionalism from a critical reformist perspective.

Critical reformists argue that it's not just that markets aren't spontaneous (as liberal institutionalists accept) but that even with the 'right' institutional support such as private property rights they are not self-equilibrating. This insight is primarily drawn from Keynes (1936), and extended by Harrod (1939), Minsky (1982), Davidson (2003) and Thirwall (2007). Under such circumstances there is a compelling case for the state to take a strong role to ensure that the paradox of thrift does not mean that countries get stuck in an equilibrium of low investment.

Radicals – chiefly Marxists, but also post-development scholars, who reject the benefits of financial integration and development, respectively – offer a different take, a more macro and more political lens on the role of global finance and development. Critical reformists recognise the fundamental problems with the way in which financial markets work, are sceptical of the benefits of FfD, but are keen to fix them and to civilise capitalism. Radicals reject capitalism, arguing that trying to civilise capitalism misses the point. For many people this makes radical critiques frustrating and unworkable. This is to miss the point as well. Radical scholars offer the clearest and sharpest analysis of the role of power, inequalities and the structure of the global system in perpetuating underdevelopment. Regardless of whether delinking is feasible or not, this doesn't make the analysis and the critique wrong.

Where radical critiques do appear to struggle is on the question of how certain states can escape the structural traps of underdevelopment, such as Brazil, India, Botswana, China, etc. It is only a number of states to be sure. And capitalist social relations within these countries have not been abolished, i.e. exploitation and inequalities of wealth characterise their societies too. So the radical critique remains important. But explaining variation, exceptions and outliers is just as much a blind spot for radicals as it is for liberal institutionalists.

The political economy of reform

Access to credit can be a wonderful thing. But the flip side is of course indebtedness, which at the right levels is not a problem. However, sudden changes in the economic environment, such as a jump in borrowing costs because of interest rate or exchange rate changes, can mean that borrowers can quickly find themselves unable to repay the amount owed. Moreover, debt flows tend to be pro-cyclical and volatile in the short term. Over the medium term, debt flows tend to be very pro-cyclical, meaning that they surge during periods of economic growth, and stop and reverse in downturns. As such, they do not help to smooth business cycles, as they are intended to, but actually accentuate them. The problem is that these sudden stops are usually exactly when states, firms and individuals most need access to funds. As such, external borrowing is a major source of vulnerability for developing countries, revealed cruelly and consistently in financial crises such as the one that struck East Asia in the 1990s. But the inequalities within, and caused by, international debt flows have been most graphically exposed by the long-term debt crisis which engulfed most developing countries from the 1980s onwards. Hence these risks have to be balanced against the need to borrow from foreign lenders. For even mainstream liberals this means the liberalisation of capital should be approached much more cautiously and with a much more critical eye than the liberalisation of trade flows (Bhagwati 1998).

The experiences of the past 30 years mean that there are few genuinely unapologetic champions of capital liberalisation. Most recognise the dark side to capital flows. Proponents of financial globalisation and integration tend to portray the negative aspects of capital flows, volatility and crises, as inevitable 'growing pains' (see World Bank 2001c). They argue that the question is not whether or not capital flows bring benefits, but that increased integration is inevitable and policy should be focused on how to minimise the costs of capital flow reversals and manage inflows so as to harness and maximise their development benefits. Critics are somewhat less sanguine, arguing that it is not simply a case of developing countries being better governors and managers of capital flows, but that financial markets are inherently unstable and speculative, and no amount of the 'right policies' will remove the risks.

These problems are inherent to financial markets: their speculative and unstable nature is a natural function of their role in predicting the future. As we have seen, debt flows tend to stop or even reverse very quickly if market sentiment changes. This causes immediate refinancing problems. But, even during non-crisis periods, it also means that economic agents have to operate under the shadow of uncertainty. This acts as a drag on long-term investment, planning, and ultimately growth and development. The small size of domestic financial markets with respect to global financial flows means that developing economies are easily drowned and overheat with large inflows. And when capital is suddenly withdrawn it causes a complete liquidity drought. Critical reformists argue that these can be managed and regulated in order to be minimised, but probably never removed (Warwick Commission 2009). In addition to the problems of accessing financial flows and managing their volatility, governments are also keen to balance their diet of inflows and maximise the proportion of long-term finance, and there are overhanging systemic issues such as global imbalances. As such more interventionist and tight regulation of flows is necessary.

How to manage crises in developing countries

Domestically, developing countries can do a number of things to minimise the possibility of crisis. First, they can concentrate on improving macroeconomic stability. This would include maintaining steady and low rates of inflation, reducing fiscal imbalances and reducing current account deficits. These are all things that international investors tend to punish developing countries for. And these are all orthodox recommendations. In addition, many countries have also moved away from fixed exchange rates. This has tended to reduce the frequency of crises, but the trade-off has been increased volatility which impacts upon growth rates.

Second, many developing countries have built up reserves by way of self-insurance against sudden stops in funding and attacks on their currency. In 2005 the aggregated current and capital accounts of the developing world amounted to $710 billion. Of this, $392 billion was invested in reserves. While this is thought to be effective, it is also costly in terms of forgone resources for domestic investment. This was covered in Chapter 4.

Third, domestic prudential regulation is important. Prudential regulation refers to the banking and financial standards that require banks and other financial institutions to hold adequate capital and control the risks they expose themselves to as well as regulating the use of financial products and innovation. Prudential regulation is designed to protect the health of the financial system and depositors. Developing countries are particularly susceptible to financial crises because many lack effective systems of prudential regulation. Or, even if prudential regulation is in place, there are often problems with the enforcement of the rules with, for example, insider trading regulations being flouted and/or political interference (Barth *et al.* 2001).

Relatedly, recognising that external short-term borrowing, especially through the banking system, has been a key factor in developing country financial crises has led to support for measures which help limit developing countries' reliance on this kind of financing. Though note, this assumes that short-term debt is the cause of crises. It is certainly correlated, but a reliance on short-term debt may well be a symptom rather than a cause of crisis-prone countries (Eichengreen and Hausmann 1999). One common tool – to supplement domestic prudential regulation – has been the use of capital controls to manage capital flows. Examples of types of capital controls include taxes, minimum maturity requirements and marginal reserve requirements on foreign bank financing. These controls are used to limit the surges of inflows and sudden reversals, support macroeconomic stability and encourage longer-term capital flows. Interestingly, the IMF has conducted something of a U-turn on capital controls. Despite being a fundamental part of the post-war international monetary system, support for capital controls had become economic heresy within the IMF during the 1980s and 1990s (Chwieroth 2009). But in a recent IMF paper, it is argued that 'capital controls are a legitimate part of the toolkit to manage capital inflows' (Ostry *et al.* 2010: 15). Research has shown that countries with capital controls have increased their monetary policy autonomy and also managed to change the composition of capital inflows away from short-term debt flows to longer-term inflows (Ariyoshi *et al.* 2000). And significantly, countries with capital controls in place during the 2000s suffered smaller output falls from the financial crisis (Ostry *et al.* 2010).

Fifth, a further solution to relying on international capital is for developing countries to grow their domestic bond markets (Goldstein and Turner 2004). There is some evidence that certain emerging markets – such as Chile, Hungary, India and Thailand – are managing to borrow longer-term domestically, thereby solving the maturity mismatch (i.e. having to borrow short-term finance to fund long-term investment) (Eichengreen *et al.* 2003). Local currency bond markets grew from $1.3 trillion in 1997 to $3.5 trillion in September 2005 (World Bank 2006a). But for most developing countries this is a very long-term strategy and does not offer much by way of an immediate fix. Plus, even for these countries, while the maturity mismatch problem is addressed, the ability to borrow abroad in domestic currency remains limited. In addition, there is little evidence that local capital markets can act as a buffer against international interest rate volatility. Local currency bond spreads still respond to and tend to be correlated with changes in yield spreads in international markets (UN DESA 2005).

Finally, another solution available to domestic authorities to protect themselves from the vicissitudes of global finance is dollarisation (Eichengreen and Hausmann 1999). Dollarisation is the process of abandoning a weaker national currency for a stronger international one in which debts are denominated; so typically this is the US dollar, but it could also be the euro. As such the currency mismatch is resolved in a stroke. The problem is that it completely removes the possibility of an independent monetary policy. However, advocates argue that the increased access to international lending reduces the need for active monetary policy, as the increased capital flows will help attenuate business cycles. Yet this clearly flies in the face of a notion of development as increased autonomy and independence.

At the international level, a number of initiatives are important. The role of the IMF as an early warning system, identifying potential problems, has been revived after the global financial crisis. There are a number of ways that the IMF conducts this role, for example through its Article IV consultations with countries and the Financial Sector Stability Assessments. While individual countries have tended to prefer to self-insure rather than have to rely on the IMF and the intrusive policy conditionalities which borrowing entails, there are a number of lines of emergency funding offered by the IMF in its role as a quasi lender of last resort. For instance, in a capital account crisis countries can turn to the Supplemental Reserve Facility, and for trade financing there are the Compensatory Financing Facilities.

Also at the international level, the Basel Committee on Banking Supervision at the BIS sets the standards on capital requirements. While the Basel Committee has no legal jurisdiction to impose the standards on developing countries, the Basel Accord has been widely adopted because governments fear that they will be punished if they do not embrace international best practices, and evidence suggests that they are gradually doing so (Gottschalk 2010). The latest standard (Basel III) now requires a common equity cushion of 4.5 per cent (with another 2.5 per cent to be phased in by 2019). The idea is to ensure that banks maintain a capital cushion in order to absorb losses during periods of financial crisis. Basel III also includes a counter-cyclical buffer which is designed to reduce the build-up of excess risk if credit is growing too quickly.

The capital adequacy rules may well help the stability of the system, but the trade-off is that they will reduce credit access for the poor. For example, research into the implications of adopting Basel standards in Brazil revealed that it led to a drop in the amount of credit available (Gottschalk and Sodré 2007). It also reduced the number of banks and increased banking concentration which had negative implications for SMEs' access to finance.

Box 10.1

The Tobin Tax and the Robin Hood tax

In 1978, James Tobin, in his presidential address to the Eastern Economic Association, set out an idea for a currency transactions tax (Tobin 1978). This wasn't the first time he'd made the argument. Although it is now most commonly thought of as a revenue-raising device associated with the work of the 'anti-globalisation' protestors, James Tobin's tax proposal was first and foremost a response to the 'globalisation' of financial markets in the 1970s (*Der Spiegel* 2001). Tobin noted, with dismay, the increasing difficulty with which governments were able to exercise an autonomous macroeconomic policy. His solution was to impose a small tax to penalise short-horizon round trips, i.e. speculation (Tobin 1978). However, while discouraging speculation the tax would negligibly affect commodity trade and long-term investments. Hence the tax would also succeed in restoring the link between exchange rates and long-term fundamentals (Tobin 1996).

The major criticisms of the tax have tended to focus on it feasibility (Eichengreen 1999). First, that because it was intended to be small, then in times of crisis currency traders would simply ignore the tax in their decision-making. Second, there was the continuing availability of offshore, the problem of 'market migration'; moreover, because of the dependency that the leading financial centres had on offshore there was unlikely to be the political will to close this loophole. Third, there was the likelihood that traders would simply engage in asset substitution, that is, rather than buy currencies they could still speculate on currencies through, say, treasury bills instead. For an excellent even-handed survey of the literature, see McCulloch and Pacillo (2011).

Nevertheless, the financial crisis has seen a revival of Tobin's ideas for a financial transactions tax, to throw sand in the wheels of finance. By 2013, the European Parliament had voted 3 to 1 to support 11 countries who are planning on implementing a European financial transactions tax. The countries are Germany, France, Spain, Italy, Austria, Belgium, Portugal, Greece, Slovenia, Slovakia and Estonia, and notably don't include the UK. And France has already had such a tax in place (at 0.2 per cent), since August 2012.

Much of this lobbying success and policy change has been down to the work of the Robin Hood Tax. A coalition of NGOs and popular figures – such as Bill Nighy and Richard Curtis – launched the initiative as a deliberately 'popular' campaign, trying to put pressure on politicians via the public. A 10-metre high projection on the side of the Bank of England saying 'Be part of the world's greatest bank job' accompanied campaign ads and heavy use of social media websites. The campaign has argued that a small levy on all transactions in financial markets (0.05 per cent, on average) could raise up to $400 billion globally every year, depending upon the exact rate and number of jurisdictions it is set up in. Comparing this to aid figures gives a sense of the size. The campaign, cleverly, sought to appeal to a number of interests in how the revenue would be dispersed: 50 per cent retained domestically, so focusing on domestic concerns, education, health; 25 per cent to environment; and 25 per cent to go on development. The Robin Hood Tax campaign's strapline is 'turning a global crisis into a global opportunity', and Richard Curtis's view at the launch was telling:

> I think it is a great time for a big, heroic, interesting step forward . . . This campaign is a glimpse into the future . . . We do move towards humanity . . . I bet that in 100 years from now we do have the equivalent of an international welfare arrangement. And this is a glimpse of such a thing.

To back up his argument he points to an analogy to slavery and the slave trade, in that people 'in this country were happy about slavery 200 years ago'. Thus grabbing the window of opportunity, along with active lobbying, scheming, campaigning and persuasion, plus a clever reframing of the debate in terms of 'nurses, polar bears and development' has been successful in building support among the domestic public and making Tobin's original plan as close to fruition as ever. See 'Useful websites' for further information.

The politics of political economy

Put crudely, the policy implications from the discussion above are twofold. Governments that rely less on international capital tend to do better. Thus they need to protect and steer their economies, and they need to increase domestic resources – through increasing productive investment, savings, taxation and so forth. Internationally, the community needs to stop harms emanating from the international monetary and financial system as well as fulfilling the more common positive duties of giving aid. Taken together, this underlines the importance of better governance: domestic and global.

However, to state that developing countries need to raise more taxes, spend them better, and that the international community needs to reform the international monetary system or the aid regime, is less of an answer than to ask a further question: how? The bald claim that domestic and international finance needs to be better governed is *not* a very satisfactory conclusion. It merely begs the questions, how to do this, how to bring it about?

The debate between Justin Lin and Ha-Joon Chang (see Box 3.9) starts to identify a much more interesting set of proposals than simply improving governance across the board. This is very much in line with the growing school of thought that it is *good enough* governance rather than good governance that should be the focus (Grindle 2004, 2007). Grindle's argument is that the list of 'best practice' institutional reforms is too long and too demanding and presents an unrealistic burden on governments to do everything at once. Instead there should be a prioritisation of reforms in terms of which are more essential, in terms of which will work in the specific country context, and finally in terms of which are politically feasible. A related approach is Dani Rodrik's (2008) notion of a second-best mindset. Instead of trying to transplant 'best practice' institutions and policies from overseas – essentially following the old blueprint one-size-fits-all model – it is better to understand what will work in each individual context. Best practice may actually be harmful. Finally, institutional copying and pasting is likely to fail if it doesn't 'fit' with existing institutions – informal, cultural and formal. It is not the case that poor countries *lack* institutions, but according to the logic of liberal institutionalism these are not the right *sort* of institutions to encourage market efficiency. 'Throughout most of history, the experience of the agents and the ideologies of the actors do not combine to lead to efficient outcomes' (North 1990: 96). This is because institutions reflect the interests of the powerful as they are created, and then through their institutionalisation become locked in, creating a path dependency whereby they are difficult to change. A second-best mindset is more pragmatic and it often leads to more interventionist policies associated with the developmental state model as it was developed in East Asia and Peter Evans's (1995) work on 'embedded autonomy'.

To recap – an autonomous state has two distinctive features. First, it has the 'prowess and perspicacity of technocrats within the state apparatus' and second, 'an institutional structure that is durable and effective' (Evans 1995: 141). The rationalised bureaucracy is therefore free from manipulation by rent-seeking groups. Yet, it needs to retain some flexibility to help aggregate preferences and change course. Basically, the state elites are in touch with important players within business and civil society. So, the question becomes how is it that the right blend of embeddedness and autonomy was achieved in South Korea (the prototypical developmental state), but the delicate balance was less successful in India and Brazil (intermediate states in Evans's terminology), and in countries like Zaire the state became downright 'predatory'? Evans discusses the importance of generating an *esprit de corps* among policymakers – that they share a cultural background, set of beliefs and assumptions, and commitment to national development. But, at the risk of infinite regress, how is *this* made possible? Recent efforts to develop a more political economy approach to governance reforms in developing countries point in the right direction. To illustrate, compare the good governance with political economy approaches to financial access.

Understanding financial access: from governance to political economy

It is widely accepted that limited access to financial services – savings, loans and insurance – is a binding constraint on economic growth and limits the freedoms of individuals (Collins *et al.* 2009; Demirgüç-Kunt *et al.* 2008). If financial development is indeed welfare-enhancing why don't we see more of it? What explains the barriers to financial development? The standard liberal institutionalist explanation is that the legal and institutional inheritance that countries are left with determines whether or not they grow (La Porta 1997). However, given that institutions change and there are advances and setbacks in liberalisation over time, and across countries, and there is considerable variation in financial development even between countries with the same legal origins, suggests that the argument is incomplete (Becerra *et al.* 2012). We need to reinsert agency and politics.

Why are financial services undersupplied in LICs? A political explanation is that well-connected elites abuse their influence to gain preferential treatment from banks. Politically connected interest groups (either from industry or banking) oppose financial deepening because widening access to credit would increase competition and erode profits for incumbent firms (Rajan and Zingales 2003). For example, in Pakistan, firms with a director who had participated in an election borrowed 45 per cent more and had 50 per cent higher default rates (Khwaja and Mian 2005). Significantly, the preferential treatment was exclusively from government banks, not private ones. Moreover, the strength of the political preference effect increased if the director was in power. Such political lending stifles growth, to an extent that has been estimated to be between 0.3–1.9 per cent of GDP every year in the case of Pakistan. Others go further, arguing that an underdeveloped financial system allows governments to abuse the existing banking system for their own credit needs (Haber *et al.* 2008).

Good governance-style arguments note that without sound legal systems to guarantee property rights (Beck *et al.* 2001) and/or in the absence of sufficient checks and balances to limit government expropriation (North and Weingast 1989; Keefer 2007), depositors limit their savings, banks limit their lending and investors withhold their capital. A political economy approach differs insofar as it begins by acknowledging the problem of vested interests. Political elites fear financial sector development because it undermines their own advantages, increases competition and potentially empowers others. The question, then, is not what the ideal reforms would be, but given the configuration of political interests and institutions, which reforms are possible and how might they be achieved?

A political economy approach begins from the premise that the promotion or blocking of political reforms to expand access to credit is sensitive to the incentives facing political decision-makers and the interests of their constituents (Keefer 2007). The policy conclusion to be drawn is that financial sector development will improve when the interests of a politician (assumed to be staying in office) are more aligned with the economic interests of voters than with the special interests of the banking or industrial sector. Constituents have an interest in voting for the politician who promises to expand access to credit and secure property rights and so forth. But when voters don't know what politicians do when they get elected or where there is an inability to hold them to account when they renege on their promises, there are no incentives for politicians to push for increased financial access. The trick, from a political economy perspective, is to shift the incentives facing politicians by holding them to account for their pre-electoral promises to voters. Two common solutions are proposed (Keefer 2007). First, competitive elections increase the likelihood that politicians will be held to account, making their pre-election promises more credible. Second, an enfranchised citizenry increases the focus on what governments and elected representatives have and haven't done during office.

The same type of analysis can reveal where and why taxation reforms meet political logjams. Yet, there are real limits to this type of analysis. First, analytically, they are too economistic. They tend to build on neoclassical assumptions about rationality and utility maximisation typically found in the public choice school. This means they miss the vitally important and dynamic role of ideas (Blyth 2003). Agents have to interpret the opportunities and risks facing them in a situation of uncertainty. As such, agents' beliefs, values and other cognitive filters are central in understanding how they act (Hay and Wincott 1998). See the discussion of the Tobin Tax in Box 10.1.

The political economy of global reform

Global imbalances

Is the current financial system fit for purpose? Does it efficiently and effectively allocate capital to maximise sustainable economic development? Standard economic theory suggests that surplus capital will be invested in capital-poor regions, a function of the relative scarcity of capital to labour and guided by the greater growth prospects. It also suggests that capital account convertibility allows capital to travel to where it will be most productive and thus provide an engine of growth for developing countries to start to 'catch up'. Thus, investment should flow from the capital-rich countries of the core to the capital-poor periphery. However, this is not the case (Lucas 1990).

Globally, capital runs uphill: from the high-risk-reward-capital-poor countries to the low-risk-reward-capital-rich countries. Plus, it seems to be becoming more pronounced over time. Studies show that the result holds even when controlling for variables which would reduce investors' risk-adjusted returns, such as political instability, poor infrastructure, corruption, etc. (Prasad *et al.* 2007).

As Mervyn King (2006), the then governor of the Bank of England, noted, 'The invisible hand of international capital markets has not successfully coordinated monetary and exchange rate policies'. This has led to the bizarre situation of the rest of the world financing the US. Essentially the US deficit, i.e. its external debt, which has helped to pay for everything from personal consumption to arms expenditure, is being financed by poor countries (Teunissen 2006). The Commission of Experts, established by the president of the United Nations General Assembly, estimated that during 2007 the developing countries loaned $3.7 trillion to the developed countries (UN Commission of Experts 2009). The system not only runs counter to established economic theory, but it is also highly inequitable and effectively redistributively regressive.

The large US current account deficit grew from zero in 1991 to reach 6.5 per cent of GDP in 2006, and then fell back to 3 per cent. This is the other side of the coin to the current account surpluses in the developing world and the holding of reserves in US Treasuries and other securities. These global imbalances – the pattern of current account surpluses and deficits built up in the world economy – have formed around two key poles: the US and China; but also include other deficit economies such as the UK and Southern and Eastern Europe and other surplus economies such as Japan and the other East Asian economies, Germany, and the oil exporters.

There are competing explanations of the proximate cause of the imbalances. Ben Bernanke (2005) famously identified the high savings rates in China, and East Asia more generally, as the source of the problem. However, whether one agrees with Bernanke's account of a global savings glut, or the counter argument that it is excessive

US consumption that lies behind the imbalances, the underlying point is that the imbalances are a function of the fundamental asymmetries which characterise the international monetary system (Costabile 2007). The issuer of the international currency will always tend to end up being a net debtor because of natural recycling. The dollars tend to be invested back into the markets of the issuing country – thus producing the uphill flow of capital (as we saw in Chapter 4). The absolutely crucial point is that under such a system, the creation of international liquidity (i.e. creation of the international currency) is done through the creation of debt in the issuing country (i.e. the US) (Costabile 2007). Which means that international liquidity completely depends on the policies of the issuing country, i.e. the US. Hence the actual provision of international liquidity does not necessarily follow the true logic of a public good but, as Charles Kindleberger (1981) noted, is provided as the side-effect of the private good of seneiorage. That is, the provision of international liquidity, which rose from 2000 and suddenly contracted following the financial crisis, is fundamentally tied to the interests and policy needs of the US, not development.

Global monetary reform

So what is to be done to reduce volatility, to decrease the deflationary consequences of reserve accumulation, and to delink the provision of international liquidity and effective global aggregate demand from domestic US economic policy?

Keynes had a plan for an institutionalised and rules-based international monetary system that would avoid deflationary pressures for the world economy. All international transactions would clear through an international monetary union and because the international currency is now credit money the need for reserves would disappear. Keynes's system not only addresses the problem of speculation, but it also prevents liquidity constraints from hindering economic growth and development. The international preference for liquidity rather than investment spending is the nub of the problem in the world economy. It has been demonstrated that Keynes's insights into the role which uncertainty plays in producing an antisocial fetish of liquidity is just as true internationally (in the form of foreign exchange reserves) as it is domestically.

Box 10.2

The ideas and politics around a new global reserve currency

The idea of a new global reserve currency is one that has received some important backing of late. The propositions for a new reserve currency are clear responses to the financial crisis and were building through 2009–10. In March 2009, the recommendations of the Commission of Experts on Reforms of the International Monetary and Financial System, chaired by Joseph Stiglitz, proposed a new global reserve system (UN Commission of Experts 2009). Also in March 2009, Zhou Xiaochuan, the governor of the People's Bank of China, called for an international reserve currency that delinked it from sovereign nations. And on 6 June 2009, John Lipsky, the IMF's first deputy managing director, observed that while revolutionary, it is entirely conceivable that the IMF could create a new global reserve currency to replace the US dollar. Lispky noted that there would be 'many, many attractions in the long run to such an outcome' (Nicholson 2009). While in February 2010, Dominique Strauss-Kahn, managing director of the IMF, also addressed the issue, saying:

> A longer-term question is whether a new global reserve asset is needed . . . And one day, the Fund might even be called upon to provide a globally issued reserve asset, similar to – but in important respects different from – the SDR. That day has not yet come. But I think it is intellectually healthy to explore these kinds of ideas now – with a view to what the global system might need at some time in the future.
>
> (Strauss-Kahn 2010)

And most graphically, at the G8 meeting in Italy, July 2009, Russian president Dmitry Medvedev paraded a sample minted coin of a new world currency to the press. UNCTAD's *Trade and Development Report* 2009 also called for major reforms of the international financial architecture, specifically backing the idea of using SDRs as the main form of international liquidity as opposed to the US dollar.

The last time significant political investment was placed into discussions of a global reserve currency was during the 1960s as a response to the breaking down of the original Bretton Woods system, and before that in the 1940s with the dislocation of the World War II. So it is of little surprise to see these discussions now. But will this time be any different? Frankly, it would seem as though the immediate window of opportunity opened by the 2008 financial crisis has probably now closed. However, the interest in Keynes's original plan for a 'super-sovereign reserve currency' by the Chinese (Zhou 2009) suggests that it would be short-sighted to rule out international monetary reform as utopian.

Keynes wanted a multilateral institution that would contain and manage conflicting national interests – an international clearing house which would manage an international currency (the bancor). By creating an expansionist (and stable) international monetary system, Keynes's 'Clearing Union Plan' was designed to fulfil the same full-employment objectives as laid out in the *General Theory* but on a global scale (Costabile 2007). Keynes sought to do this by effectively modernising the international monetary system through the introduction of an international currency as credit money and using the principles of the banking system (Costabile 2009). These are the keys to his plan. International liquidity would be elastic and adjusted to economic need (trade etc.), regulated in the collective interest of member countries solving the problems of (1) international financial volatility, (2) the hoarding of international reserves and global deflationary tendencies, and (3) providing an automatic mechanism to resolve persistent balance of payments problems.

Keynes's original plan has been most consistently and clearly elaborated by Paul Davidson (2000, 2002, 2009) in his proposals for an International Clearing Union (ICU). The ICU would set up an institutionalised international monetary order which would manage global (im)balances through a rules-based system. But, within it, domestic governments would retain autonomy over monetary and fiscal policy and capital controls in order to control net international flows. The system would have at its centre an international reserve currency: the International Money Clearing Unit (IMCU). There would be one-way convertibility between the IMCU and domestic currencies, so all public and private international transactions would clear through it. Crucially, only central banks would be able to use IMCUs, ruling out the possibility of private speculation. Exchange rates would be stable: they would be fixed, but not unchanging, reflecting trade relations which relate to underlying production costs. As such, exchange rates would be set (and changed) to reflect permanent increases in efficient wages. This means that central banks would be assured that the purchasing power of their currency would remain constant, even in the situation of other governments permitting wage and price inflation in their economies.

The system would resolve international payments imbalances through the use of a closed, double-entry bookkeeping clearing institution. Balance would be the responsibility of surplus countries as much as debtor ones, i.e. China and Germany now, the US in the 1940s. Keynes always argued that the adjustment must come from the surplus countries, not the debtors. Davidson, following Keynes, argues that it is the surplus countries who are better placed to resolve persistent balance of payments problems. Keynes was certain that there had to be a built-in mechanism which ensured that surplus countries bore the burden of adjustment. So, in Davidson's proposal there is a series of automatic triggers which encourage creditor countries with accrued excessive surpluses to spend them. First, surplus countries would be required to expand domestic demand, reduce barriers to imports and FDI and increase lending to developing countries through the direct transfer of resources such as ODI, or pay surpluses into a reserve fund, or have their currency appreciated. If these actions were not forthcoming, then, as a last resort, the international clearing agency would be mandated to confiscate the surplus credits and redistribute them unilaterally to debtor members of the union. On the other hand, for debtor countries, bank money could be created as overdraft facilities, and individual members (countries) could overdraw based on their 'index quota', say, one half of their average trade for the previous five years. And in order to avoid imbalances, enforced quotas and surpluses would be charged subject to rising interest rates, and excess surpluses would be transferred to the union. A final – and important – element of Davidson's interpretation of Keynes's plan is that if governments do not wish to limit the free movement of capital then they are perfectly at liberty to set up a currency union with other like-minded countries.

Conclusions

This chapter has reflected on the contribution of the theoretical perspectives given the empirical chapters on aid, debt and bank lending, equity and FDI, remittances and microfinance. First, it became clear that the mainstream view – the UN MDG view as well as the World Bank – is characterised by an apolitical capital fundamentalism. Second, there is clearly a hierarchy of flows, and more interestingly there is an appropriate 'fit' between a country's financial development and the costs and benefits of different flows. Third, the benefits of capital mobility have been greatly oversold; there are other areas of policy that could and should take priority over attracting external flows of capital. Fourth, development can happen without external capital but happens best when countries regulate and actively manage financial inflows. Fifth, the mainstream has mistakenly viewed finance as just a flow, making it blind to the institutional and structural aspects of finance as a system. Sixth, the study of global finance has underplayed and missed the centrality of politics.

To conclude, I believe that the really productive work is to be done at the margins of the critical reformist position; so for example, a conversation with more critical liberal institutionalists such as Dani Rodrik on the one hand, but informed by the more critical politics of the likes of Susanne Soederberg and David Harvey on the other. There is also a need to strike a productive balance between keeping one eye on the global and structural elements of global finance and the international system more generally, but to remain sensitive to the specificities and context of countries and regions. Without understanding the institutional and political barriers and opportunities that exist in particular contexts it is no good having a blueprint for reform, no matter how progressive. But on the other hand, a descent into globally decontextualised national analysis is to miss the significant causes of poverty, underdevelopment and the structural barriers to achieving progressive reforms. One thing which is clear is that good governance is not a problem of the poor and developing countries, but a problem of the financial system. In the context of the global financial crisis, the distinction between Western finance, with its trappings of science and respectability, and the assumed riskiness of banking the poor, does not hold up to scrutiny (Patten *et al.* 2001; Arun and Hulme 2008). Analysis and reform must retain a global sensitivity.

It's a tough act. But it is in the contingencies and agency of politics that this can be done best. And it is specifically politics, as the messy and universal activity, and not the technocratic notion of governance, that future research needs to address. Politics is the missing part of the puzzle that allows us to understand the limits to institutional reform, understand how to bring about developmental change and avoid predatory states, and understand how struggle and resistance are central. The challenge for future political economy analysis of global finance and development is to bring together theoretically informed analysis of the empirical drivers and impact of capital on development outcomes alongside a keen sense of what is politically feasible and/or the way in which policy recommendations can be brought about through coalitions and political leadership.

Discussion questions

1 Which of the theoretical approaches is, in your view, most persuasive? Or which combination?

2 How would you advise the government of a developing country of your choice to proceed with financial liberalisation and regulation?

3 Is it desirable or feasible for countries to de-globalise?

4 In your view, is the fact that the systemic nature of finance is ignored in mainstream debates a matter of ideology, interests or something else?

5 Has the window for global reforms now passed? If so, what might open it again?

Further reading

- The edited collection by Jose Antonio Ocampo, Jan Kregel and Stephany Griffith-Jones (2007), *International Finance and Development*, offers a good critical take on reforming global finance, with an emphasis on the systemic issues that need addressing, from imbalances and IMF surveillance and conditionality to debt and FDI.

- Another good collection which is focused on global financial reform is Jomo Kwame Sundaram's (2011) *Reforming the International Financial System*. With chapters on Basel, cross-border tax evasion, food speculation and the IMF system of SDRs, from leading critical authors, it's well worth a look.

Useful websites

- The Robin Hood Tax. The campaign was launched in London, February 2010, and has since gone global. For more information and lots of resources see the Robin Hood Tax website. The website includes a useful timeline documenting key decisions and developments as well as a large repository of working papers and research pertaining to a financial transactions tax. Available at: http://robinhoodtax.org.uk/.

- Post-2015 High-Level Panel. On 12 July 2012 the UN Secretary-General Ban Ki-moon launched a High-Level Panel to advise and consult on the post-MDG development framework. The details and reports of the Panel can be found on its website: http://www.post2015hlp.org/. And an interesting partner website holds the 'MY World' global survey asking for your development priorities for a post-2015 world http://www.myworld2015.org/.

Bibliography

Abdelal, R. (2007) *Capital Rules: The Contruction of Global Finance*, Cambridge, MA: Harvard University Press.

Abouharb, M.R. and Cingranelli, D. (2007) *Human Rights and Structural Adjustment*, Cambridge: Cambridge University Press.

Abramovitz, M. (1956) 'Resource and Output Trends in the United States Since 1870', *American Economic Review*, 46 (2): 5–23.

Abramovitz, M. (1986) 'Catching Up, Forging Ahead, and Falling Behind', *Journal of Economic History*, 46 (2): 385–406.

Abramovitz, M. (1989) *Thinking About Growth And Other Essays on Economic Growth and Welfare*, Cambridge: Cambridge University Press.

Acemoglu, D., Johnson, S. and Robinson J.A. (2001) 'The Colonial Origins of Comparative Development: An Empirical Investigation', *American Economic Review*, 91 (5): 1369–1401.

Acemoglu, D. and Robinson, J.A. (2012), *Why Nations Fail*, London: Profile Books.

Adam, C. (2005), 'Exogenous Inflows and Real Exchange Rates: Theoretical Quirk or Empirical Reality?' paper presented at IMF Seminar on Foreign Aid and Macroeconomic Management, Maputo, Mozambique.

Adams, D.W., Graham, D.H. and von Pischke, J.D. (1984) *Undermining Rural Development with Cheap Credit*, Boulder, CO: Westview Press.

Adams, D., and Von Pischke, J.D. (1992) 'Microenterprise credit programs: déjà vu', *World Development*, 20 (10): 1463–1470.

Adams, R. and John P. (2005) 'Do International Migration and Remittances Reduce Poverty in Developing Countries?', *World Development*, 33: 1645–1669.

Adams, R.H. Jr. (2005) 'Remittances, Household Expenditure and Investment in Guatemala', Policy Research Working Paper Series, No. 3532, Washington, DC: World Bank.

Adelman, I. and Morris, C.T. (1965) 'Factor Analysis of the Interrelationship between Social and Political Variables and Per Capita Gross National Product', *Quarterly Journal of Economics*, 79 (4): 555–578.

African Development Bank (2002) *Achieving the Millennium Development Goals in Africa: Progress, Prospects and Policy Implications*. Global Poverty Report, African Development Bank, Abidjan.

Agarwal, B. (1994), *A Field of One's Own: Gender and Land Rights in South Asia*, Cambridge: Cambridge University Press.

Aggarwal, R., Erel, I., Stulz, R. and Williamson, R. (2009) 'Differences in Governance Practice Between U.S. and Foreign Firms: Measurement, Causes, and Consequences', *Review of Financial Studies*, 22: 3131–3169.

Agrawal, S. (2007) 'Emerging Donors in International Development: The India Case', Research Report Prepared for IDRC. Available at: www.idrc.ca/uploads/user-S/12441474461Case_of_India.pdf> (accessed 24 October 2010).

Aitken, B.J. and Harrison, A.E. (1999) 'Do Domestic Firms Benefit from Direct Foreign Investment? Evidence from Venezuela', *American Economic Review*, 89: 605–618.

Aizenman, J. and Lee, J. (2005) 'International Reserves: Precautionary vs. Mercantilist Views, Theory, and Evidence', *National Bureau of Economic Research Working Papers*, No. 11366, Cambridge, MA: NBER.

Aizenman, J. and Ito, H. (forthcoming) 'The "impossible trinity", the international monetary framework, and the Pacific Rim', in Kaur, I.N. and Singh, N. (eds) *Handbook of the Economics of the Pacific Rim*, Oxford: Oxford University Press.

Alesina, A. and Dollar, D. (2000) 'Who Gives Foreign Aid to Whom and Why?', *Journal of Economic Growth*, 5 (1): 33–63.

Alfaro, L., Chanda, A., Kalemli-Ozcan, S. and Sayek, S. (2004) 'FDI and economic growth: the role of local financial markets', *Journal of International Economics*, 64 (1): 89–112.

Alkire, S. and Santos, M.E. (2010) 'Acute Multidimensional Poverty: A New Index for Developing Countries', OPHI Working Paper 38, Oxford: Oxford Poverty and Human Development Initiative.

Allison, G.T. (1971) *The Essence of Decision: Explaining the Cuban Missile Crisis*, Glenview, IL: Scott Foresman.

Amann, E., Aslanidis, N., Nixson, F. and Walters, B. (2006) 'Economic Growth and Poverty Alleviation: A Reconsideration of Dollar and Kraay', *The European Journal of Development Research*, 18 (1): 22–44.

Amin, S. (1974) *Accumulation on a World Scale: Critique of the Theory of Underdevelopment*, London: Monthly Review Press.

Amin, S. (1990) *Delinking*, London: Zed Books.

Amsden, A.H. (1989) *Asia's Next Giant: South Korea and Late Industrialization*, Oxford: Oxford University Press.

Amsden, A.H. (2001) *The Rise of 'the Rest': Challenges to the West from Late-Industrializing Economies*, Oxford: Oxford University Press.

Anderson, K. and Martin, W. (2005) 'Agricultural Trade Reform and the Doha Development Agenda', *World Economy*, 28 (9): 1301–1327.

Anderson, K., Francois, J., Hertel, T., Hoekman, B., and Martin, W. (2000) 'Potential Gains from Trade Reform in the New Millennium', paper presented at the Third Annual Conference on Global Economic Analysis, held at Monash University, June 27–30.

Anderson, S. and Baland, J.-M. (2002) 'The Economics of Roscas and intra-household resource allocation', *Quarterly Journal of Economics*, 117: 963–995.

Andrews, D.M. (ed.) (2006) *International Monetary Power*, Ithaca, NY: Cornell University Press.

Annan, K. (2006). 'Address of Mr. Kofi Annan, Secretary-General, to the High-Level Dialogue of the United Nations General Assembly on International Migration and Development', New York, 14 September, *International Migration Review*, 963–965.

APPG DAT (2010) *A Parliamentary Inquiry into Aid Effectiveness*, London: Houses of Parliament, All Party Parliamentary Group for Debt, Aid & Trade.

Ariyoshi, A., Habermeier, K., Laurens, B., Ötker-Robe, I., Canales-Kriljenko, J.I. and Kirilenko, A. (2000) 'Capital Controls: Country Experiences with Their Use and Liberalization', IMF Occasional Paper 190, Washington, DC: International Monetary Fund.

Armendáriz, B. and Morduch, J. (2000) 'Microfinance Beyond Group Lending', *The Economics of Transition*, 8 (2): 401–420.

Armendáriz, B. and Morduch, J. (2010) *The Economics of Microfinance*, 2nd edn, Cambridge, MA: MIT Press.

Arrow, K.J. (1962) 'The Economic Implications of Learning by Doing', *The Review of Economic Studies*, 29 (3): 155–173.

Arrow, K.J. and Debreu, G. (1954) 'Existence of an Equilibrium for a Competitive Economy', *Econometrica*, 22 (3): 265–269.

Arun, T. (2005) 'Regulating for Development: the Case of Microfinance', *The Quarterly Review of Economics and Finance*, 45: 346–357.

Arun, T. and Hulme, D. (2008) 'Microfinance – A Way Forward', BWPI Working Paper 54, available at: www.bwpi.manchester.ac.uk/resources/Working-Papers/bwpi-wp-5408.pdf (accessed 11 August 2010).

Aryeetey, E. and Steel, W.F. (1995) 'Savings Collectors and Financial Intermediation in Ghana', *Savings and Development*, 19 (2): 191–211.

Ashraf, N., Karlan, D. and Yin, W. (2010) 'Female Empowerment: Further Evidence from a Commitment Savings Product in the Philippines', *World Development*, 28 (3): 333–344.

Attanasio, O.P. and Székely, M. (2000) 'Household Saving in Developing Countries – Inequality, Demographics and All That: How Different are Latin America and South East Asia?', Working Paper 427, Washington, DC: Inter-American Development Bank.

Azariadis, C. and Stachurski, J. (2005) 'Poverty Traps', in Aghion, P. and Durlauf, S.N. (eds) *Handbook of Economic Growth Volume 1A*, Oxford: North-Holland.

Backhouse, R.E. and Bateman, B.W. (2011) *Capitalist Revolutionary: John Maynard Keynes*, Cambridge, MA: Harvard University Press.

Baird, V. (2005) 'Out of Africa: A Migrant's Story', *New Internationalist*, 379, 1 June, available at: www.newint.org/issues/2005/06/01/ (accessed 3 September 2010).

Baker, A. (2000) 'The G7 as a Global Ginger Group: Plurilateralism and Four-Dimensional Diplomacy', *Global Governance*, 6 (2): 165–189.

Baker, A., Hudson, D. and Woodward, R. (eds) (2005) *Governing Financial Globalization: IPE and Multi-Level Governance*, London: Routledge.

Balasubramanyam, V.N., Salisu, M. and Sapsford, D. (1996) 'Foreign Direct Investment and Growth in EP and IS Countries', *The Economic Journal*, 92–105.

Balasubramanyam, V. N. Salisu, M. and Sapsford, D. (1999) Foreign Direct Investment as an Engine of Growth, *Journal of International Trade & Economic Development*, 8 (1): 27–40.

Bandiera, O., Caprio, G., Honohan, P. and Schiantarelli, F. (2000) 'Does Financial Reform Raise or Reduce Saving?', *Review of Economics and Statistics*, 82 (2): 239–263.

Banerjee, A.V. and Duflo, E. (2007) 'The Economic Lives of the Poor', *Journal of Economic Perspectives*, 21: 141–167.

Banerjee, A. V. and Duflo, E. (2011) *Poor Economics: A Radical Rethinking of the way to Fight Global Poverty*, London: Penguin.

Banerjee, A.V. and Newman, A.F. (1993) 'Occupational Choice and the Process of Development', *Journal of Political Economy*, 101 (2): 274–298.

Banerjee, A.V., Duflo, E., Glennerster, R. and Kinnan, C. (2009) 'The Miracle of Microfinance? Evidence from a Randomized Evaluation', MIT Department of Economics and Abdul Latif Jameel Poverty Action Lab (J-PAL) working paper.

Banerjee, K. (2010) 'Interview with Dilip Ratha', *New Global Indian*, January, available at: https://blogs.worldbank.org/files/peoplemove/Dilip-NGI_Jan10.pdf (accessed 13 March 2012).

Bank of China Online (2010) available at: www.pbc.gov.cn/english/detail.asp?col=6500&id=178 (accessed 9 May 2010).

Barajas, A. Chami, R., Fullenkamp, C. and Garg, A. (2010) 'The Global Financial Crisis and Workers' Remittances to Africa: What's the Damage?', IMF Working Paper 10/24, Washington DC: International Monetary Fund.

Baran, P.A. ([1957] 1973) *The Political Economy of Growth*, Harmondsworth, Penguin Books, p. 402 (originally published by Monthly Review Press).

Baran, P.A. and Sweezy, P.M. (1966) *Monopoly Capital: An Essay on the American Economic and Social Order*, New York: Monthly Review Press.

Barder, O. (2005a) 'Reforming Development Assistance: Lessons from the UK Experience', Center for Global Development Working Paper 70.

Barder, O. (2005b) 'Technical Shortcomings with the IMF Analysis', 30 June, available at: www.owen. org/blog/194 (accessed 25 July 2010).

Barder, O. (2009) 'Review of "Dead Aid" by Dambiso Moyo', 31 March, available at: www.owen.org/ wp-content/uploads/review-of-dead-aid.pdf (accessed 26 July 2010).

Bardhan, P. (2005) 'Institutions Matter, but Which Ones?', *Economics of Transition*, 13: 499–532.

Bardhan, P.K. (2005) 'History, Institutions, and Underdevelopment', in Bardhan, P.K. *Scarcity, Conflicts, and Cooperation: Essays in the Political and Institutional Economics of Development*, Cambridge, MA: MIT Press.

Barker, A., Parker, B. and Lamont, J. (2011) 'UK to Give £1bn to India in Spite of Cuts', *Financial Times*, 13 February.

Barth, J.R., Caprio, G. Jr and Levine, R. (2001) 'Banking Systems Around the Globe: Do Regulations and Ownership Affect Performance and Stability?', in Mishkin, F.S. (ed.) *Prudential Supervision: What Works and What Doesn't*, Chicago: University of Chicago Press, pp. 31–96.

Bateman, M. (2010) *Why Doesn't Microfinance Work? The Destructive Rise of Local Neoliberalism*, London: Zed.

Bateman, M. (2011) (ed.) *Confronting Microfinance: Undermining Sustainable Development*, Sterling, VA: Kumarian Press.

Bateman, M. and Chang, H. (2009) 'The Microfinance Illusion', available at: www.microfinancegateway. org/p/site/m/template.rc/1.9.40987/ (accessed 16 March 2012).

Bauer, P. T. (1971) 'Economic History as Theory', *Economica*, 38: 163–179.

Bauer, P.T. (1981) *Equality, the Third World, and Economic Delusion*, Cambridge, MA: Harvard University Press.

Becerra, O., Cavallo, E. and Scartascini, C. (2012) 'The Politics of Financial Development: The Role of Interest Groups and Government Capabilities', *Journal of Banking & Finance*, 36 (3): 626–643.

Beck, T., Demirgüç-Kunt, A. and Levine, R. (2001) 'Legal Theories of Financial Development', *Oxford Review of Economic Policy*, 17 (4): 483–501.

Beck, T., Demirgüç-Kunt, A., Laeven, L. and Levine, R. (2005) 'Finance, Firm Size and Growth', *World Bank Policy Research Working Paper*, No. 3485, Washington, DC: World Bank.

Beck, T., Demirgüç-Kunt, A., Laeven, L. and Maksimovic, V. (2006) 'The Determinants of Financing Obstacles', *Journal of International Money and Finance*, 25: 932–952.

Beck, T., Demirgüç-Kunt, A. and Levine, R. (2007) 'Finance, Inequality, and the Poor', *Journal of Economic Growth*, 12 (1): 27–49.

Beer, L. (1999) 'Income inequality and Transnational Corporate Penetration', *Journal of World Systems Research*, 5 (1): 1–25.

Beer, L. and Boswell, T. (2002) 'The Resilience of Dependency Effects in Explaining Income Inequality in the Global Economy: A Cross-national analysis, 1975–1995', *Journal of World-systems Research*, 8 (1): 30–59.

Beeson, M. (2004) 'The Rise and Fall (?) of the Developmental State: The Vicissitudes and Implications of East Asian Interventionism', in Low, L. (ed.) *Developmental States: Relevant, Redundant or Reconfigured?*, New York: Nova Science Publishers.

Bekaert, G. (1995) 'Market Integration and Investment Barriers in Emerging Equity Markets', *World Bank Economic Review*, 9 (1): 75–107.

Bello, W., Bullard, N. Malhotra, K. (eds) (2000) *Global Finance: New Thinking on Regulating Speculative Capital Markets*, London: Zed.

Benería, L. (1999) 'The Enduring Debate Over Unpaid Labour', *International Labour Review*, 138: 287–309.

Benhabib, J. and Spiegel, M.M. (2000) 'The Role of Financial Development in Growth and Investment', *Journal of Economic Growth*, 5: 341–360.

Benn, J., Rogerson, A. and Steensen, S. (2010) 'Getting Closer to the Core –Measuring Country Programmable Aid', *OECD Development Brief Issue 1 2010*, Paris: OECD, available at: www.oecd.org/dataoecd/32/51/45564447.pdf (accessed 26 July 2010).

Berger, P. and Dore, S. (1996) *National Diversity and Global Capitalism*, London: Cornell University Press.

Bernanke, B. (2005) The Global Saving Glut and the U.S. Current Account Deficit. Speech at the Sandridge Lecture, Virginia Association of Economics, 10 March, Richmond, VA.

Betts, H. (2004) 'Interview: Noreena Hertz', *The Times*, 18 September.

Bhagwati, J. (1998) 'The Capital Myth. The Difference between Trade in Widgets and Dollars', *Foreign Affairs*, 77 (3): 7–12.

Bhattacharyya, S. (2011) *Growth Miracles and Growth Debacles: Exploring Root Causes*, Cheltenham: Edward Elgar.

Bibow, J. (2009) *Keynes on Monetary Policy, Finance and Uncertainty: Liquidity Preference Theory and the Global Financial Crisis*, London: Routledge.

Bidwai, P. (2011) 'Why India Needs Aid', *The Guardian*, 7 February 2012, available at: www.guardian.co.uk/commentisfree/2012/feb/07/why-india-needs-aid (accessed 20 March 2012).

Bird, G. and Rajan, R. (2003) 'Too Much of a Good Thing? The Adequacy of International Reserves in the Aftermath of Crises', *World Economy*, 86: 873–879.

Birdsall, N. and de la Torre, A. with Menezes, R. (2001) *Washington Contentious: Economic Policies for Social Equity in Latin America*, Washington, DC: Carnegie Endowment for International Peace and Inter-American Dialogue.

BIS (2010) *Triennial Central Bank Survey Foreign Exchange and Derivatives Market Activity in April 2010*, Basel: Bank of International Settlements.

Black, R.E., Allen, L.H., Bhutta, Z.A., Caulfield, L.E., de Onis, M. Ezzati, M., Mathers, C. and Rivera, J. (2008) 'Maternal and Child Undernutrition: Global and Regional Exposures and Health Consequences', *The Lancet*, 371 (9608): 243–260.

Blaug, M. (1997/1962) *Economic Theory in Retrospect*, Cambridge: Cambridge University Press.

Block, F.L. (1977) *The Origins of International Economic Disorder: A Study of United States International Monetary Policy From World War II to the Present*, Berkeley, CA: University of California Press.

Blomstrom, M., Lipsey, R.E. and Zejan, M. (1994) 'What Explains Developing Country Growth?', *NBER Working Papers*, 4132, Cambridge, MA: NBER.

Bloomberg (2007) 'Yunus Blasts Compartamos', *Bloomberg Businessweek Magazine*, 12 December, avalable at: www.businessweek.com/stories/2007-12-12/online-extra-yunus-blasts-compartamos (accessed 1 May 2013).

Boone, P. (1996) 'Politics and the Effectiveness of Foreign Aid', *European Economic Review*, 40 (2): 289–29.

Bordo, M., Eichengreen, B., Klingebiel, D. and Soledad Martinez-Peria, M. (2001) 'Is the Crisis Problem Growing More Severe?', *Economic Policy*, 32.

Bordo, M.D. and Eichengreen, B. (eds) (1993) *A Retrospective on the Bretton Woods System: Lessons for International Monetary Reform*, Chicago IL: University of Chicago Press.

Borensztein, E., De Gregorio, J. and Lee, J-W. (1998) 'How Does Foreign Direct Investment Affect Economic Growth?', *Journal of International Economics*, 45 (1): 115–135.

Bornschier, V. (1980) 'Multinational Corporations and Economic Growth: A Cross-National Test of the Decapitalization Thesis', *Journal of Development Economics*, 7: 191–10.

Bornschier, V. and Chase-Dunn, C. (1985) *Transnational Corporations and Underdevelopment*, New York: Praeger.

Boserup, E. (1970) *Women's Role in Economic Development*, London: George Allen & Unwin.

Bosworth, B. and Collins, S.M. (1999) 'Capital Flows to Developing Economies: Implications for Saving and Investment', *Brookings Papers on Economic Activity*, 1: 143–169.

Bouab, A.H. (2004) 'Financing for Development, the Monterrey Consensus: Achievements and Prospects', *Michigan Journal of International Law*, 26: 359–369.

Bovard, J. (1986) 'The Continuing Failure of Foreign Aid', *Cato Policy Analysis*, 65, available at: www.cato.org/pubs/pas/pa065.html> (accessed 5 February 2012).

Box, G.E.P. (1979) 'Robustness in the Strategy of Scientific Model Building', in *Robustness in Statistics*, New York: Academic Press, pp. 202–236.

Boyer, R. and Drache, D. (eds) (1996) *States Against Markets: The Limits of Globalization*, London: Routledge.

BRAC (2010) 'Meet BRAC Uganda's Model Farmers', available at: www.brac.net/content/meet-brac-uganda's-model-farmers (accessed 11 August 2010).

Brautigum, D. (2009) *The Dragon's Gift: The Real Story of China in Africa*, Oxford: Oxford University Press.

Brautigam, D.A. and Knack, S. (2004) 'Foreign Aid, Institutions, and Governance in Sub-Saharan Africa', *Economic Development & Cultural Change*, 52 (2): 255–285.

Brenner, R. (1977), 'The Origins of Capitalist Development: A Critique of Neo-Smithian Marxism', *New Left Review*, 104: 25–92.

Brewer, A. (1990) *Marxist Theories of Imperialism: A Critical Survey*, London: Routledge.

Brossard, M. and Gacougnolle, L.C. (2001) *Financing Primary Education for All: Yesterday, Today and Tomorrow*, Paris: UNESCO.

Brown, Foster, Norton and Fozzard (2001) 'The Status of Sector Wide Approaches', ODI Working Paper 142, January.

Browne, S. (2007) 'Target the MDGs – Not Aid Amounts', in Ehrenpreis, D. (ed.) *Does Aid Work – for the MDGs, Poverty in Focus*, Brasilia: International Poverty Centre (IPC).

Bruns, B., Mingat, A. and Rakotomalala, R. (2003) *Achieving Universal Primary Education by 2015 – A Chance for Every Child*, Washington, DC: World Bank.

Bulir, A. and Lane, T.D. (2002) 'Aid and Fiscal Management', IMF Working Paper, No. 02/112, Washington, DC: International Monetary Fund.

Bullard, N. (1998) 'Miracle to Meltdown', *New Internationalist*, 306.

Bunting, M. (2011) 'Is India ready to refuse UK aid?', *The Guardian*, 10 January.

Burgess, R. and Pande, R. (2005) 'Do Rural Banks Matter? Evidence from the Indian Social Banking Experiment', *American Economic Review*, 95 (3): 780–795.

Burnside, C. and Dollar, D. (2000) 'Aid, Policies, and Growth', *American Economic Review*, 90 (4): 847–868.

Busch, A. (2000) 'Unpacking the Globalisation Debate: Approaches, Evidence and Data', in Hay, C. and Marsh, D. (eds) *Demystifying Globalisation*, London: Macmillan.

Calderisi, R. (2006) *The Trouble with Africa: Why Foreign Aid Isn't Working*, Houndmills: Palgrave Macmillan.

Calvo, G.A., Leiderman, L. and Reinhart, C.M. (1996) 'Inflows of Capital to the Developing Countries in the 1990s', *Journal of Economic Perspectives*, 10 (2): 123–139.

Caprio, G. and Klingebiel, D. (2002) 'Episodes of Systemic and Borderline Banking Crises', Managing the Real and Fiscal Effects of Banking Crises, World Bank discussion paper 428, Washington, DC: World Bank.

Card, D., Dustmann, C. and Preston, I. (2012) 'Immigration, Wages, and Compositional Amenities', *Journal of the European Economic Association*, 10 (1): 78–119.

Cardoso, F.H. (1972) 'Dependent Capitalist Development in Latin America', *New Left Review*, 1 (74).

Carkovic, M. and Levine, R. (2000) 'Does Foreign Direct Investment Accelerate Economic Growth?', University of Minnesota Department of Finance working paper.

Carrier, J.G. and Miller, D. (eds) (1998) *Virtualism: A New Political Economy*, Oxford: Berg.

Caselli, F. and Feyrer, J. (2007) 'The Marginal Product of Capital', *The Quarterly Journal of Economics*, 122: 535–568.

Cassen, R. *et al.* (1986) *Does Aid Work?: Report to an Intergovernmental Task Force*, Oxford: Clarendon Press.

Cater, N. (2005) 'The Global Remittance Rip-off', *The Guardian*, 31 March.

Cerny, P. (1994) 'The Dynamics of Financial Globalization: Technology, Market Structure, and Policy Response', *Policy Sciences*, 27 (4): 319–342.

Cerny, P. (1997) 'Paradoxes of the Competition State: The Dynamics of Political Globalization', *Government and Opposition*, 32: 251–274.

CGAP (2004) 'The Key Principles of Microfinance', available at: http://www.cgap.org/gm/document-1.9.2747/KeyPrincMicrofinance_CG_eng.pdf (accessed 11 August 2010).

CGAP (2006) *Good Practice Guidelines for Funders of Microfinance*, 2nd edn, Washington, DC: Consultative Group to Assist the Poor/The World Bank.

Chambers, R. (1997) *Whose Reality Counts? Putting the First Last*, London: Intermediate Technology Publications.

Chandler, D. (2003) 'Rhetoric Without Responsibility: The Attraction of "Ethical" Foreign Policy', *British Journal of Politics and International Relations*, 5: 295–316.

Chang, H-J. (1994) *The Political Economy of Industrial Policy*, London: Macmillan.

Chang, H-J. (1999) 'The Economic Theory of the Developmental State', in Woo-Cumings, M. (ed.) *The Developmental State*, Ithaca, NY: Cornell University Press.

Chang, H-J. (2002), *Kicking Away the Ladder: Development Strategy in Historical Perspective*, London: Anthem Press.

Chang, H-J. (2003) 'The Market, the State and Institutions in Economic Development', in Chang, H-J. (ed.) *Rethinking Development Economics*, London: Anthem Press.

Chang, H-J. (2007) *Bad Samaritans: The Guilty Secrets of Rich Nations and the Threat to Global Prosperity*, London: Random House.

Chang, H-J. and Grabel, I. (2004) *Reclaiming Development: An Alternative Economic Policy Manual*, London: Zed.

Chase-Dunn, C. (1975) 'The Effects of International Economic Dependence on Development and Inequality: A Cross-National Study', *American Sociological Review*, 40 (6): 720–738.

Chauvet, L. and Guillaumont, P. (2004) 'Aid and Growth Revisited: Policy, Economic Vulnerability and Political Instability', in Tungodden, B., Stern, N. and Kolstad, I. (eds) *Toward Pro-Poor Policies – Aid, Institutions and Globalisation*, World Bank/Oxford University Press, pp. 95–109.

Chen, S. and Ravallion, M. (2008) 'The Developing World is Poorer than we Thought, But No Less Successful in the Fight against Poverty', Policy Research Working Paper 4703, Washington, DC: World Bank.

Chen, S. and Ravallion, M. (2012) 'More Relatively-poor People in a Less Absolutely-poor World', Policy Research Working Paper 6114, Washington, DC: World Bank.

Chenery, H.B. (1960) 'Patterns of Industrial Growth', *American Economic Review*, 50: 624–654.

Chenery, H.B. and Strout, A.M. (1966) 'Foreign Assistance and Economic Development', *American Economic Review*, 56, (4): 679–733.

Chwieroth, J. (2007) 'Neoliberal Economists and Capital Account Liberalization in Emerging Markets', *International Organization*, 61 (2): 443–463.

Chwieroth, J. (2009) *Capital Ideas: The IMF and the Rise of Financial Liberalization*, Princeton, NJ: Princeton University Press.

Claessens, S. (1995) 'The Emergence of Equity Investment in Developing Countries: Overview', *World Bank Economic Review*, 9 (1): 1–17.

Claessens, S., Dasgupta, S. and Glen, J. (1995) 'Return Behavior in Emerging Stock Markets', *World Bank Economic Review*, 9 (1): 131–152.

Clarke, G., Cull, R. and Martinez Peria, S.M. (2001a) 'Does Foreign Bank Penetration Reduce Access to Credit in Developing Countries? Evidence from Asking Borrowers', World Bank Policy Research Working Paper, No. 2716, Washington, DC: World Bank.

Clarke, G., Cull, R., Martinez Peria, M.S. and Sanchez, S.M. (2001b) 'Foreign Bank Entry: Experience, Implications for Developing Countries and Agenda for Further Research', World Bank Policy Research Working Paper, No. 2698, Washington, DC: World Bank.

Clarke, G., Cull, R., Martinez-Peria, M.S. and Sanchez, S.M. (2002) 'Bank Lending to Small Businesses in Latin America: Does Bank Origin Matter?', Policy Research Working Paper Series, No. 2760, Washington, DC: World Bank.

Clay, E. J., Geddes, M. and Natali, L. (2009) *Untying Aid: Is it Working? An Evaluation of the Implementation of the Paris Declaration and of the 2001 DAC Recommendation of Untying ODA to the LDCs*, Copenhagen: Danish Institute for International Studies.

Cleaver, H. (1989) 'Close the IMF, Abolish Debt and End Development: a Class Analysis of the International Debt Crisis', *Capital & Class*, 13 (3): 17–50.

Clemens, M.A. (2011) 'Economics and Emigration: Trillion-Dollar Bills on the Sidewalk?' *Journal of Economic Perspectives*, 25 (3): 83–106.

Clemens, M.A. and Moss, T.J. (2005) 'Ghost of 0.7%: Origins and Relevance of the International Aid Target', Center for Global Development Working Paper 68.

Clemens, M.A., Radelet, S. and Bhavnani, R. (2004) 'Counting Chickens when they Hatch: The Short Term Effect of Aid on Growth', Center for Global Development Working Paper, No. 44, Washington, DC: CGDEV.

Clemens, M.A., Kenny, C.J. and Moss, T.J. (2007) 'The Trouble with the MDGs: Confronting Expectations of Aid and Development Success', *World Development*, 35 (5): 735–751.

Clements, B., Bhattacharya, R. and Nguyen, T.Q. (2003) 'External Debt, Public Investment, and Growth in Low-Income Countries', IMF Working Paper 03/249, Washington, DC: International Monetary Fund.

Clift, J. (2003) 'Beyond the Washington Consensus', *Finance and Development*, 40 (3): 9.

Coase, R. (1937) 'The Nature of the Firm', *Economica*, 4: 386–405.

Coase, R. (1960) 'The Problem of Social Cost', *Journal of Law and Economics*, 3: 1–44.

Coase, R. (1998) 'The New Institutional Economics', *American Economic Review*, 88 (2).

Cohen, B.J. (1977) *Organizing the World's Money: The Political Economy of International Monetary Relations*, New York: Basic Books.

Cohen, B.J. (1998) *The Geography of Money*, Ithaca, NY: Cornell University Press.

Cohen, B.J. (2002) 'Bretton Woods System', in Jones, R.J.B. (ed.) *Routledge Encyclopaedia of International Political Economy*, London: Routledge.

Cohen, B.J. (2008) *Global Monetary Governance*, London: Routledge.

Cohen, D. (1993) 'Convergence in the Closed and in the Open Economy', in Giovannini, A. (ed.) *Finance and Development: Issues and Experience*, Cambridge: Cambridge University Press.

Coleman, B. (1999) 'The Impact of Group Lending in Northeast Thailand', *Journal of Development Economics*, 60: 105–141.

Coleman, J. (2010) *Employment of Foreign Workers: 2007–2009*, Office for National Statistics, available at: /www.ons.gov.uk/ons/rel/lmac/employment-of-foreign-workers/employment-of-foreign-workers--2007-2009/index.html.

Collier, P. (2007) *The Bottom Billion*, London: Penguin.

Collier, P. and Dollar, D. (2001) 'Can the World Cut Poverty in Half? How Policy Reform and Effective Aid Can Meet International Development Goals', *World Development*, 29 (11): 1787–1802.

Collier, P. and Dollar, D. (2002) 'Aid Allocation and Poverty Reduction', *European Economic Review*, 46 (8): 1475–1500.

Collier, P. and Dollar. D. (2004) 'Development Effectiveness: What Have we Learnt?', *Economic Journal*, 114: 244–271.

Collins, D., Morduch, J., Rutherford, S. and Ruthven, O. (2009) *Portfolios of the Poor: How the World's Poor Live on $2 a Day*, Princeton, NJ: Princeton University Press.

Collins, S.M. and Bosworth, B. (2003) 'The Empirics of Growth: An Update', *Brookings Papers on Economic Activity*, 2: 113–179.

Commission on the Private Sector & Development (2004) *Unleashing Entrepreneurship: Making Business Work for the Poor*, New York: United Nations Development Programme.

Commission on Growth and Development (2008) *The Growth Report: Strategies for Sustained Growth and Inclusive Development*, Washington, DC: World Bank.

Confino, J. (2012) 'Children's Rights Take Centre Stage', *The Guardian*, 12 March, available at: www.guardian.co.uk/sustainable-business/childrens-rights-business-principles-guidance> (accessed 18 March 2012).

Conning, J. (1999) 'Outreach, Sustainability and Leverage in Monitored and Peer-monitored Lending', *Journal of Development Economics*, 60 (1): 51–77.

Cooksey, B. (2003) 'Marketing Reform? The Rise and Fall of Agricultural Liberalisation in Tanzania', *Development Policy Review*, 21 (1): 67–91.

Costabile, L. (2007) 'Current Global Imbalances and the Keynes Plan', Political Economy Research Institute Working Paper Series, No. 156, Amherst, MA: University of Massachusetts.

Costabile, L. (2009) 'Current Global Imbalances and the Keynes Plan: A Keynesian Approach for Reforming the International Monetary System', *Structural Change and Economic Dynamics*, 20 (2): 79–89.

Counts, A. (2008) 'Reimagining Microfinance', *Stanford Social Innovation Review*, Summer, 46–53.

Cowen, M.P. and Shenton, R.W. (1996) *Doctrines of Development*, London: Routledge.

Cox, R.W. (1981) 'Social Forces, States and World Orders: Beyond International Relations Theory', *Millennium: Journal of International Studies*, 10 (2): 126–155.

Cox-Edwards, A. and Ureta, M. (2003) 'International Migration, Remittances and Schooling: Evidence from El Salvador', *Journal of Development Economics*, 72 (2): 429–461.

CPSS/World Bank (2007) *General Principles for International Remittance Services*, Basel: Bank for International Settlements and The World Bank, available at: www.bis.org/publ/cpss76.pdf (accessed 3 September 2010).

Cruz, M. and Walters, B. (2008) 'Is the Accumulation of International Reserves Good for Development?', *Cambridge Journal of Economics*, 32: 665–681.

Cull, R., Demirgüç-Kunt, A. and Morduch, J. (2007) 'Financial Performance and Outreach: A Global Analysis of Leading Microbanks', *Economic Journal*, 117 (517): F107–F133.

Cull, R., Demirgüç-Kunt, A. and Morduch, J. (2009) 'Microfinance Meets the Market', *Journal of Economic Perspectives*, 23: 167–192.

Cutler, C.A., Haufler, V. and Porter, T. (eds) (1999) *Private Authority and International Affairs*, Albany, NY: State University of New York Press.

Cypher, J.M. and Dietz, J.L. (2009) *The Process of Economic Development*, 3rd edn, London: Routledge.

D'Espallier, B., Guérin, I. and Mersland, R. (2011) 'Women and Repayment in Microfinance: A Global Analysis', *World Development*, 39 (5): 758–772.

Dafe, F. (2011) 'The Potential of Pro-Market Activism as a Tool for Making Finance Work for Africa: A Political Economy Perspective', DIE discussion paper, 2/2011, Bonn: Deutsches Institut für Entwicklungspolitik.

Daley-Harris (2009) *State of the Microcredit Summit Campaign Report 2009*, Washington, DC: Microcredit Summit Campaign.

Darnton, A. and Kirk, M. (2011) *Finding Frames: New Ways to Engage the Public in Global Poverty*, London: BOND.

Das, N.C. and Misha, F.A. (2010) 'Addressing Extreme Poverty in a Sustainable Manner: Evidence from CFPR Programme', CFPR Working paper No. 19, Dhaka: BRAC.

Dasandi, N. (2009) 'Poverty Reductionism: The Exclusion of History, Politics, and Global Factors from Mainstream Poverty Analysis', IPEG 'Papers in Global Political Economy Series', available at: www.bisa-ipeg.org/papers/39_dasandi.pdf (accessed 26 March 2012).

Davidson, P. (1972) 'Money and the Real World', *The Economic Journal*, 82 (325): 101–115.

Davidson, P. (1992) 'Reforming the World's Money', *Journal of Post Keynesian Economics*, 15 (2): 153–179.

Davidson, P. (2000) 'Is a Plumber or a New Financial Architect Needed to End Global International Liquidity Problems?', *World Development*, 28 (6): 1117–1131.

Davidson, P. (2002) *Financial Markets, Money and the Real World*, Cheltenham: Edward Elgar.

Davidson, P. (2007) *John Maynard Keynes*, Basingstoke: Palgrave Macmillan.

Davidson, P. (2009) *The Keynes Solution: The Path to Global Economic Prosperity*, Basingstoke: Palgrave Macmillan.

Davies, G. (1994) *A History of Money: From Ancient Times to the Present Day*, Cardiff: University of Wales Press.

Davies, M. and Ryner, R. (eds) (2006) *Poverty and the Production of World Politics: Unprotected Workers in the Global Economy*, Basingstoke: Palgrave Macmillan.

De Haas, H. (2007) 'Turning the Tide? Why Development will not Stop Migration', *Development and Change*, 38 (5): 819–841.

De Haas, H. (2010) 'Migration and Development: A Theoretical Perspective', *International Migration Review*, 44 (1): 227–264.

de Onis, M., Blossne, M. and Borghi, E. (2011) 'Prevalence of Stunting Among Pre-school Children 1990–2020', *Public Health Nutrition*, 14: 1–7.

De Parle, J. (2008) 'World Banker and His Cash Return Home', *New York Times*, 17 March, available at: www.nytimes.com/2008/03/17/world/asia/17remit.html?_r=1&ref=world_bank&pagewanted=all (accessed 3 September 2010).

Deans, F., Lonnqvist, L. and Sen, K. (2006) 'Remittances and Migration: Some Policy Considerations for NGOs', *INTRAC Policy Briefing 8*, Oxford: International NGO Training and Research Centre.

Delamonica, E., Mehrotra, S. and Vandemoortele, J. (2001) 'Is EFA Affordable? Estimating the Global Minimum Cost of "Education for All"', Innocenti Working Paper, No. 87, Florence: UNICEF Innocenti Research Centre.

Demirgüç-Kunt, A. and Detragiache, E. (1998) 'The Determinants of Banking Crises: Evidence from Developing and Developed Countries', IMF Staff Papers, 45: 81–109.

Demirgüç-Kunt, A. and Levine, R. (2008) 'Finance and Economic Opportunity', Policy Research Working Paper 4468, Washington, DC: World Bank.

Demirgüç-Kunt, A., Beck, T. and Honohan, P. (2008) *Finance for All?: Policies and Pitfalls in Expanding Access*, World Bank Policy Research Report, Washington, DC: World Bank.

Depetris Chauvin, N. and Kraay, A. (2005) 'What has 100 Billion Dollars Worth of Debt Relief Done for Low-income Countries?', available at SSRN 818504.

Der Spiegel (2001) 'They are Misusing My Name', interview with James Tobin, 2 September.

Dercon, S. (2004) 'Risk, Insurance, and Poverty: A Review', in Dercon, S. (ed.) *Insurance Against Poverty*, Oxford: Oxford University Press.

Dercon, S. and Hoddinott, J. (2004) 'Health, Shocks, and Poverty Persistence', in Dercon, S. (ed.) *Insurance Against Poverty*, Oxford: Oxford University Press.

Deshingkar, P. and Aheeyar, M.M.M. (2006) 'Remittances in Crisis: Sri Lanka After the Tsunami', HPG Background Paper, London: Overseas Development Institute.

Dessus, S., Fukasaku, K. and Safadi, R. (1999) 'Multilateral Tariff Liberalisation and the Developing Countries', OECD Development Centre Policy Brief, No. 18, Paris: OECD.

Detragiache, E. and Spilimbergo, A. (2002) 'Empirical Models of Short-term Debt and Crises: Do they Test the Creditor Run Hypothesis?', IMF Working Paper, No. 01/2, Washington, DC: International Monetary Fund.

Devarajan, S., Miller, M.J. and Swanson, E.V. (2002) 'Goals for Development: History, Prospects, and Costs', World Bank Policy Research Working Paper, No. 2819, Washington, DC: World Bank.

Devlin, R., Ffrench-Davis, R. and Griffith-Jones, S. (1994) 'Surges in Capital Flows and Development: An Overview of Policy Issues', in Ffrench-Davis, R. and Griffith-Jones, S. (eds) *Coping with Capital Surges: Latin American Macroeconomics and Investment*, London: Lynne Rienner.

DFID (2011) 'India Operational Plan 2011–2012', London: DFID, available at: www.dfid.gov.uk/Documents/publications1/op/india-2011.pdf (accessed 20 March 2012).

Diamond, D. (1984) 'Financial Intermediation and Delegated Monitoring', *The Review of Economic Studies*, 51 (3): 393–414.

Diamond, D. and Dybvig, P. (1983) 'Bank Runs, Deposit Insurance, and Liquidity', *Journal of Political Economy*, 91: 401–419.

Diamond, D. and Verrecchia, R.E (1982) 'Optimal Managerial Contracts and Equilibrium Security Prices', *Journal of Finance*, 37 (2): 275–287.

Diamond, S. (1974) *In Search of the Primitive: A Critique of Civilisation*, Transaction Books.

Dijkstra, G. (2005) 'The PRSP Approach and the Illusion of Improved Aid Effectiveness: Lessons from Bolivia, Honduras and Nicaragua', *Development Policy Review*, 23 (4): 443–464.

Dijkstra, G. (2008) *The Impact of International Debt Relief*, London: Routledge.

Ditcher, T. (2007) 'The Chicken and Egg Dilemma in Microfinance: An Historical Analysis of the Sequence of Growth and Credit in the Economic Development of the "North', in Dichter, T. and Harper, M. (eds) *What's Wrong with Microfinance?*, Warwickshire: Practical Action Publishing.

Dixon, W.J. and Boswell, T. (1996) 'Dependency, Disarticulation, and Denominator Effects: Another Look at Foreign Capital Penetration', *The American Journal of Sociology*, 102: 543–562.

Dobson, L. and Hufbauer, G. (2001) *World Capital Markets: Challenge to the G-10*, Washington, DC: Institute for International Economics.

Dodd, N. (1994) *The Sociology of Money*, Cambridge: Polity Press.

Dodd, R. and Munck, L. (2002) *Dying for Change: Poor People's Experience of Health and Ill-health*, Geneva: World Health Organization and World Bank.

Dohn, A.L., Chávez, A., Dohn M.N., Saturria L. and Pimentel C. (2004) 'Changes in Health Indicators Related to Health Promotion and Microcredit Programs in the Dominican Republic', *Revista Panamericana de Salud Pública*, 15 (3): 185–193.

Doidge, C., Karolyi, G.A. and Stulz, R.M. (2004) 'Why are Foreign Firms Listed in the U.S. Worth More?', *Journal of Financial Economics*, 71: 205–238.

Dollar, D. and Kraay, A. (2002) 'Growth is Good for the Poor', *Journal of Economic Growth*, 7 (3): 195–225.

Domar, E.D. (1946) 'Capital Expansion, Rate of Growth, and Employment', *Econometrica*, 14 (2): 137–147.

Dooley, M., Fernandez-Arias, E. and Kletzer, K. (1994) 'Is the Debt Crisis History? Recent Private Capital Inflows to Developing Countries', Policy Research Working Paper Series, No. 1327, Washington, DC: World Bank.

Doornbos, M. (2001) 'Good Governance: The Rise and Decline of a Policy Metaphor', *Journal of Development Studies*, 37 (6): 93–108.

Dos Santos, T. (1970), 'The Structure of Dependence', *American Economic Review*, 60 231–236.

Dow, A. and Dow, S.C. (2011) 'Animal Spirits Revisited', *Capitalism and Society*, 6 (2): 1–23.

Dowd, K. (1996) 'The Case for Financial *Laissez-Faire*', *Economic Journal*, 106: 679–687.

Duflo, E., Glennerster, R. and Kremer, M. (2008) 'Using Randomization in Development Economics Research: A Toolkit', in Schultz, T.P. and Strauss, J. (eds) *Handbook of Development Economics Volume 4*, Oxford: North-Holland.

Duhigg, C. and Barboza, D. (2012) 'In China, Human Costs Are Built Into an iPad', *New York Times*, 25 January, available at: www.nytimes.com/2012/01/26/business/ieconomy-apples-ipad-and-the-human-costs-for-workers-in-china.html.

Duhigg, C. and Bradsher, K. (2012) 'How the U.S. Lost Out on iPhone Work', *New York Times*, 21 January, available at: http://www.nytimes.com/2012/01/22/business/apple-america-and-a-squeezed-middle-class.html.

Dunning, J. (2008) *Multinational Enterprises and the Global Economy*, 2nd edn, Cheltenham: Edward Elgar.

Dustmann, C., Frattini, T. and Halls, C. (2010) 'Assessing the Fiscal Costs and Benefits of A8 Migration to the UK', *Fiscal Studies*, 31: 1–41.

Duval Smith, A. (2009) 'Critics Fear all is not well in the Darling of Africa on Eve of election', *The Independent*, 16 October.

Easterly, W. (1999) 'The Ghost of Financing Gap: Testing the Growth Model Used in the International Financial Institutions', *Journal of Development Economics*, 60 (2): 423–438.

Easterly, W. (2001) 'The Lost Decades: Developing Countries' Stagnation in Spite of Policy Reform 1980–1998', *Journal of Economic Growth*, 6 (2): 135–157.

Easterly, W. (2003) 'Can Foreign Aid Buy Growth?', *Journal of Economic Perspectives*, 17 (3): 23–48.

Easterly, W. (2006) *The White Man's Burden: Why the West's Efforts to Aid the Rest Have Done So Much Ill and So Little Good*, Oxford: Oxford University Press.

Easterly, W., and Levine, R. (2003) 'Tropics, Germs, and Crops: How Endowments Influence Economic Development', *Journal of Monetary Economics*, 50 (1): 3–39.

Easterly, W., Levine, R. and Roodman, D. (2004) 'Aid, Policies and Growth: A Comment', *American Economic Review*, 94 (3): 774–780.

Ebrahim-Zadeh, C. (2003) 'Dutch Disease: Too Much Wealth Managed Unwisely', *Finance and Development*, 40 (1).

Edison, H.J. (2000) 'Do Indicators of Financial Crises Work? An Evaluation of An Early Warning System', International Finance Discussion Papers, No. 675 (July), Federal Reserve Board of Governors.

Edison, H. and Warnock, F. (2001) 'A Simple Measure of the Intensity of the Capital Controls', IMF Working Papers, No. 01/180, Washington, DC: International Monetary Fund.

Edison H. and Warnock F. (2003) 'A Simple Measure of the Intensity of Capital Controls', *Journal of Empirical Finance*, 10: 83–105.

Edwards, M. (2011) 'The Role and Limitations of Philanthropy', commissioned paper for the Bellagio Initiative, November, Institute of Development Studies, the Resource Alliance and the Rockefeller Foundation.

Edwards, S. (2000) *Capital Flows and the Emerging Economies: Theory, Evidence, and Controversies*, Chicago: University of Chicago Press.

Eichengreen, B. (1985) 'International Policy Coordination in Historical Perspective: A View from the Interwar Years', NBER Working Paper No. 1440, Cambridge, MA: NBER.

Eichengreen, B. (1992) *Golden Fetters: The Gold Standard and the Great Depression, 1919–39*, Oxford: Oxford University Press.

Eichengreen, B. (1996) *Globalizing Capital: A History of the International Monetary System*, Princeton, NJ: Princeton University Press.

Eichengreen, B. (1999) *Toward a New International Financial Architecture: A Practical Post-Asia Agenda,* Washington DC: Institute of International Economics.

Eichengreen, B. (2001) 'Capital Account Liberalization: What Do Cross-Country Studies Tell Us?', *World Bank Economic Review*, 15 (3): 341–365.

Eichengreen, B. (2004) 'Financial Instability', in Lomborg, B. (ed.) *Global Crises, Global Solutions*, Cambridge: Cambridge University Press.

Eichengreen, B. (2008) *Globalizing Capital: A History of the International Monetary System*, 2nd edn, Princeton, NJ: Princeton University Press.

Eichengreen, B. (2011) *Exorbitant Privilege: The Rise and Fall of the Dollar and the Future of the International Monetary System*, Oxford: Oxford University Press.

Eichengreen, B. and Hausmann, R. (1999), 'Exchange Rate and Financial Fragility', NBER Working Paper 7418, Cambridge MA: NBER.

Eichengreen, B. and Mody, A. (1998) 'Interest Rates in the North and Capital Flows to the South: Is There a Missing Link?', *International Finance*, 1 (1): 35–57.

Eichengreen, B. and Mussa with Dell'Arriccia, G., Detragiache, E., Milesi-Ferretti, G.M. and Tweedie, A. (1998) 'Capital Account Liberalization: Theoretical and Practical Aspects', IMF Occasional Paper 172, Washington, DC: IMF.

Eichengreen, B. and Rose, A.K. (1998) 'Staying Afloat When the Wind Shifts: External Factors and Emerging-Market Banking Crises', NBER Working Paper No. 6370. Cambridge, MA: NBER.

Eichengreen, B., Tobin, J. and Wyplosz, (1995) 'Two Cases for Sand in the Wheels of International Finance', *Economic Policy*, 105 (428): 162–172.

Eichengreen, B., Hausmann, R. and Panizza, U. (2003) 'Currency Mis-matches, Debt Intolerance and Original Sin: Why They are Not the Same and Why it Matters', NBER Working Paper 10036, Cambridge, MA: NBER.

EMTA (2009) 'The Brady Plan', available at: www.emta.org/template.aspx?id=35 (accessed on 17 March 2010).

Engel, J. (2011) *Ethiopia's Progress in Education: A Rapid and Equitable Expansion of Access*, London: Overseas Development Institute.

England, A. (2004) 'Africa's Poorest should refuse to repay debt', *Financial Times*, 6 July.

Escobar, A. (1995) *Encountering Development: The Making and Unmaking of the Third World*, Princeton, NJ: Princeton University Press.

Eurobarometer. (2005) 'Europeans and Development Aid, 222 EB62.2', Brussels: European Commission.

Evans, P. (1995) *Embedded Autonomy: States and Industrial Transformation*, Princeton, NJ: Princeton University Press.

Faal, G. (2006) 'Introduction to RemitAid™: Remittance Tax Relief for International Development', briefing document, available at: www.remitaid.org/downloads/Introduction%20to%20RemitAid%20-%20GKP%2006.pdf (accessed on 17 March 2010).

Faini, R. (2002) 'Migration, Remittances and Growth,' paper presented at the World Institute for Development Economics, New York: United Nations University.

Fajnzylber, P. and López, J.H. (2007) *Close to Home: The Development Impact of Remittances in Latin America*, Washington, DC: World Bank.

Fajnzylber, P. and López, J.H. (2008) 'The Development Impact of Remittances in Latin America', in Fajnzylber, P. and López, J.H. (eds) *Remittances and Development: Lessons from Latin America*, Washington, DC: World Bank.

Fama, E. (1970) 'Efficient Capital Markets: A Review of Theory and Empirical Work', *Journal of Finance*, 25 (2): 383–417.

Faria, A., Mauro, P., Minnoni, M. and Zaklan, A. (2006) 'The External Financing of Emerging Market Countries: Evidence from Two Waves of Financial Globalization', IMF Working Paper 06/205, Washington, DC: International Monetary Fund.

FATF (2001) 'Special Recommendations on Terrorist Financing', 31 October, Paris: OECD.

FATF (2002) *Annual Report 2001–2002*, Paris: OECD/Financial Action Task Force, available at: www.fatf-gafi.org/dataoecd/13/1/34328160.pdf (accessed 1 September 2010).

Federal Ministry of Finance (2007) 'The 7 Recommendations of the G8 Outreach Meeting on Remittances', 30 November, available at: http://siteresources.worldbank.org/INTPROSPECTS/Resources/334934-1215617363964/7recommendations.pdf (accessed 1 September 2010).

Feenstra, R.C. and Hanson, G.H. (1997) 'Foreign Direct Investment and Relative Wages: Evidence from Mexico's Maquiladoras', *Journal of International Economics*, 42: 371–393.

Fei, J.C., and Ranis, G. (1969) 'Economic Development in Historical Perspective', *American Economic Review*, 59 (2): 386–400.

Feldstein, M. (1999) 'A Self-Help Guide for Emerging Markets', *Foreign Affairs*, 78 (2): 93–109.

Ferguson, J. (1994) *The Anti-Politics Machine: Development, Depoliticization, and Bureaucratic Power in Lesotho*, Minnesota, MN: Minnesota University Press.

Fernando, J. (2006) *Microfinance: Perils and Prospects*, London: Routledge.

Ffrench-Davis, R. and Reisen, H. (eds) (1998) *Capital Flows and Investment Performance: Lessons from Latin America*, Paris: OECD Development Centre/ECLAC.

Filmer, D. (2002) 'Fever and its Treatment Among the More and Less Poor in Sub-Saharan Africa', World Bank Development Working Paper, No. 2789, Washington, DC: World Bank.

Fine, B. (2007) 'Financialisation, Poverty, and Marxist Political Economy', paper prepared for Poverty and Capital Conference 2–4 July, Manchester: University of Manchester.

Fine, B. (2010) 'Economics and Scarcity: With Amartya Sen as Point of Departure?' in Mehta, L. (ed.) *The Limits to Scarcity: Contesting the Politics of Allocation*, London: Earthscan.

Fine, B. and Lapavitsas, C. (2000) 'Markets and Money in Social Theory: What Role for Economics?', *Economy and Society*, 29 (3): 357–382.

Firebaugh, G. (1996) 'Does Foreign Capital Harm Poor Nations? New Estimates Based on Dixon and Boswell's Measures of Capital Penetration', *American Journal of Sociology*, 102 (2): 563–575.

Fischer, S. (1997) 'Capital Account Liberalization and the Role of the IMF', speech at the 'Asia and the IMF' seminar, 19 September, available at: www.imf.org/external/np/apd/asia/fischer.htm (accessed 20 September 2010).

Fiszbein, A. and Schady, N. (2009) 'Conditional Cash Transfers: Reducing Present and Future Poverty', Washington, DC: World Bank.

Fitzgibbons, A. (1988) *Keynes's Vision: A New Political Economy*, Oxford: Clarendon Press.

Fleming, J. Marcus (1962) 'Domestic Financial Policies Under Fixed and Floating Exchange Rates', *IMF Staff Papers*, 9: 369–379.

Ford, L. (2011) 'UK minister defends policy as aid to India comes under fire', *The Guardian*, 16 February.

Foster, J., Greer, J. and Thorbecke, E. (1984) 'A Class of Decomposable Poverty Measures', *Econometrica*, 52: 761–766.

Foster, M. and Leavy, J. (2001) 'The Choice of Financial Aid Instruments', ODI Working Paper 158, London: ODI, Centre for Aid and Public Expenditure.

Frank, A.G. (1967) *Capitalism and Underdevelopment in Latin America*, New York: Monthly Review Press.

Frank, A.G. (1991) 'The Underdevelopment of Development', *Scandinavian Journal of Development Alternatives*, 10 (3), September.

Frankel, J.A. and Rose, A.K. (1997) 'Is EMU more justifiable ex post than ex ante?', *European Economic Review*, 41 (3–5): 753–760.

Frei, M. (1998) 'When Crisis Hits Home', BBC Online, 30 September, available at: http://news.bbc.co.uk/1/hi/world/asia-pacific/182463.stm (accessed 24 October 2010).

Friedman, M. (1953) 'The Case for Flexible Exchange Rates', in *Essays in Positive Economics*, Chicago: University of Chicago Press.

Friedman, T.L. (2000) *The Lexus and the Olive Tree*, New York: Anchor Books.

Fritz, V. and Kolstad, I. (2008) 'Corruption and Aid Modalities', *U4 Issue*, 4, Bergen: Chr. Michelsen Institute.

Fukuda-Parr, S. (2010) 'Reducing Inequality: The Missing MDG', *IDS Bulletin*, 41 (1): 26–35.

G8 (2008) *Final Declaration on Responsible Leadership for a Sustainable Future*, available at: http://www.wcoomd.org/en/topics/key-issues/~/media/21E3FE6008454F9D8EC0B73A8BB7F5E2.ashx (accessed 16 December 2013).

Gallagher, K.P., Griffith-Jones, S. and Ocampo, J.A. (2011) 'Capital Account Regulations for Stability and Development: A New Approach', *Issues in Brief*, 22, November, Boston, MA: The Frederick S. Pardee Center for the Study of the Longer-Range Future, available at: http://www.bu.edu/pardee/publications-library/iib-22-capital-account-regulations.

Gallagher, K.P., Griffith-Jones, S. and Ocampo, J.A. (2012) 'Capital Account Regulations for Stability and Development: A New Approach', in *Regulating Global Capital Flows for Long-Run Development*, Boston, MA: The Frederick S. Pardee Center for the Study of the Longer-Range Future.

Gallie, W.B. (1956) 'Essentially Contested Concepts', *Proceedings of the Aristotelian Society*, 56: 167–198.

Galtung, J. (1971) 'A Structural Theory of Imperialism', *Journal of Peace Research*, 8 (2): 81–117.

Gansmann, H. (1988) 'Money – A Symbolically Generalised Medium of Communication? On the Concept of Money in Recent Sociology', *Economy and Society*, 17: 285–315.

Garrett, G. (2001) 'Globalization and Government Spending Around the World', *Studies in Comparative International Development*, 35 (4): 3–29.

GCIM (2005) *Migration in an Interconnected World: New Directions for Action*, Geneva: Global Commission on International Migration.

Gelos, R.G., Sahay, R. and Sandleris, G. (2011) 'Sovereign Borrowing by Developing Countries: What Determines Market Access?', *Journal of International Economics*, 83 (2): 243–254.

Gerlach, M.L. (1992) *Alliance Capitalism: The Social Organization of Japanese Business*, Berkley, CA: University of California Press.

Germain, R. (1997) *The International Organisation of Credit: States and Global Finance in the World-Economy*, Cambridge: Cambridge University Press.

Germain, R. (2010) *Global Politics and Financial Governance*, London: Palgrave.

Ghatak, M. (1999) 'Group Lending, Local Information and Peer Selection', *Journal of Development Economics*, 60: 27–50.

Ghose, A.K. (2004) 'Capital Flows and Investment in Developing Countries', ILO Employment Strategy Papers 2004/11, available at: www.ilo.org/wcmsp5/groups/public/---ed_emp/---emp_elm/documents/publication/wcms_114304.pdf (accessed 1 November 2010).

Giddens, A. (1990) *The Consequences of Modernity*, Cambridge: Polity.

Giddens, A. (1999) *Globalisation*, 1999 BBC Reith Lecture, available at: www.lse.ac.uk/Giddens/reith_99/week1/week1.htm, and in Giddens, A. (1999) *Runaway World: How Globalisation is Reshaping our Lives*, London: Profile Books.

Gill, S.R. and Law, D. (1989) 'Global Hegemony and the Structural Power of Capital', *International Studies Quarterly*, 33 (4): 475–499.

Gilligan, A. (2012) 'India tells Britain: We don't want your aid' *The Daily Telegraph*, 4 February.

Giné, X., Jakiela, P., Karlan, D. and Morduch, J. (2010) 'Microfinance Games', *American Economic Journal: Applied Economics*, 2 (3): 60–95.

Giuliano, P. and Ruiz-Arranz, M. (2005) 'Remittances, Financial Development, and Growth', IMF Working Paper, Washington, DC: International Monetary Fund.

Glennie, J. (2008) *The Trouble with Aid: Why Less Could Mean More for Africa*, London: Zed.

GlobeScan (2012) Multi-Country Nutrition Poll 2011, available at: www.savethechildren.org.uk/sites/default/files/docs/Nutrition%20survey.pdf (accessed 26 March 2012).

Goetz, A.M. and Gupta, R.S. (1996) 'Who Takes the Credit? Gender, Power and Control Over Loan Use in Rural Credit Programs in Bagladesh', *World Development*, 24 (1): 45–63.

Goldin, I., Knudsen, O. and van der Mensbrugghe, D. (1993) *Trade Liberalisation: Global Economic Implications*, Paris: OECD.

Goldsmith, R.W. (1969) *Financial Structure and Economic Development*, New Haven, CT: Yale University Press.

Goldstein, M. and Turner, P. (2004) *Controlling Currency Mismatches in Emerging Market Economies: An Alternative to the Original Sin Hypothesis*, Washington, DC: Institute for International Economics.

Goldstein, M., Mathieson, D.J., Folkerts-Landau, D.F.I., Lane, T.D., Lizondo, J.S. and Rojas-Suárez, L. (1991) *Determinants and Systemic Consequences of International Capital Flows*, occasional paper No. 77, Washington, DC: International Monetary Fund.

Gomanee, K., Girma, S. and Morrissey, O. (2005a) 'Aid, Public Spending and Human Welfare: Evidence from Quantile Regressions', *Journal of International Development*, 17.

Gomanee, K., Morrissey, O., Mosley, P. and Verschoor, A. (2005b) 'Aid, Government Expenditure and Aggregate Welfare', *World Development*, 33 (3).

Gonzalez, A. and Rosenberg, R. (2006) 'The State of Microfinance – Outreach, Profitability and Poverty: Findings from a Database of 2300 Microfinance Institutions', manuscript, May, available at: http://ssrn.com/abstract=1400253.

Goodhart, C.A.E. (1975) *Money, Information and Uncertainty*, London: Macmillan.

Gordon, R.J. (1990) 'What Is New-Keynesian Economics?', *Journal of Economic Literature*, 28 (3): 1115–1171.

Gottschalk, R. (ed.) (2010) *The Basel Capital Accords in Developing Countries: Challenges for Development Finance*, Basingstoke: Palgrave Macmillan.

Gottschalk, R. and Sodré, C.A. (2007) 'Implementation of Basel Rules in Brazil: What are the Implications for Development Finance?' IDS Working Papers 273, Brighton: Institute of Development Studies.

Gourinchas, P.O. and Jeanne, O. (2006) 'The Elusive Gains From International Financial Integration', *Review of Economic Studies*, 73 (3): 715–741.

Graeber, D. (2011) *Debt: The First 5,000 Years*, New York: Melville House Publishing.

Grahl, J. and Teague, P. (2000) 'The Regulation School, the Employment Relation and Financialization', *Economy and Society*, 29 (1): 160–178.

Grameen Bank (2011) 'Grameen Bank At A Glance', available at: www.grameen-info.org (accessed 5 March 2012).

Granovetter, M. (1985) 'Economic Action and Social Structure: the Problem of Embeddedness', *American Journal of Sociology*, 91 (1): 481–493.

Greeley, M. (2006) 'Microfinance Impact and the MDGs: The Challenge of Scaling-up', IDS Working Paper 225, Brighton: IDS.

Greenhill, R. (2002) *The Unbreakable Link: Debt Relief and the Millennium Development Goals*, report from Jubilee Research at the New Economics Foundation, London.

Griffith-Jones, S. and Leape, J. (2002) 'Capital Flows to Developing Countries: Does the Emperor Have Clothes?', QEH Working Paper Series, Working Paper No. 89.

Griffith-Jones, S. and Kimmis, J. (2003) 'International Financial Volatility', *Journal of Human Development: A Multi-Disciplinary Journal for People-Centered Development*, 4 (2): 209–225.

Grilli, V. and Milesi-Ferretti, G.M. (1995) 'Economic Effects and Structural Determinants of Capital Controls', *IMF Staff Papers*, 42 (3): 517–551.

Grindle, M.S. (2004), 'Good Enough Governance: Poverty Reduction and Reform in Developing Countries', *Governance*, 17: 525–548.

Grindle, M. (2007) 'Good Enough Governance Revisited', *Development Policy Review*, 25 (5): 533–574.

Grindle, M.S. (2010) 'Good Governance: The Inflation of an Idea', HKS Faculty Research Working Paper Series, RWP10-023, John F. Kennedy School of Government, Harvard University.

Gronemeyer, M. (1992) 'Helping', in Sachs, W. (ed.) *The Development Dictionary: A Guide to Knowledge as Power*, London: Zed.

Gugerty, M. K. (2007) 'You Can't Save Alone: Commitment in Rotating Savings and Credit Associations in Kenya', *Economic Development and Cultural Change*, 55 (2): 251–282.

Gupta, S., Powell, R. and Yang, Y. (2006) *Macroeconomic Challenges of Scaling up Aid*, Washington, DC: IMF.

Gupta, S., Pattillo, C. and Wagh, S. (2007) 'Impact of Remittances on Poverty and Financial Development in Sub-Saharan Africa', IMF Working Paper 07/38, Washington, DC: International Monetary Fund.

Gurley, J.G. and Shaw, E.S. (1955) 'Financial Aspects of Economic Development', *American Economic Review*, 45: 515–538.

Haber, S., North, D. and Weingast, B. (2008) *Political Institutions and Financial Development*, Stanford, CA: Stanford University Press.

Haddad, M. and Harrison, A. (1993) 'Are There Positive Spillovers from Direct Foreign Investment? Evidence from Panel Data for Morocco', *Journal of Development Economics*, 42: 51–74.

Hall, R.B. and Biersteker, T.J. (eds) (2002) *The Emergence of Private Authority in Global Governance*, Cambridge: Cambridge University Press.

Hall, R.E. and Jones, C.I. (1999) 'Why Do Some Countries Produce So Much More Output Per Worker than Others?', *Quarterly Journal of Economics*, 114 (1): 83–116.

Hallward-Driemeier, M. (2003) 'Do Bilateral Investment Treaties Attract Foreign Direct Investment? Only a Bit . . . And They Could Bite', World Bank Policy Research Working Paper 3121, Washington, DC: World Bank.

Hamilton, A. (1791) 'Report on the Subject of Manufactures', in *Alexander Hamilton: Writings*, New York: Literary Classics of the United States Inc., 2001.

Hamilton, B. and Whalley, J. (1984) 'Efficiency and Distributional Implications of Global Restrictions on Labour Mobility', *Journal of Development Economics*, 14: 61–75.

Hanlon, J., Barrientos, A. and Hulme, D. (2010) *Just Give Money to the Poor: The Development Revolution from the Global South*, Sterling, VA: Kumarian Press.

Hanson, G.H. and Woodruff, C. (2003) *Emigration and Educational Attainment in Mexico*, mimeo, University of California at San Diego.

Hardie, I. (2012) *Financialization and Government Borrowing Capacity in Emerging Markets*, Basingstoke: Palgrave Macmillan.

Harriss, J., Hunter, J. and Lewis, C.M. (eds) (1995) *The New Institutional Economics and Third World Development*, London: Routledge.

Harrod, R.F. (1939) 'An Essay in Dynamic Theory', *Economic Journal*, 49 (193): 14–33.

Hart, K. (1986) 'Heads or Tails? Two Sides of the Coin', *Man*, 21 (4): 637–656.

Harvey, D. (1982) *The Limits to Capital*, Oxford: Blackwell.

Harvey, D. (1989) *The Condition of Postmodernity*, Oxford: Blackwell.

Harvey, D. (2003) *The New Imperialism*, Oxford: Oxford University Press.

Harvey, D. (2005) *A Brief History of Neoliberalism*, Oxford: Oxford University Press.

Harvey, D. (2010) *The Enigma of Capital and the Crises of Capitalism*, Oxford: Oxford University Press.

Hausmann, R. and Rodrik, D. (2003) 'Economic Development as Self-Discovery', *Journal of Development Economics*, 72 (2): 603–633.

Hausmann, R., Pritchett, L. and Rodrik, D. (2005) 'Growth Accelerations', *Journal of Economic Growth*, 10 (4): 303–329.

Hausmann, R., Rodrik, D. and Velasco, A. (2008) 'Growth Diagnostics', in Serra, N. and Stiglitz, J.E. (eds) *The Washington Consensus Reconsidered: Towards a New Global Governance*, New York, Oxford University Press.

Hay, C. and Wincott, D. (1998) 'Structure, Agency and Historical Institutionalism', *Political Studies*, 46 (5): 951–957.

Hayek, F.A. (1978) 'Competition as a Discovery Procedure', in *New Studies in Philosophy, Politics, Economics and the History of Ideas*, Chicago, IL: University of Chicago Press.

Hayter, T. (1971) *Aid as Imperialism*, Harmondsworth: Penguin.

Headey, D. (2008) 'Geopolitics and the Effect of Foreign Aid on Economic Growth: 1970–2001', *Journal of International Development*, 20 (2): 161–180.

Heilbroner, R. (1990) 'Economics as Ideology', in Samuel, W.J. (ed.) *Economics as Discourse, An Analysis of the Language of Economists*, Boston, MA: Kluwer-Nijhoff.

Heilbroner, R. (1993) 'Was Schumpeter right after all?', *The Journal of Economic Perspectives*, 7 (3): 87–96.

Helleiner, E. and Kirshner, J. (2009) *Geopolitics*.

Helleiner, E. (1994) *States and the Reemergence of Global Finance: From Bretton Woods to the 1990s*, Ithaca, NY: Cornell University Press.

Helleiner, E. (1996) *States and the Reemergence of Global Finance*, Ithaca, NY: Cornell University Press.

Helleiner, E. (2003) *The Making of National Money: Territorial Currencies in Historical Perspective*, Ithaca, NY: Cornell University Press

Helleiner, E. (2011) 'The Evolution of the International Monetary and Financial System', in Ravenhill, J. (ed.) *Global Political Economy*, 3rd edn, Oxford: Oxford University Press.

Henderson, C. (1998) *Asia Falling: Making Sense of the Asia Crisis and its Aftermath*, New York: McGraw-Hill.

Henderson, J. (1989) *The Globalization of High Technology Production: Society, Space and Semiconductors in the Restructuring of the Modern World*, London: Routledge.

Henson, S., Lindstrom, J., Haddad, L. and Mulmi R. (2010) 'Public Perceptions of International Development and Support for Aid in the UK: Results of a Qualitative Enquiry', IDS Working Paper, Brighton: IDS, available at: www.ids.ac.uk/index.cfm?objectid=9EC8ACF6-91C1-CED2-7C6912061A53BC53 (accessed 1 August 2010).

Herkenrath, M. and Bornschier, V. (2003) 'Transnational Corporations in World Development – Still the Same Harmful Effects in an Increasingly Globalized World Economy', *Journal of World Systems Research*, 105–139.

Hertel, T. and Keeney, R. (2006) 'What is at Stake: The Relative Importance of Import Barriers, Export Subsidies, and Domestic Support', in Anderson, K. and Martin, W. (eds) *Agricultural Trade Reform and the Doha Development Agenda*, Washington, DC: World Bank.

Hertz, N. (2005) *IOU: The Debt Threat and Why We Must Defuse It*, London: Harper Perennial.

Hicks, J.R. (1937) 'Mr. Keynes and the "Classics": A Suggested Interpretation', *Econometrica*, 5 (2): 147–159.

Hicks, J.R. (1976) 'Some Questions of Time in Economics', in Tang, A.M., Westfield, F.M. and Worley, J.S. (eds) *Evolution, Welfare and Time*, Lexington MA: Heath.

Higgott, R. (1998) 'The Asian Economic Crisis: A Study in the Politics of Resentment', *New Political Economy*, 3 (3): 333–356.

Higgott, R. (2002) 'Taming Economics, Emboldening International Relations. The Theory and Practice of International Political Economy in an Era of Globalization', in Lawson, S. (ed.) *The New Agenda for International Relations*, Cambridge: Polity.

Higgott, R., Underhill G. and Bieler, A. (eds) (2000) *Non-State Actors and Authority in the Global System*, London: Routledge.

Hilferding, R. ([1910] 1981) *Finance Capital: A Study of the Latest Phase of Capitalist Development*, London: Routledge & Kegan Paul.

Hirschman, A.O. (1977) *The Passions and the Interests: Political Arguments for Capitalism Before its Triumph*, Princeton, NJ: Princeton University Press.

Hirschman, A.O. (1982) 'The Rise and Decline of Development Economics', in Gersovitz, M. and Lewis, W.A. (eds) *The Theory and Experience of Economic Development*, London: Allen & Unwin.

Hirst, P. and Thompson, G. (1999) *Globalization in Question: The International Economy and the Possibilities of Governance*, 2nd edn, Cambridge: Polity.

Hoeffler, A. and Outram, V. (2011) 'Need, Merit, or Self-Interest – What Determines the Allocation of Aid?', *Review of Development Economics*, 15: 237–250.

Hoff, K. and Stiglitz, J.E. (2001) 'Modern Economic Theory and Development', in Meier, G.M. and Stiglitz, J.E. (eds) *Frontiers of Development Economics: The Future in Perspective*, Wanshington, DC: World Bank.

Holmes, R., Hagen-Zanker, J. and Vandemoortele, M. (2011) *Social Protection in Brazil: Impacts on Poverty, Inequality and Growth*, London: ODI.

Honohan, P. (2004) 'Financial Sector Policy and the Poor', Working Paper 43, Washington, DC: World Bank.

Honohan, P. (2008) 'Cross-country Variation in Household Access to Financial Services', *Journal of Banking and Finance*, 32: 2493–2500.

Honohan, P. and Klingebiel, D. (2000) 'Controlling the Fiscal Costs of Banking Crises', *World Bank Policy Research Working Paper Series*, No. 2441, Washington, DC: World Bank.

Hook, S.W. (1995) *National Interest and Foreign Aid*, Boulder, CO: Lynne Rienner Publishers.

Hopwood, A.G. and Miller, P. (eds) (1994) *Accounting as Social and Institutional Practice*, Cambridge: Cambridge University Press.

Horton, S., Shekar, M., McDonald, C., Mahal, A. and Brook, J.K. (2010) *Scaling Up Nutrition: What Will it Cost?*, Washington, DC: World Bank.

Howse, R. (2007) 'The Concept of Odious Debt in Public International Law', Geneva: United Nations Conference on Trade and Development (UNCTAD) Discussion Paper, No. 185.

Hudson, D. (2008) 'Developing Geographies of Financialisation: Banking the Poor and Remittance Securitisation', *Contemporary Politics*, 14 (3): 315–333.

Hudson, D. (2010) 'Financing for Development and the Post Keynesian Case for a New Global Reserve Currency', *Journal of International Development*, 22 (6): 772–787.

Hudson, D. and van Heerde, J. (2010) ' "The Righteous Considereth the Cause of the Poor"? Public Attitudes Towards Poverty in Developing Countries', *Political Studies*, 58 (3): 389–409.

Hudson, D. and van Heerde-Hudson, J. (2012) ' "A Mile Wide and an Inch Deep": Surveys of Public Attitudes towards Development Aid', *International Journal of Development Education and Global Learning*, 4 (1).

Hughes, N. and Lonie, S. (2007) 'M-PESA: Mobile Money for the "Unbanked": Turning Cellphones into 24-Hour Tellers in Kenya', *Innovations: Technology, Governance, Globalization*, 2 (1–2): 63–81.

Hulme, D. (2009) 'The Millennium Development Goals (MDGs): A Short History of the World's Biggest Promise', BWPI Working Paper 100, Manchester: Brooks World Poverty Institute.

Hulme, D. and Mosley, P. (1996) *Finance Against Poverty*, London: Routledge.

Hulten, C.R. and Isaksson, A. (2007) 'Why Development Levels Differ: The Sources of Differential Economic Growth in a Panel of High and Low Income Countries', NBER Working Paper No. 13469.

Hyden, G. (2008) 'After the Paris Declaration: Taking on the Issue of Power', *Development Policy Review*, 26: 259–274.

Hymer, S. (1976) *The International Operations of National Firms: A Study of Foreign Direct Investment*, Cambridge, MA: MIT Press.

IDA (2008) *Aid Architecture: An Overview of the Main Trends in Official Development Assistance Flows*, Washington, DC: World Bank.

IDA (2013) 'The World Bank's Fund for the Poorest', Washington, DC: World Bank, available at: www.worldbank.org/ida/what-is-ida/fund-for-the-poorest.pdf (accessed 15 December 2013).

IDB (2006) *Sending Money Home: Leveraging the Development Impact of Remittances*, Washington, DC: Inter-American Development Bank, Multilateral Investment Fund.

IFAD (2007) *Sending Money Home: Worldwide Remittance Flows to Developing and Transition Countries*, Rome: IFAD.

IIF (2010) 'Capital Flows to Emerging Markets', *IFF Research Note*, 15 April.

ILO (1998) *Labour and Social Issues Relating to Export Processing Zones*, report for discussion at the Tripartite Meeting of the Export Processing Zones-Operating Countries, Geneva: International Labour Organization.

IMF (2003) *Foreign Direct Investment Statistics: How Countries Measure FDI*, Washington, DC: IMF/OECD.

IMF (2005) *World Economic Outlook: Globalization and External Balances*, Washington, DC: The International Monetary Fund.

IMF (2010a) 'Chile: 2010 Article IV Consultation – Staff Report', IMF Country Report No. 10/298, available at: www.imf.org/external/pubs/ft/scr/2010/cr10298.pdf (accessed 11 October 2010).

IMF (2010b) 'Debt Relief Under the Heavily Indebted Poor Countries (HIPC) Initiative', IMF Factsheet, available at: www.imf.org/external/np/exr/facts/hipc.htm (accessed 11 October 2010).

IMF (2011) *Recent Experiences in Managing Capital Inflows – CrossCutting Themes and Possible Policy Framework*, Washington, DC: International Monetary Fund.

Ingham, G. (1996) 'Money is a Social Relation', *Review of Social Economy*, 54 (4): 243–275.

Ingham, G (1999) 'Money is a Social Relation', in Fleetwood, S. (ed.) *Critical Realism in Economics*, London: Routledge.

Ingham, G. (2000) ' "Babylonian Madness": On the Historical and Sociological Origins of Money', in Smithin, J. (ed.) *What is Money?* London: Routledge.

Ingham, G. (2001) 'Fundamentals of a Theory of Money: Untangling Fine, Lapavitsas and Zelizer', *Economy and Society*, 30 (3): 304–323.

Innes, A.M. (1913) 'What is Money?', *Banking Law Journal*, May: 377–408.

Iregui, A.M. (2005) 'Efficiency Gains from the Elimination of Global Restrictions on Labour Mobility', in Borjas, G.J. and Crisp, J. (eds) *Poverty*,

Iskander, N. (2005) 'Social Learning as a Productive Project: The Tres por Uno (Three for One) Experience at Zacatecas, Mexico', in OECD (ed.) *Remittances, Migration and Development*, Paris: OECD.

Jacques, M. (2009) *When China Rules the World: The End of the Western World and the Birth of a New Global Order*, London: Allen Lane.

Jensen, M.C. (1978) 'Some Anomalous Evidence Regarding Market Efficiency', *Journal of Financial Economics*, 6 (2/3): 95–101.

Jevons, W.S. (1871) *The Theory of Political Economy*, London: Macmillan, 1888 version.

Johnson, C. (1982) *MITI and the Japanese Miracle: The Growth of Industrial Policy, 1925–1975*, Stanford, CA: Stanford University Press.

Johnson, C. (1999) 'The Developmental State: Odyssey of a Concept', in Woo-Cummings, M. (ed.) *The Developmental State*, Ithaca, NY: Cornell University Press.

Johnston, D. and Morduch, J. (2008) 'The Unbanked: Evidence from Indonesia', *World Bank Economic Review*, 22 (3): 517–537.

Jomo, K. S. (ed.) (1998) *Tigers in Trouble: Financial Governance, Liberalisation and Crises in East Asia*, London: Zed.

Jung, W.S. (1986) 'Financial Development and Economic Growth: International Evidence', *Economic Development and Cultural Change*, 35: 333–46.

Kaminsky, G.L. and Reinhart, C.M. (1999) 'The Twin Crises: The Causes of Banking and Balance-of-Payments Problems', *American Economic Review*, 89 (3): 473–500.

Kapoor, M., Ravi, S. and Morduch, J. (2007) 'From Microfinance to M-Finance', *Innovations: Technology, Governance, Globalization*, 2 (1–2): 82–90.

Kapur, D. (2005) 'Remittances: The New Development Mantra?', in Maimbo, S.M. and Ratha, D. (eds) *Remittances: Development Impact and Future Prospects*, Washington, DC: World Bank.

Karlan, D. (2007) 'Social Connections and Group Banking', *Economic Journal*, 117 (517): F52-F84, 02.

Karlan, D. and Appel, J. (2011) *More than Good Intentions*, New York: Dutton.

Karlan, D. and Morduch, J. (2010) 'Access to Finance', in Rodrik, D. and Rosenzeig, M. (eds) *Handbook of Development Economics*, Amsterdam: Elsevier.

Karlan, D. and Valdivia, M. (2011) 'Teaching Entrepreneurship: Impact of Business Training on Microfinance Clients and Institutions', *Review of Economics and Statistics*, 93 (2): 510–527.

Karlan, D. and Zinman, J. (2009) 'Observing Unobservables: Identifying Information Asymmetries with a Consumer Credit Field Experiment', *Econometrica*, 77 (6): 1993–2008.

Karlan, D. and Zinman, J. (2010) 'Expanding Credit Access: Using Randomized Supply Decisions to Estimate the Impacts', *Review of Financial Studies*, 23 (1): 433–464.

Karlan, D., Goldberg, N. and Copestake, J. (2009) 'Randomized Control Trials are the Best Way to Measure Impact of Microfinance Programmes and Improve Microfinance Product Designs', *Enterprise Development and Microfinance*, 20 (3): 167–176.

Karnani, A. (2007) 'Microfinance Misses its Mark', *Stanford Social Innovation Review*, Summer: 34–40.

Karp, D.J. (2009) 'Transnational Corporations in "Bad States": Human Rights Duties, Legitimate Authority and the Rule of Law in International Political Theory', *International Theory*, 1 (1): 87–118.

Kaufmann, D. (1997) 'Corruption: The Facts', *Foreign Policy*, 114–131.

Kaufmann, D., Kraay, A. and Mastruzzi, M. (2005) 'Governance Matters IV: Governance Indicators for 1996–2004', available at: www.worldbank.org/wbi/governance/govdata/ (accessed 29 October 2010).

Keefer, P. (2007) 'Beyond Legal Origin and Checks and Balances: Political Credibility, Citizen Information, and Financial Sector Development', World Bank Policy Research Working Paper Series 4154, Washington, DC: World Bank.

Keen, S. (2011) *Debunking Economics: The Naked Emperor of the Social Sciences*, 2nd edn, London: Zed Books.

Kenny, C. (2005) 'Why are we Worried about Income? Nearly Everything that Matters is Converging', *World Development*, 33 (1): 1–19.

Kenny, C. (2011) *Getting Better: Why Global Development Is Succeeding – and How We Can Improve the World Even More*, New York: Basic Books.

Kenny, C. and Sumner, A. (2011) 'How 28 poor countries escaped the poverty trap', *The Guardian*, 21 July.

Kentor, J. (2001) 'The Long Term Effects of Globalization on Income Inequality, Population Growth, and Economic Development', *Social Problems*, 48 (4): 435–455.

Kentor, J. and Boswell, T. (2003) 'Foreign Capital Dependence and Development: A New Direction', *American Sociological Review*, 68: 301–313.

Kektar, S. and Ratha, D. (2001) 'Securitization of Future Flow Receivables: A Useful Tool for Developing Countries', *Finance & Development*, 38 (1).

Kektar, S. and Ratha, D. (2004–2005) 'Recent Advances in Future-Flow Secuiritization', *The Financier*, 11/12: 1–14.

Ketkar, S.L. and Ratha, D. (2007) 'Development Finance via Diaspora Bonds: Track Record and Potential', Washington, DC: World Bank, available at: http://siteresources.worldbank.org/INTPROSPECTS/Resources/334934-1100792545130/Diasporabonds.pdf (accessed 3 September 2010).

Kevane, M. and Wydick, B. (2001) 'Microenterprise Lending to Female Entrepreneurs: Sacrificing Economic Growth for Poverty Alleviation?' *World Development*, 29 (7): 1225–1236.

Keynes, J.M. (1933) 'National Self-Sufficiency', *The Yale Review*, 22 (4): 755–769.

Keynes J.M. (1936) *The General Theory of Employment, Interest, and Money*, New York: Prometheus Books, 1997 edition.

Keynes J.M. (1937) 'The general theory of employment', *The Quarterly Journal of Economics*, 51 (2): 209–223.

Keynes, J.M. (1939) 'The Process of Capital Formation', *Economic Journal*, 49 (195): 569–574.

Khandker, S.R. (1998) *Fighting Poverty With Microcredit: Experience in Bangladesh*, Oxford: Oxford University Press.

Khandker, S.R., Koolwal,G.B. and Samad, H.A. (2010) *Handbook on Impact Evaluation: Quantitative Methods and Practices*, Washington, DC: World Bank.

Kharas, H. (2007) *The New Reality of Aid*, Wolfensohn Center for Development at Brookings.

Kharas, H. (2008) *Measuring the Cost of Aid Volatility*, Wolfensohn Center for Development, Brookings Institution, July, paper 3.

Khwaja, A. and Mian, A. (2005) 'Do Lenders Favor Politically Connected Firms? Rent Provision in an Emerging Financial Market', *Quarterly Journal of Economics*, 120: 1371–1411.

Kilburn, P.M. (1969) 'Excerpt from International Bank Notes', December issue, reproduced by the World Bank 'Pages from World Bank History: The Pearson Commission', available at: http://go.worldbank.org/JYCU8GEWA0 (accessed 17 December 2013).

Killen, B. and Rogerson, A. (2010) 'Global Governance of International Development: Who's in Charge?', OECD Development Brief, Issue 2, 2010, Paris: OECD, available at: www.oecd.org/dataoecd/34/63/45569897.pdf (accessed 26 July 2010).

Killick T. (1998) *Aid and the Political Economy of Policy Change*, London: Routledge.

Kim, J.C., Watts, C.H., Hargreaves, J.R., Ndhlovu, L.X., Phetla, G., Morison, L.A., Busza, J., Porter, J.D.H. and Pronyk, P. (2007) 'Understanding the Impact of a Microfinance-Based Intervention on Women's Empowerment and the Reduction of Intimate Partner Violence in South Africa', *American Journal of Public Health*, 97 (10): 1–9.

Kindleberger, C. (1978) *Manias, Panics and Crashes*, Basingstoke, Palgrave Macmillan.

Kindleberger, C. (1981) 'Dominance and Leadership in the International Economy: Exploitation, Public Goods, and Free Rides', *International Studies Quarterly*, 25 (2): 242–254.

King M. (2006) 'Reform of the International Monetary Fund', speech at the Indian Council for Research on International Economic Relations, New Dehli, India; 20 February, available at: www. bankofengland.co.uk/archive/Documents/historicpubs/speeches/2006/speech267.pdf (accessed 24 October 2010).

King, R.G. and Levine, R. (1993a) 'Finance and Growth: Schumpeter Might be Right', *Quarterly Journal of Economics*, 108 (3): 717–737.

King, R.G. and Levine, R. (1993b) 'Finance, Entrepreneurship, and Growth: Theory and Evidence', *Journal of Monetary Economics*, 32 (3): 513–542.

Kirshner, J. (1995) *Currency and Coercion: The Political Economy of International Monetary Power*, Princeton, NJ: Princeton University Press.

Kitching, G. (1989) *Development and Underdevelopment in Historical Perspective: Populism, Nationalism and Industrialisation*, London: Routledge.

Kiva (2012) 'Statistics', available at: http://www.kiva.org/about/stats (accessed 31 January 2012).

Klein, P. and Ventura, G. J. (2007) 'TFP Differences and the Aggregate Effects of Labor Mobility in the Long Run', *The B.E. Journal of Macroeconomics*, 7 (1): 1–38.

Knack, S. (2001) 'Aid Dependence and the Quality of Governance: Cross-Country Empirical Tests', *Southern Economic Journal*, 68 (2): 310–329.

Koch, D.J. (2009) *Aid from International NGOs: Blind Spots on the Aid Allocation Map*, Abingdon: Routledge.

Kosack, S. (2003) 'Effective Aid: How Democracy Allows Development Aid to Improve the Quality of Life', *World Development*, 31: 1–22.

Kragelund, P. (2008) 'The Return of non-DAC Donors to Africa: New Prospects for African Development', *Development Policy Review*, 26 (5): 555–584.

Kristof, ND (2006) 'Aid Can it Work?', *New York Review of Books*, 53 (15), 5 October.

Krueger, A.O. (1974) 'The Political Economy of the Rent-Seeking Society', *American Economic Review*, 64 (3): 291–303.

Krueger, A.O. (1993) *Political Economy of Policy Reform in Developing Countries*, Cambridge, MA: MIT Press.

Krueger, A.O. (2002) 'A New Approach to Sovereign Debt Restructuring', Washington, DC: International Monetary Fund, available at: www.imf.org/external/pubs/ft/exrp/sdrm/eng/sdrm.pdf (accessed 10 October 2010).

Krueger, A.O. (2004) 'Meant Well, Tried Little, Failed Much: Policy Reforms in Emerging Market Economies', Roundtable lecture, Economic Honors Society, New York University, 23 March, available at: www.imf.org/external/np/speeches/2004/032304a.htm (accessed 19 December 2011).

Krugman, P. (1988) 'Financing vs. Forgiving a Debt Overhang', *Journal of Development Economics*, 29 (3): 253–268.

Krugman, P. (1994) 'The Myth of Asia's Miracle', *Foreign Affairs*, 73 (6): 62–78.

Krugman, P. (1995) 'Dutch Tulips and Emerging Markets', *Foreign Affairs*, 74 (4): 28–4.

Krugman, P. (2000) 'Fire-sale FDI', in Edwards, S. (ed.) *Capital Flows and the Emerging Economies: Theory, Evidence, and Controversies*, Chicago, IL: University of Chicago Press.

Kuczynski, P-P. and Williamson, J. (eds) (2003) *After the Washington Consensus: Restarting Growth and Reform in Latin America*, Washington, DC: Institute for International Economics.

Kunz, R. (2011) *The Political Economy of Global Remittances: Gender, Governmentality and Neoliberalism*, London: Routledge.

Kuznets, S. (1973) 'Modern Economic Growth: Findings and Reflections', *American Economic Review*, 63: 247–258.

Kyrili, K. and Martin, M. (2010) 'The Impact of the Global Economic Crisis on the Budgets of Low-Income Countries', available at: www.oxfam.org.uk/resources/policy/economic_crisis/ downloads/rr_gec_impact_budget_lics_200710.pdf (accessed 13 October 2010).

La Porta, R.F. *et al.* (1997) 'Legal determinants of external finance', *Journal of Finance*, 52 (3): 1131–1150.

Lall, S. (1978) 'Transnationals, Domestic Enterprises and Industrial Structure in Host LDCs: A Survey, *Oxford Economic Papers*, 30 (2): 217–248.

Lamont, J. and Barker, A. (2010) 'Future of UK aid to India in question', *Financial Times*, 15 September.

Lancaster, C. (1983) 'Africa's Economic Policy', *Foreign Policy*, fall.

Lancaster, C. (1999) 'Aid Effectiveness: The Problem of Africa', *Development Outreach*, 1 (2): 22–24.

Lancaster, C. (2007) *Foreign Aid: Diplomacy, Development, Domestic Politics*, Chicago, IL: University of Chicago Press.

Landes, D. (1969) *The Unbound Prometheus: Technological Change and Industrial Development in Western Europe from 1750 to the Present*, Cambridge: Cambridge University Press.

Lane, G.M. and Milesi-Ferretti, P.R. (2004) 'International Investment Patterns', IMF Working Paper 134, Washington, DC: International Monetary Fund.

Langley, P. (2002) *World Financial Orders: An Historical International Political Economy*, London: Routledge.

Lapavitsas, C. (2008) 'Financialised Capitalism: Direct Exploitation and Periodic Bubbles', SOAS mimeo, available at: www.soas.ac.uk/economics/events/crisis/43939.pdf (accessed 20 December 2011).

Lasswell, H.D. (1936) *Politics: Who Gets What, When, How*, New York: McGraw-Hill.

Lauridsen, L.S. (1995) 'The Developmental State and the Asian Miracles: An Introduction to the Debate', in *Institutions and Industrial Development: Asian Experiences*, Occasional Paper no. 16, Roskilde: International Development Studies Roskilde University.

Leatherman, S. and Dunford, C. (2010) 'Linking Health to Microfinance to Reduce Poverty', *Bulletin of the World Health Organization*, 88: 470–471.

Lee, J. (1993), 'International Trade, Distortions and Long-Run Economic Growth', *IMF Staff Papers*, 40: 299–328.

Leftwich, A. (1993) 'States of Underdevelopment. The Third World State in Theoretical Perspective', *Journal of Theoretical Politics*, 6 (1): 55–74.

Leftwich, A. (1995) 'Bringing Politics Back in: Towards a Model of the Developmental State', *Journal of Development Studies*, 31 (3): 400–427.

Leibenstein, H. (1957) *Economic Backwardness and Economic Growth*, Chichester: Wiley.

Lenin, V.I. (1916) *Imperialism: The Highest Stage of Capitalism*, London: Penguin, 2010.

Leo, B. (2010) *Will World Bank and IMF Lending Lead to HIPC IV? Debt Deja-Vu All Over Again*,

Leo, B. and Thuotte, R. (2011) 'MDG Progress Index 2011 The Good (Country Progress), the Bad (Slippage), and the Ugly (Fickle Data)', *CGD Notes*, Washington DC: Center for Global Development.

Levine, R. (1997) 'Financial Development and Economic Growth', *Journal of Economic Literature*, 35 (2): 688–727.

Levine, R. and Kinder, M. (2004) *Millions Saved: Proven Successes in Global Health*, Washington, DC: Center for Global Development, What Works Working Group.

Levine, R. and Zervos, S.J. (1993) 'What we Have Learned About Policy and Growth from Cross-country Regressions?', *The American Economic Review*, 83 (2): 426–430.

Levine, R., Loayza, N. and Beck, T. (2000) 'Financial Intermediation and Growth: Causality and Causes', *Journal of Monetary Economics*, 46: 31–77.

Levitt, P. (2001) *The Transnational Villagers*, Berkeley, CA: University of California Press.

Lewis, A.W. (1954) 'Economic Development with Unlimited Supplies of Labor', *Manchester School of Economic and Social Studies*, 22 (2): 139–191.

Lewis, W.A. (1954) 'Economic Development with Unlimited Supplies of Labour', *Manchester School of Economic and Social Studies*, 22: 139–191.

Lewis, W.A. (1955) *The Theory of Economic Growth*, London: Allen & Unwin.

Leyshon, A. and Thrift, N. (1997) *Money/Space*, London: Routledge.

Lin, J.Y. (2012) *New Structural Economics: A Framework for Rethinking Development and Policy*, Washington, DC: World Bank.

Lin, J.Y. and Chang H-J. (2009) 'Should Industrial Policy in Developing Countries Conform to Comparative Advantage or Defy it? A Debate Between Justin Lin and Ha-Joon Chang', *Development Policy Review*, 27 (5): 483–502.

Lin, J.Y. and Monga, C. (2010) 'Growth Identification and Facilitation: The Role of the State in the Dynamics of Structural Change', World Bank Policy Research Working Paper 5313, Washington, DC: World Bank

List, F. (1841) *The National System of Political Economy*, New Jersey: Augustus M. Kelly, 1991.

Lindblom, C.E. (1959) 'The science of "muddling through" ', *Public Administration Review*, 19 (2): 78–88.

Littlefield, E., Morduch, J. and Hashemi, S. (2003) 'Is Microfinance an Effective Strategy to Reach the Millennium Development Goals?', Focus Note 24, Washington, DC: CGAP.

Lübker, M., Smith, G. and Weeks, J. (2002), 'Growth and the Poor: A Comment on Dollar and Kraay', *Journal of International Development*, 14: 555–571.

Lucas, R.E. (1988) 'On the Mechanics of Economic Development', *Journal of Monetary Economics*, 22: 3–42.

Lucas, R.E. (1990) 'Why Doesn't Capital Flow from Rich to Poor Countries?', *The American Economic Review*, 80 (2): 92–96.

Lucas, R.E. (2004) 'The Industrial Revolution: Past and Future, 2003 Annual Report Essay', *The Region*, May, available at: www.minneapolisfed.org/publications_papers/pub_display.cfm?id=3333 (accessed 20 March 2012).

Lumsdaine, D. (1993) *Moral Vision in International Politics: The Foreign Aid Regime, 1949–1989*, Princeton, NJ: Princeton University Press.

Luxemburg R. ([1913] 1951) *The Accumulation of Capital*, London: Routledge & Kegan Paul.

MacLaughlin, A. and Ricaurte, M. (2010) Appendix to IMF *Chile: 2010 Article IV Consultation-Staff Report; Staff Supplement; Public Information Notice on the Executive Board Discussion; and Statement by the Executive Director for Chile*, IMF Country Report No. 10/298, Washington, DC: International Monetary Fund, available at: www.imf.org/external/pubs/ft/scr/2010/cr10298.pdf (accessed 11 October 2010).

Macrae, J. and Leader, N. (2001) 'Apples, Pears and Porridge: The Origins and Impact of the Search for "Coherence" Between Humanitarian and Political Responses to Chronic Political Emergencies', *Disasters*, 25 (4): 290–307.

Maddison, A. (2007) *Contours of the World Economy 1–2030 AD: Essays in Macro-Economic History*, Oxford: Oxford University Press.

Madeley, J. (2008) *Big Business, Poor Peoples: How Transnational Corporations Damage the World's Poor*, 2nd edn, London: Zed.

Maes, J.P. and Reed, L.R. (2012) *State of the Microcredit Summit Campaign Report 2012*, Washington, DC: Microcredit Summit Campaign.

Maimbo, S.M. and Ratha, D. (eds) (2005a) *Remittances: Development Impact and Future Prospects*, Washington, DC: World Bank.

Maimbo, S.M. and Ratha, D. (2005b) 'Remittances: An Overview', in Maimbo, S.M. and Ratha D. (eds) *Remittances: Development Impact and Future Prospects*, Washington, DC: World Bank.

Mallick, R. (2002) 'Implementing and Evaluation of Microcredit in Bangladesh', *Development in Practice*, 12: 153–163.

Malthus, T. (1820) *Principles of Political Economy*, 2nd edn, published 1836, London: W. Pickering.

Mankiw, N.G. (1989) 'Real Business Cycle: A New Keynesian Perspective', *Journal of Economic Perspectives*, 3 (3): 79–90.

Mankiw, N.G., Romer, D. and Weil, D.N. (1992) 'A Contribution to the Empirics of Economic Growth', *The Quarterly Journal of Economics*, 107 (2): 407–437.

Manning, R. (2006) 'Will "Emerging Donors" Change the Face of International Co-operation?', *Development Policy Review*, 24 (4): 371–385.

Marshall, A. (1907) *Principles of Economics*, London: Macmillan.

Marston, R. (2013) 'What is a Rating Agency?', BBC website, 25 February 2013, available at: www. bbc.co.uk/news/10108284 (accessed 1 March 2013).

Martens, B (ed.) (2004) *The Institutional Economics of Foreign Aid*, Cambridge: Cambridge University Press.

Marx, K. (1844) 'On the Jewish Question', in Tucker, R.C. (ed.) *The Marx-Engels Reader*, New York: W.W. Norton & Co., 1978.

Marx, K. (1857–1858) *Grundrisse: Foundations of the Critique of Political Economy*, London: Penguin, 1973.

Marx, K. (1867) *Capital: A Critique of Political Economy, Volume I*, Harmondsworth: Penguin, 1976.

Marx, K. (1886) *Capital, Volume II*, London: Penguin, 1978.

Marx, K. (1894) *Capital, Volume III*, Harmondsworth: Penguin, 1981.

Marx, K. (1970) *A Contribution to the Critique of Political Economy*, Moscow: Progress Publishers, 1859.

Marx, K. (1975) 'On the Jewish Question' in Marx, K. and Engels, F., *Collected Works*, Vol. 3, *Marx and Engels 1843–44*, London: Lawrence & Wishart, 1843, pp. 146–174.

Marx, K. (1986) *Karl Marx: A Reader*, ed. J. Elster. Cambridge: Cambridge University Press.

Maslow, A.H. (1943) 'A Theory of Human Motivation', *Psychological Review*, 50 (4): 370–396.

Matin, I., Sulaiman, M. and Rabbani, M. (2008) 'Crafting a Graduation Pathway for the Ultra Poor: Lessons and Evidence from a BRAC programme', Chronic Poverty Research Centre Working Paper No. 109, available at: www.chronicpoverty.org/uploads/publication_files/WP109_Martin.pdf (accessed 11 August 2010).

McCulloch, N. and Pacillo, G. (2011) 'The Tobin Tax: A Review of the Evidence', *IDS Research Reports*, 68: 1–77.

McDonnell, I., Lecomte, H. and Wegimont, L. (eds) (2003) *Public Opinion and the Fight Against Poverty*, Paris: OECD/North-South Centre.

McGillivray, M. and Morrissey, O. (2000): 'Aid Fungibility in Assessing Aid: Red Herring or True Concern?', *Journal of International Development*, 12 (3): 413–428.

McGillivray, M. and Morrissey, O. (2004) 'Fiscal Effects of Aid', in Addison, T. and Roe, A. (eds) *Fiscal Policy for Development: Poverty, Reconstruction, and Growth*, Basingstoke: Palgrave Macmillan in association with UNU-WIDER.

McKenzie, D. and Woodruff, C. (2008) 'Experimental Evidence on Returns to Capital and Access to Finance in Mexico', *World Bank Economic Review*, 22 (3): 457–482.

McKinlay, R.D. and Little, R. (1979) 'The US Aid Relationship: A Test of the Recipient Need and Donor Interest Models', *Political Studies*, XXVII (2): 236–250.

McKinnon, R. (1973) *Money and Capital in Economic Development*, Washington, DC: The Brookings Institution.

McKinnon, R. (1991) *The Order of Economic Liberalization: Financial Control in the Transition to a Market Economy*, Baltimore, MD: Johns Hopkins University Press.

McKinnon, R. and Pill, H. (1997) 'Credible Economic Liberalizations and Overborrowing', *American Economic Review*, 87 (2): 189–193.

McMichael, P. (1998) 'Development and Structural Adjustment', in Carrier, J.G. and Miller, D. (eds) *Virtualism: A New Political Economy*, Oxford: Berg.

MDG Gap Task Force (2011) *Millennium Development Goal 8: The Global Partnership for Development: Time to Deliver. MDG Gap Task Force Report 2011*, New York: United Nations, available at: www. un.org/en/development/desa/policy/mdg_gap/mdg_gap2011/mdg8report2011_engw.pdf (accessed 16 September 2011).

Meka, S. and Johnson, D. (2010) 'Access to Finance in Andhra Pradesh', Chennai: IFMRC MFC and CFMRB IRD.

Mellor, M. (2012) 'Money as a Public Resource for Development', *Development*, 55 (1): 45–5.

Menger, C. (1871) *Principles of Economics*, London: The Free Press, 1959.

Menger, C. (1892) 'On the Origin of Money', *Economic Journal*, 2 (6): 239–255.

Merton, R.K. (1968) 'The Matthew Effect in Science', *Science*, 159: 56–63.

Merz, B.J., Chen, L.C. and Geithner, P.F. (eds) (2007) *Diasporas and Development*, Cambridge, MA: Harvard University Press.

Mihalache-O'keef, A. and Li, Q. (2011) 'Modernization vs. Dependency Revisited: Effects of Foreign Direct Investment on Food Security in Less Developed Countries', *International Studies Quarterly*, 55: 71–93.

Milanovic, B. (2005) *World's Apart: Measuring International and Global Inequality*, Princeton, NJ: Princeton University Press.

Milanovic, B. (2011) *The Haves and the Have-Nots*, New York: Basic Books.

Mill, J.S. (1848) *Principles of Political Economy with some of their Applications to Social Philosophy*, 7th edn, London: Longmans, Green & Co., 1909.

Millet, D. and Toussaint, E. (2004) *Who Owes Who?: 50 Questions about World Debt*, London: Zed.

Milner and Tingley (2010) 'The Political Economy of US Foreign Aid: American Legislators and the Domestic Politics of Aid', *Economics & Politics*, 22 (2): 200–232.

Mingat, A., Rakotomalala, R. and Tan, J.P. (2002) *Financing Education for All by 2015: Simulations for 33 African Countries. Africa Region*, Washington, DC: World Bank.

Minsky, H. (1982) *Can 'It' Happen Again?*, New York: M.E. Sharpe.

Minsky, H. (1986) *Stabilizing an Unstable Economy*, New Haven, CT: Yale University Press.

Mishra, D., Mody, A. and Murshid, A.P. (2001) 'Private Capital Flows and Growth', *Finance & Development*, 38 (2).

Mitchell, A. (2011) 'The Future of DFID's Programme in India', witness statement for the International Development Committee, Monday 28 March.

Mody, A. and Taylor, M.P. (2002) 'International Capital Crunches: The Time-Varying Role of Informational Asymmetries', IMF Working Papers, No. 02/43, Washington, DC: International Monetary Fund.

Mohapatra, S., Ratha, D. and Silwal, A. (2011) 'Outlook for Remittance Flows 2012–14', Migration and Development Brief 17, Washington, DC: World Bank Migration and Remittances Unit.

Montes, M. (1998) *The Currency Crisis in Southeast Asia*, Singapore: Institute of Southeast Asian Studies.

Moran, T.H. (1998) *FDI and Development: The New Policy Agenda for Developing Countries and Economics in Transition*, Washington, DC: Institute for International Economics.

Moran, T.H. (2002) *Beyond Sweatshops: Foreign Direct Investment and Globalization in Developing Nations*, Washington, DC: Brookings Institution Press.

Morduch, J. (1999a) 'The Microfinance Schism', *World Development*, 28 (4): 617–629.

Morduch, J. (1999b) 'Between the Market and State: Can Informal Insurance Patch the Safety Net?', *World Bank Research Observer*, 14: 187–207.

Morduch, J. (2006) 'Micro-insurance: The Next Revolution? What have we learned about poverty?', in Banerjee, A., Benabou, R. and Mookherjee, D. (eds) *Understanding Poverty*, Oxford: Oxford University Press.

Morduch, J. and Haley, B. (2002) 'Analysis of the Effects of Microfinance on Poverty Reduction', Wagner Working Paper 1014, New York: New York University.

Morgenthau, H. (1960) 'Foreword', in Liska, G. *The New Statecraft: Foreign Aid in American Foreign Policy*, Chicago: University of Chicago Press.

Morgenthau, H. (1962) 'A Political Theory of Foreign Aid', *American Political Science Review*, 56 (2): 301–309.

Morris, M.D. (1979) *Measuring the Condition of the World's Poor, The Physical Quality of Life Index*, New York: Pergamon Press.

Morris, M.D. (1980) 'Physical Quality of Life Index', *Development Digest*, 18 (1): 95–109.

Morrissey, O. (2002) 'Aid Effectiveness for Growth and Development', ODI Opinions, February, available at: www.odi.org.uk/resources/download/446.pdf (accessed 16 December 2013).

Moses, J.W., and Letnes, B. (2005) 'If People Were Money: Estimating the Gains and Scope of Free Migration', in Borjas G.J. and Crisp, J. (eds) *Poverty, International Migration and Asylum*, New York: Palgrave Macmillan.

Mosley, L. (2000) 'Room to Move: International Financial Markets and National Welfare States', *International Organization*, 54: 4.

Mosley, L. (2003) *Global Capital and National Governments*, Cambridge: Cambridge University Press.

Mosley, L. (2005) 'Globalization and the State: Still Room to Move?' *New Political Economy*, 10 (3): 355–362.

Mosley, P. and Toye, J. (1988) 'The Design of Structural Adjustment Programmes', *Development Policy Review*, 6 (4): 395–413.

Mosley, P., Harringan, J. and Toye, J. (1991) *Aid as Power: The World Bank and Policy-Based Lending*, Vol. 1, London: Routledge.

Moss, T., Pettersson, G. and van de Walle, N. (2008) 'An Aid-Institutions Paradox? A Review Essay on Aid Dependency and State Building in Sub-Saharan Africa', in W. Easterly (ed.) *Reinventing Foreign Aid*, Cambridge, MA: MIT Press, pp. 255–228.

Moyo, D. (2009) *Dead Aid: Why Aid Is Not Working and How There Is Another Way for Africa*, London: Penguin.

Mundell, R.A. (1963) 'Capital Mobility and Stabilization Policy Under Fixed and Flexible Exchange Rates', *Canadian Journal of Economic and Political Science*, 29 (4): 475–485.

Myrdal, G. (1989) 'The Equality Issue in World Development', *American Economic Review*, 79 (6): 8–17.

Naím, M. (1995) 'Latin America: The Morning After', *Foreign Affairs*, 74 (4): 45–61.

Naím, M. (2000a) 'Washington Consensus or Washington Confusion?' *Foreign Policy*, spring: 86–103.

Naím, M. (2000b) 'Fads and Fashions in Economic Reform: Washington Consensus or Washington Confusion?', *Third World Quarterly*, 21 (3): 514–515.

Naím, M. (2002) 'Washington Consensus: A Damaged Brand', *Financial Times*, 28 October.

Narayan, D. with Patel, R., Schafft, K., Rademacher, A. and Koch-Schulte, S. (2000) *Voices of the Poor: Can Anyone Hear Us?*, New York: Oxford University Press.

Naschold, F. (2002) 'Why Inequality Matters for Poverty', Inequality Briefing Paper, No. 2 (March), London: Department for International Development.

National Executive Committee (1944) 'Full Employment and Financial Policy', British Labour Party, June.

Nelson, A. (2001) 'Marx's Theory of the Money Commodity', *History of Economic Review*, 33: 44–63.

Nelson, R.R. (1956) 'A Theory of the Low-Level Equilibrium Trap in Underdeveloped Economies', *The American Economic Review*, 46 (5): 894–908.

Nesvetailova, A. (2007) *Fragile Finance: Debt, Speculation and Crisis in the Age of Global Credit*, Basingstoke: Palgrave Macmillan.

Nesvetailova, A. (2010) *Financial Alchemy in Crisis: The Great Liquidity Illusion*, London: Pluto.

Neumayer, E. and Spess, L. (2005) 'Do Bilateral Investment Treaties Increase Foreign Direct Investment to Developing Countries?', *World Development*, 33 (10): 1567–1585.

Nguyen, T.Q., Clements, B.J. and Bhattacharya, R. (2005) 'Can Debt Relief Boost Growth in Poor Countries?', *Economic Issues*, 34.

Noel, A. and Therien, J-P. (1995) 'From Domestic to International Justice: The Welfare State and Foreign Aid', *International Organization*, 49 (3): 523–553.

Nicholson, A. (2009) 'IMF Says New Reserve Currency to Replace Dollar is Possible', *Bloomberg*, 6 June, available at: www.bloomberg.com/apps/news?pid=newsarchive&sid=aUYeJEwZaQrw (accessed 15 December 2013).

North, D.C. (1981) *Structure and Change in Economic History*, New York: W.W. Norton & Co.

North, D.C. (1990) *Institutions, Institutional Change and Economic Performance*, Cambridge: Cambridge University Press.

North, D.C. (1995) 'The New Institutional Economics and Third World Development', in Harriss, J., Hunter, J. and Lewis, C.M. (eds) *The New Institutional Economics and Third World Development*, London: Routledge.

North, D.C. and Weingast, B.R. (1989) 'Constitutions and Commitment: The Evolution of Institutions Governing Public Choice in Seventeenth-Century England', *Journal of Economic History*, XLIX (4): 803–832.

Nurske, R. (1944) *International Currency Experience*, Geneva: League of Nations.

Nurske, R. (1953) *Problems of Capital-Formation in Underdeveloped Countries*.

O'Brien, R. and Williams, M. (2010) *Global Political Economy*, Basingstoke: Palgrave Macmillan.

O'Meally, S. (2011) *Unsung Progress in Rural Sanitation: Building the Foundations in Lao PDR*, London: ODI.

Ocampo, J.A., Kregel, J. and Griffith-Jones, S. (eds) (2007) *International Finance and Development*, London: Zed.

Odedokun, M. (1996) 'Alternative Econometric Approaches for Analysing the Role of the Financial Sector in Economic Growth: Time Series Evidence from LDCs', *Journal of Development Economics*, 50 (1): 119–146.

OECD (1972) *Development Co-operation Report 1972*, Paris: OECD.

OECD (2002) 'History of the 0.7% ODA Target', *The DAC Journal*, 3 (4): III-9-III-11.

OECD (2005a) *Migration, Remittances and Development*, Paris: OECD.

OECD (2005b) *The Paris Declaration on Aid Effectiveness*, Paris: OECD, available at: www.oecd.org/development/effectiveness/34428351.pdf.

OECD (2008a) 'Is it ODA?', *OECD Factsheet*, available at: www.oecd.org/dataoecd/21/21/34086975.pdf (accessed 1 August 2010).

OECD (2008b) *OECD Benchmark Definition of Foreign Direct Investment*, 4th edn, Paris: OECD.

OECD (2008c) *The Accra Agenda for Action*, Paris: OECD, available at: www.oecd.org/development/effectiveness/34428351.pdf.

OECD (2009a) *Development Brief Issue 1 2009*, Paris: OECD, available at: www.oecd.org/dataoecd/14/34/43853485.pdf (accessed 26 July 2010).

OECD (2009b) *International Development Statistics 2009*, Paris: OECD.

OECD (2010a) *Geographical Distribution of Financial Flows to Developing Countries 2010*, Paris: OECD.

OECD (2010b), *Development Co-operation Report 2010*, Paris: OECD.

OECD (2010c) 'Development Aid Rose in 2009 and Most Donors will Meet 2010 Aid Targets', OECD Aid Statistics Website, 14 April, available at: www.oecd.org/document/11/0,334 3,en_2649_34447_44981579_1_1_1_37413,00.html (accessed 1 August 2010).

OECD (2011a) *Aid Effectiveness 2005–10: Progress in Implementing the Paris Declaration*, Paris: OECD, available at: http://www.oecd.org/dataoecd/25/30/48742718.pdf (accessed 21 December 2011).

OECD (2011b) *Busan Partnership for Effective Development Co-operation*, available at: http://www.aideffectiveness.org/busanhlf4/images/stories/hlf4/OUTCOME_DOCUMENT_-_FINAL_EN.pdf (accessed 21 December 2011).

OECD (2011c) 'Getting the Most out of International Capital Flows', OECD Economics Department Policy Notes 6, Paris: OECD.

OECD (2013) 'Aid to Poor Countries Slips Further as Governments Tighten Budgets', 3 April, available at: http://www.oecd.org/dac/stats/aidtopoorcountriesslipsfurtherasgovernmentstightenbudgets.htm.

OECD/DAC (2013) *Geographical Distribution of Financial Flows to Developing Countries 2012: Disbursements, Commitments, Country Indicators*, Paris: OECD.

OECD DCD/DAC Development Co-Operation Directorate/Development Assistance Committee (2007a) *DAC Statistical Reporting Directives*, DCD/DAC(2007)34, Paris: OECD, available at: www.oecd.org/dataoecd/28/62/38429349.pdf (accessed 1 August 2010).

OECD DCD/DAC Development Co-Operation Directorate/Development Assistance Committee (2007b) *Reporting Directives for the Creditor Reporting System*, DCD/DAC(2007)39/FINAL, Paris: OECD, available at: www.oecd.org/dataoecd/28/62/38429349.pdf (accessed 1 August 2010).

OECD.Stat (2013) Creditor Reporting System, available at: www.oecd.org/development/stats/internationaldevelopmentstatisticsidsonlinedatabasesonaidandotherresourceflows.htm#dac.

Olson, M. (1996) 'Big Bills Left on the Sidewalk: Why Some Nations are Rich and Others Poor', *Journal of Economic Perspectives*, 10 (2): 3–24.

ONS (2010) *Labour Force Survey 2010*, London: Office for National Statistics.

Orozco, M. (2003) *Worker Remittances: An International Comparison*, Washington, DC: Inter-American Development Bank.

Orozco, M. (2004) 'The Remittance Marketplace: Prices, Policy and Financial Institutions', Pew Hispanic Center Report, Washington, DC: Georgetown University and Institute for the Study of International Migration.

Orozco, M. (2005) 'Transnationalism and Development: Trends and Opportunities in Latin America', in Maimbo, S.M. and Ratha, D. (eds) *Remittances. Development Impact and Future Prospects*, Washington, DC: World Bank.

Orozco, M. and Fedewa, R. (2006) 'Leveraging Efforts on Remittances and Financial Intermediation', INTAL-ITD Working Paper 24, Buenos Aires: Inter-American Development Bank.

Orozco, M. with Lapointe, M. (2004) 'Mexican Hometown Associations and Development Opportunities', *Journal of International Affairs*, 57 (2): 1–21.

Ostry, J.D., Ghosh, A.R., Habermeier, K., Chamon, M., Qureshi, M.S. and Reinhardt, D.B.S. (2010) 'Capital Inflows: The Role of Controls', IMF Staff Position Note 10/04, Washington, DC: IMF, available at: www.imf.org/external/pubs/ft/spn/2010/spn1004.pdf (accessed 10 October 2010).

Ostry, J., Atish, D., Ghosh, R., Habermeier, K., Chamon, M., Qureshi, M.S. and Reinhardt, D.B.S. (2010) 'Capital Inflows: The Role of Controls', IMF Staff Position Note, Washington, DC: International Monetary Fund.

Oxfam (2002) 'Last Chance at Monterrey: Meeting the Challenge of Poverty Reduction', Briefing Paper No. 17, Oxford: Oxfam.

Oxfam (2010) *21st Century Aid: Recognising Success and Tackling Failure*, Oxfam Briefing Paper 137, Oxford: Oxfam International.

Özden, Ç. and Schiff, M. (eds) (2006) *International Migration, Remittances and the Brain Drain*, Basingstoke: Palgrave/World Bank.

Packenham, R.A. (1966) 'Foreign Aid and the National Interest', *Midwest Journal of Political Science*, 10 (2): 214–221.

Page, B. (2007) 'Slow Going: The Mortuary, Modernity and the Hometown Association in Bali-Nyonga, Cameroon', *Africa*, 77 (3): 419–441.

Palan, R., Murphy, R. and Chavagneux, C. (2011) *Tax Havens: How Globalization Really Works*, Ithaca, NY: Cornell University Press.

Patrick, H.T. (1966) 'Financial Development and Economic Growth in Underdeveloped Countries', *Economic Development and Cultural Change*, 14: 174–189.

Pascaline, D. and Robinson, J. (2013) 'Savings Constraints and Microenterprise Development: Evidence from a Field Experiment in Kenya', *American Economic Journal: Applied Economics*, 5 (1): 163–192.

Patten, R., Rosengard, J. and Johnston, D. (2001) 'Microfinance Success Amidst Macroeconomic Failure: The Experience of Bank Rakyat Indonesia During the East Asian Crisis', *World Development*, 29 (6): 1057–1069.

Pattillo, C., Poirson, H. and Ricci, L. (2002) 'External Debt and Growth', IMF Working Paper 02/69, Washington, DC: International Monetary Fund.

Pauly, L. (1997) *Who Elected the Bankers? Surveillance and Control in the World Economy*, Ithaca, NY: Cornell University Press.

Pauly, L. (2000) 'Capital Mobility and the New Global Order', in Stubbs, R. and Underbill, G.R.D. (eds) *Political Economy and the Changing Global Order*, Oxford: Oxford

Payne, A. (2005) *The Global Politics of Unequal Development*, Basingstoke: Palgrave Macmillan.

Payne, A. and Phillips, N. (2010) *Development*, Cambridge: Polity.

Peachey, S. and Roe, A. (2004) *Access to Finance: A Study for the World Savings Bank Institute*, Oxford: Oxford Policy Management.

Pearson, L.B. (1969), *Partners in Development: Report of the Commission on International Development*, London: Pall Mall Press.

Peet, R. (2003) *Unholy Trinity: The IMF, World Bank, and WTO*, London: Zed.

Peet, R. (2007) *Geography of Power: The Making of Global Economic Policy*, London: Zed.

Perlman, M. (1987) 'Introduction' in Schumpeter, J.A. *History of Economic Analysis*,

Perry, G. (2007) 'Foreword', in Fajnzylber, P. and López, J.H. (2007) *Close to Home: The Development Impact of Remittances in Latin America*, Washington, DC: World Bank.

Petras, R./OECD/DGF (Rudolphe Petras) (2009) *Comparative Study of Data Reported to the OECD Creditor Reporting System (CRS) and to the Aid Management Platform (AMP)*.

Pitt, M. and Kandker, S. (1998) 'The Impact of Group-Based Credit Programs on Poor Households in Bangladesh: Does the Gender of Participants Matter?', *Journal of Political Economy*, 106 (5): 958–996.

Pogge, T. (2008) *Human Rights and World Poverty*, 2nd edn, Cambridge: Polity.

Polanyi, K. (1944) *The Great Transformation: The Political and Economic Origins of Our Time*, Boston, MA: Beacon Books.

Porter, T. (2005) *Globalization and Finance*, Cambridge: Polity Press.

Portes, R. and Rey, H. (2003) 'The Determinants of Cross-Border Equity Flows: The Geography of Information',

Povcal (2013) 'Povcal: the on-line tool for poverty measurement developed by the Development Research Group of the World Bank', available at: http://iresearch.worldbank.org/PovcalNet/index. htm (accessed 1 May 2013).

Prahalad, C.K. (2004) *The Fortune at the Bottom of the Pyramid: Eradicating Poverty through Profits*, Philadelphia, PA: Wharton School Press.

Prasad, E., Rogoff, K., Wei, S-J. and Kose, M.A. (2003) 'Effects of Financial Globalization on Developing Countries: Some Empirical Evidence', IMF Occasional Paper 220, Washington, DC: International Monetary Fund.

Prasad, E., Rogoff, K., Wei, S.J. and Kose, M.A. (2004) 'Financial Globalization, Growth and Volatility in Developing Countries', National Bureau of Economic Research Working Paper, No. 10942, Cambridge, MA: NBER.

Prasad, E., Rajan, J. and Subramanian, A. (2007) 'Foreign Capital and Economic Growth', *Brookings Papers on Economic Activity*, 1: 153–230.

Prati, A., Sahay, R. and Tressel, T. (2003) 'Is There a Case for Sterilizing Foreign Aid Inflows?' Paper presented at IMF Seminar at European Economic Association/Econometric Society European Meeting, Stockholm, Sweden.

Prebisch, R. (1950) 'The Economic Development of Latin America and its Principal Problems', *Economic Bulletin for Latin America*, 7 (1): 1–12.

Prebisch, R. (1981) *Capitalismo periférico. Crisis y transformación*, Mexico City: Fondo de Cultura Económica.

Pritchett, L. (1997) 'Divergence, Big Time', *Journal of Economic Perspectives*, 11 (3): 3–17.

Pritchett, L. (2006) 'Who is Not Poor? Dreaming of a World Truly Free of Poverty', *World Bank Research Observer*, 21 (1): 1–23.

PIPA (Program on International Policy Attitudes) (2001) *Americans on Foreign Aid and World Hunger: A Study of US Public Attitudes*, available at: www.pipa.org/OnlineReports/ForeignAid/ForeignAid_ Feb01/ForeignAid_Feb01_rpt.pdf (accessed 1 August 2010).

Pronyk, P.M., Hargreaves, J.R., Kim, J.C., Morison, L.A., Phetla G., Watts, C., Busza, J. and Porter, J.D. (2006) 'Effect of a Structural Intervention for the Prevention of Intimate-partner Violence and HIV in Rural South Africa: A Cluster Randomised Trial', *Lancet*, 368 (9551): 1973–1983.

Pronyk, P.M., Hargreaves, J.R. and Morduch, J. (2007) 'Microfinance Programs and Better Health: Prospects for Sub-Saharan Africa', *JAMA*, 298 (16): 1925–1927.

Przeworski, A. (1985) *Capitalism and Social Democracy*, Cambridge: Cambridge University Press.

Radelet, S. (2006) 'A Primer on Foreign Aid', Center for Global Development, Working Paper Number 92, July.

Radelet, S., Clemens, M. and Bhavnani, R. (2005) 'Aid and Growth', *Finance and Development*, 42 (3).

Rahman A. (1999) 'Micro-credit Initiatives for Equitable and Sustainable Development: Who Pays?', *World Development*, 27: 67–82.

Rahnema, M. and Bawtree, V. (eds) (1997) *The Post-development Reader*, London: Zed.

Rajan and Zingales (1998) 'Financial Dependence and Growth', *American Economic Review*, 88 (3): 559–586.

Rajan, R. and Subramanian, A. (2005) 'What Undermines Aids Impact on Growth?', IMF Working Paper, No. 05/127, Washington, DC: International Monetary Fund.

Rajan, R. and Zingales, L. (2003) 'The Great Reversals: The Politics of Financial Development in the Twentieth Century', *Journal of Financial Economics*, 69: 5–50.

Randewich, N. (2007) 'Mexican Microlending Bank Surges in Market Debut', *Reuters*, 20 April, available at: www.reuters.com/article/2007/04/20/mexico-compartamos-idUSN2025193920070420 (accessed 1 May 2013).

Rao, P.K. (2003) *Development Finance*, London: Springer.

Ratha, D. (2005) 'Remittances: A Lifeline for Development', *Finance & Development*, 42 (2), available at: www.imf.org/external/pubs/ft/fandd/2005/12/basics.htm (accessed 16 December 2013).

Ratha, D. (2010) 'Mobilize the Diaspora for the Reconstruction of Haiti', SSRC Features: Haiti: Now and Next, 11 February, available at: www.ssrc.org/features/pages/haiti-now-and-next/1338/1438/ (accessed 30 August 2010).

Ratha, D. (2012) 'Remittances: Funds for the Folks Back Home', *Finance & Development Back to Basics*, available at: www.imf.org/external/pubs/ft/fandd/basics/remitt.htm (accessed 13 March 2012).

Ratha, D., Mohapatra, S. and Silwal, A. (2010) 'Outlook for Remittance Flows 2010–11', *Migration and Development Brief 12*, Washington, DC: World Bank.

Ravallion, M. (1998) 'Poverty Lines in Theory and Practice', *Living Standards Measurement Study Working Paper 133*, Washington, DC: World Bank.

Ravallion, M. (2008a) 'How *Not* to Count the Poor? A Reply to Reddy and Pogge', in Anand, S., Segal, P. and Stiglitz, J.E. (eds) *Debates on the Measurement of Global Poverty*, Oxford: Oxford University Press.

Ravallion, M. (2008b) 'Evaluating Anti-Poverty Programs', in Schultz, T.P. and Strauss, J. (eds) *Handbook of Development Economics Volume 4*, Oxford: North-Holland.

Ravallion, M. (2009) 'Should the Randomistas Rule?', *The Economists' Voice*, 6 (2): Article 6.

Ravallion, M., Datt G. and van de Walle, D. (1991) 'Quantifying Absolute Poverty in the Developing World', *Review of Income and Wealth*, 37: 345–361.

Ravallion, M., Chen, S., Sangraula, P. (2008) 'Dollar a Day Revisited', *Policy Research Working Paper 4620*, Washington, DC: World Bank.

Reddy, S.G. and Pogge, T. (2010) 'How Not to Count the Poor', in Anand, S., Segal, P. and Stiglitz, J.E. (eds) *Debates on the Measurement of Global Poverty*, Oxford: Oxford University Press.

Reille, X. and Forster, S. (2008) 'Foreign Capital Investment in Microfinance: Balancing Social and Financial Returns', CGAP Focus Note 44, Washington, DC: CGAP.

Reinhart, C.M. and Reinhart, V.R. (2001) 'What Hurts Most? G-3 Exchange Rate or Interest Rate Volatility', in Edwards, S. and Frankel, J. (eds) *Preventing Currency Crises in Emerging Markets*, Chicago: University of Chicago Press.

Reinhart, C.M. and Rogoff, K.S. (2004) 'Serial Default and the "Paradox" of Rich to Poor Capital Flows', *American Economic Review Papers and Proceedings*, 94 (2): 53–58.

Reinhart, C.M., and Rogoff, K.S. (2009) *This Time is Different: Eight Centuries of Financial Folly*, Princeton, NJ: Princeton University Press.

Reinhart, C.M. and Rogoff, K.S. (2011) 'From Financial Crash to Debt Crisis', *American Economic Review*, 101: 1676–1706.

Reisen, H. and Soto, M. (2001) 'Which Types of Capital Inflows Foster Developing-Country Growth?', *International Finance*, 4 (1): 1–14.

Rhyne, E. (1999) 'The Yin and Yang of Microfinance: Reaching the Poor and Sustainability', *Microfinance Bulletin*, July: 6–9.

Ricardo, D. (1817) *On the Principles of Political Economy and Taxation*, ed. P. Sraffa, Cambridge: Cambridge University Press.

Riddell, R.C. (2007) *Does Foreign Aid Really Work?*, Oxford: Oxford University Press.

Roberge, I. (2011) 'Financial Action Task Force', in Hale, T. and Held, D. (eds) *Handbook of Transnational Governance: New Institutions and Innovations*, Cambridge: Polity.

Robinson, J. (1952) 'The Generalization of the General Theory', in *The Rate of Interest and Other Essays*, London: Macmillan.

Robinson, J. (1953–1954) 'The Production Function and the Theory of Capital', *Review of Economic Studies*, 21 (2): 81–106.

Robinson, J. (1956) *The Accumulation of Capital*, London: Macmillan.

Robinson, J. (1962) *Economic Philosophy*, Harmondsworth: Penguin.

Robinson, J. (1974) 'History Versus Equilibrium', in *Collected Economic Papers, Volume 5* (1980), Cambridge, MA: MIT Press.

Robinson, M.S. (2001) *The Microfinance Revolution: Sustainable Finance for the Poor*, Washington, DC: World Bank.

Robinson, W.I. (2011) 'Globalization and the Sociology of Immanuel Wallerstein: A Critical Appraisal', *International Sociology*, 26 (6): 723–745.

Rodriguez, E.R. (1998) 'International Migration and Income Distribution in the Philippines', *Economic Development and Cultural Change*, 46 (2): 329–350.

Rodrik, D. (1997) *Has Globalization Gone Too Far?* Washington, DC: Institute for International Economics.

Rodrik, D. (1998) 'Who Needs Capital Account Convertibility?', in Kenen P. (ed.) *Should the IMF Pursue Capital Account Convertibility?*, Princeton Essays in International Finance no. 207.

Rodrik, D. (2000) 'Saving Transitions', *World Bank Economic Review*, 14 (3): 481–507.

Rodrik, D. (2005) 'Growth Strategies', in Aghion, P. and Durlauf, S.N. (eds) *Handbook of Economic Growth*, Vol. 1A, Oxford: Elsevier.

Rodrik, D. (2006a) 'Goodbye Washington Consensus, Hello Washington Confusion', *Journal of Economic Literature*, 44 (4): 973–987.

Rodrik, D. (2006b) 'The Social Cost of Foreign Exchange Reserves', *International Economic Journal*, 20: 253–66.

Rodrik, D. (2006c) 'Institutions for High-quality Growth: What They Are and How to Acquire Them', in Roy, K.C. and Sideras, J. (eds) *Institutions, Globalization, and Empowerment*, Chetenham: Edward Elgar.

Rodrik, D. (2007) *One Economics, Many Recipes: Globalization, Institutions, and Economic Growth*, Princeton, NJ: Princeton University Press.

Rodrik, D. (2008) 'Second-Best Institutions', *American Economic Review*, 98 (2): 100–104.

Rodrik, D., (2009a) 'Industrial Policy: Don't Ask Why, Ask How', *Middle East Development Journal*, 1 (1): 1–29.

Rodrik, D. (2009b) 'Let Finance Sceptics Take Over', *Business Standard*, 11 August, available at: www.business-standard.com/india/news/dani-rodrik-let-finance-sceptics-take-over/366591/ (accessed 29 December 2011).

Rodrik, D. and Velasco, A. (1999) 'Short-Term Capital Flows', National Bureau of Economic Research Working Papers, No. 7364, Cambridge, MA: NBER.

Rodrik, D., Subramanian, A. and Trebbi, F. (2004) 'Institutions Rule: The Primacy of Institutions Over Geography and Integration in Economic Development', *Journal of Economic Growth*, 9 (2): 131–165.

Rogerson, A. and Steensen, S. (2009) 'Aid Orphans: Whose Responsibility?', *OECD Development Brief*, October, available at: www.oecd.org/dac/aid-architecture/43853485.pdf> (accessed 16 December 2013).

Romer, P.M. (1986) 'Increasing Returns and Long Run Growth', *Journal of Political Economy*, 94: 1002–1037.

Romer, P.M. (1990) 'Endogenous Technological Change', *Journal of Political Economy*, 98: S71–S102.

Romer, P.M. (1993) 'Idea Gaps and Object Gaps in Economic Development', *Journal of Monetary Economics*, 32 (3): 543–573.

Romer, P.M. (1994) 'The Origins of Endogenous Growth', *The Journal of Economic Perspectives*, 8 (1): 3–22.

Roodman, D. (2007) 'The Anarchy of Numbers: Aid, Development, and Cross-Country Empirics', Center for Global Development Working Paper 32, Washington, DC: CGD.

Roodman, D. (2012) *Due Diligence: An Impertinent Inquiry into Microfinance*, Washington, DC: CGD.

Rose, E. (1999) 'Consumption Smoothing and Excess Female Mortality in Rural India', *Review of Economics and Statistics*, 81: 41–49.

Rosenberg, R. (1999) 'Measuring Microcredit Delinquency: Ratios Can Be Harmful to Your Health', *CGAP Occasional Paper No. 3*, Washington, DC: CGAP.

Rosenberg, R. (2002) 'Microcredit Interest Rates', CGAP Occasional Paper No. 1, Washington, DC: CGAP.

Rosenberg, R. (2007) 'CGAP Reflections on the Compartamos Initial Public Offering: A Case Study on Microfinance Interest Rates and Profits', CGAP Focus Note No. 42, Washington, DC: Consultative Group to Assist the Poor, June.

Rosenberg, R. (2009) 'Does Microcredit Really Help Poor People?', CGAP Blog, 5 October, available at: www.cgap.org/blog/does-microcredit-really-help-poor-people (accessed 25 March 2012).

Rosenstein-Rodan, P. (1943) 'Problems of Industrialization of Eastern and South-Eastern Europe', *Economic Journal*, 53: 202–211.

Rostow, W.W. (1952) *The Process of Economic Growth*, New York: Norton.

Rostow, W.W. (1959) 'The Stages of Economic Growth', *The Economic History Review*, 12 (1): 1–16.

Rostow, W.W. (1960) *The Stages of Economic Growth: A Non-Communist Manifesto*, Cambridge: Cambridge University Press.

Royal Government of Bhutan (2000) *Bhutan National Human Development Report 2000: Gross National Happiness and Human Development – Searching for Common Ground*, Bhutan: The Planning Commission Secretariat.

Ruggie, J. G. (1982) International Regimes, Transactions, and Change: Embedded Liberalism in the Postwar Economic Order, *International Organization*, 36 (2): 379–415.

Ruggie, J.G. (2013) *Just Business: Multinational Corporations and Human Rights*, New York: W.W. Norton & Company.

Rutherford, S. (2000a) *The Poor and Their Money*, New Delhi: Oxford University Press.

Sachs, J. (1989) 'The Debt Overhang of Developing Countries', in Calvo, G., Findlay, R., Kouri, P. and de Macedo, J.B. (eds) *Debt Stabilization and Development: Essay in Memory of Carlos F. Diaz Alejandro*, Oxford: Basil Blackwell.

Sachs, J. (2002) 'Weapons of Mass Salvation', *The Economist*, 24 October.

Sachs, J. (2005) *The End of Poverty*, London: Penguin.

Sachs, J. and McArthur, J. (2009) 'Moyo's Confused Attack on Aid for Africa', *Huffington Post*, 27 May, available at: www.huffingtonpost.com/jeffrey-sachs/moyos-confused-attack-on_b_208222. html (accessed 24 June 2014).

Sachs, J. and Williamson, J. (1986) 'Managing the LDC debt crisis', *Brookings Papers on Economic Activity*, 2: 397–440.

Sachs, J.D., McArthur, J.W., Schmidt-Traub, G., Kruk, M., Bahadur, C., Faye M. and McCord G. (2004) 'Ending Africa's Poverty Trap', *Brookings Paper on Economic Activity*, 35 (1): 117–240.

Sachs, W. (ed.) (1992) *The Development Dictionary: A Guide to Knowledge as Power*, London: Zed.

Said, E.W. (1978) *Orientalism: Western Conceptions of the Orient*, London: Penguin Books.

Samuelson, P.A. (1955) *Economics*, 3rd edn, McGraw-Hill.

Sarno, L. and Taylor, M.P. (1999a) 'The Persistence of Capital Inflows and the Behaviour of Stock Prices in East Asia Emerging Markets: Some Empirical Evidence', CEPR Discussion Papers 2150, Centre for Economic Policy Research.

Sarno, L. and Taylor, M. (1999b) 'Hot Money, Accounting Labels and the Permanence of Capital Flows to Developing Countries: An Empirical Investigation', *Journal of Development Economics*, 59 (2): 337–364.

Sassen, S. (1996) *Losing Control? Sovereignty in an Age of Globalization*, New York, Columbia University Press.

Savage, K. and Harvey, P. (eds) (2007) 'Remittances During Crises: Implications for Humanitarian Response', *HPG Report 25*, London: Overseas Development Institute.

Save the Children (2012) *A Life Free from Hunger: Tackling Child Malnutrition*, London: Save the Children UK.

Savedoff, W.D., Levine R. and Birdsall, N. (2006) 'When Will We Ever Learn?: Improving Lives through Impact Evaluation', Washington, DC: CGD.

Say, J.B. ([1803] 1832) *A Treatise on Political Economy; or, The Production, Distribution, and Consumption of Wealth*, 4th edn, Philadelphia, PA: Lippincott, Grambo & Co.

Sayan, S. (2006) 'Business Cycles and Workers' Remittances: How Do Migrant Workers Respond to Cyclical Movements of GDP at Home?', IMF Working Paper 06/52, Washington, DC: International Monetary Fund.

Schneider, F. and Enste, D. (2000) 'Informal Economies: Size, Causes, and Consequences', *The Journal of Economic Literature*, 38 (1): 77–114.

Schuler, S.R., Hashemi, S.M. and Badal, S.H. (1998) 'Men's Violence Against Women in Rural Bangladesh: Undermined or Exacerbated by Microcredit Programmes?', *Development in Practice*, 8 (2): 148–157.

Schumpeter, J. (1911) *The Theory of Economic Development*, New Brunswick, NJ: Transaction Publishers, 1983.

Schumpeter, J. (1954) *History of Economic Analysis*, London: Routledge, 2006 with an introduction by M. Perlman.

Scott, J. (1998) *Seeing Like a State: How Certain Schemes to Improve the Human Condition Have Failed*, Yale University Press.

Scott, J. (2009) *The Art of Not Being Governed: An Anarchist History of Upland Southeast Asia*, New Haven, CT: Yale University Press.

Seers, D. (1969) 'The Meaning of Development', *International Development Review*, 11 (4): 3–4.

Seers, D. (1979) 'The Meaning of Development, with a Postscript', in Lehmann, D. (ed.) *Development Theory: Four Critical Studies*, London: Frank Cass.

Sen, A.K. (1984) *Resources, Values, and Development*, Cambridge, MA: Harvard University Press.

Sen, A.K. (1990) 'More than 100 Million Women are Missing', *New York Review of Books*, 20 December.

Sen, A. (1992) *Inequality Reexamined*, Cambridge, MA: Harvard University Press.

Sen, A. (1999) *Development as Freedom*, Oxford: Oxford University Press.

Shackle, G.L.S. (1983) 'The Bounds of Unknowledge', in Wiseman J. (ed.) *Beyond Positive Economics?* London: Macmillan.

Shaw, G.K. (1992) 'Policy Implications of Endogenous Growth Theory', *The Economic Journal*, 102 (412): 611–662.

Shaw, S. (1973) *Financial Deepening in Economic Development*, Oxford: Oxford University Press.

Shaxson, N. (2011) *Treasure Islands: Tax Havens and the Men who Stole the World*, London: The Bodley Head.

Shelley, T. (2007) *Exploited: Migrant Labour in the New Global Economy*, London: Zed.

Simmel, G. (1907) *The Philosophy of Money*, London: Routledge.

Sidaway, J.D. (2008) 'Post-development', in Desai, V. and Potter, R.B. (eds) *The Companion to Development Studies*, London: Routledge.

Sinclair, T. (1994) 'Passing Judgement: Credit Rating Processes as Regulatory Mechanisms of Governance in the Emerging World Order', *Review of International Political Economy*, 1 (1): 133–159.

Sinclair, T.J. (2005) *The New Masters of Capital: American Bond Rating Agencies and the Politics of Creditworthiness*, Ithaca, NY: Cornell University Press.

Singer H. W. (1950) 'The Distribution of Gains Between Investing and Borrowing Countries', *American Economic Review*, 40 (2): 473–485.

Singh, A. (1998) 'Savings, Investment and the Corporation in the East Asian Miracle', *Journal of Development Studies*, 34 (6): 112–137.

Skidelsky R. (2009) *Keynes: The Return of the Master*, London: Allen Lane.

Skocpol, T. (1995) 'Stategies of Analysis in Current Research', in Evans, P.B., Rueschemeyer, D. and Skocpol, T. (eds) *Bringing the State Back in*, Cambridge: Cambridge University Press.

Smith, A. (1776) *An Inquiry into the Nature and Causes of the Wealth of Nations*, London: Penguin.

Smithin, J. (2000a) 'What is Money? Introduction', in J. Smithin (ed.) *What is Money?* London: Routledge.

Smithin, J. (ed.) (2000b) *What is Money?*, London: Routledge.

Soederberg, S. (2004) *The Politics of the New International Financial Architecture: Reimposing Neoliberal Dominance in the Global South*, London: Zed.

Solow, R. (1956) 'A Contribution to the Theory of Economic Growth', *Quarterly Journal of Economics*, 70 (1): 65–94.

Solow, R. (1957) 'Technical Change and the Aggregate Production Function', *Review of Economics and Statistics*, 39 (3): 312–320.

Solow, R. (2001) 'Applying Growth Theory Across Countries', *The World Bank Economic Review*, 15: 283–288.

Soto, M. (2000) 'Capital Flows and Growth In Developing Countries: Recent Empirical Evidence', OECD Technical Paper 160, Paris: OECD Development Centre.

Spratt, S. (2009) *Development Finance: Debates, Dogmas and New Directions*, London: Routledge.

Spread, P. (2012) 'Science and Support: The Struggle for Mastery in Economics', *Real-world Economics Review*, 59: 39–57, available at: www.paecon.net/PAEReview/issue59/Spread59.pdf (accessed 25 March 2012).

Steger, M.B. and Roy, R.K. (2010) *Neoliberalism: A Very Short Introduction*, Oxford: Oxford University Press.

Stigler, G.J. (1971) 'The Theory of Economic Regulation', *Bell Journal of Economics and Management Science*, 2: 3–21.

Stiglitz, J. (2002) *Globalization and its Discontents*, New York: W.W. Norton.

Stiglitz, J.E. and Weiss, A. (1981) 'Credit Rationing in Markets with Imperfect Information', *The American Economic Review*, 71 (3): 393–410.

Strange, S. (1986) *Casino Capitalism*, London: Basil Blackwell.

Strange, S. (1994) *States and Markets*, 2nd edn, London: Continuum.

Strange, S. (1996) *The Retreat of the State: The Diffusion of Power in the World Economy*, Cambridge: Cambridge University Press.

Strange, S. (1997) *Casino Capitalism*, 2nd edn, Manchester: Manchester University Press.

Strauss-Kahn D. (2010) 'An IMF for the 21st century', address to the Bretton Woods Committee Annual Meeting, Washington DC, 26 February, available at: www.imf.org/external/np/speeches/2010/022610.htm (accessed 9 May 2010).

Suleri, A.Q. and Savage, K. (2006) 'Remittances in Crises: A Case Study from Pakistan', *HPG Background Paper*, London: Overseas Development Institute.

Sulla, O. (2007) 'Philanthropic Foundations and their Role in International Development Assistance', International Finance Briefing Note, Number 3, Washington, DC, World Bank/DECPG.

Sullivan, N.P. (2007) *You Can Hear Me Now: How Microloans and Cell Phones are Connecting the World's Poor to the Global Economy*, London: Wiley.

Summers, L. (2000) 'International Financial Crises: Causes, Prevention, and Cures', *American Economic Review*, 90 (2): 1–16.

Sumner, A. (2005) 'Is Foreign Direct Investment Good for the Poor?' *Development in Practice*, 15 (3–4): 269–285.

Sumner, A. (2010) 'Global Poverty and the New Bottom Billion: Three-quarters of the World's Poor Live in Middle-income Countries', *IDS Working Paper 349*, Brighton: IDS.

Sumner, A. (2011) 'Aid, India and Peanuts', *Global Dashboard*, available at: www.globaldashboard. org/2011/01/06/aid-india-and-peanuts/ (accessed 20 March 2012).

Sumner, A. and Tiwari, M. (2010) 'Global Poverty to 2015 and Beyond: What has been the Impact of the MDGs and what are the Options for a Post-2015 Global Framework?', IDS Working Paper 348, Brighton: IDS.

Sundaram, J.K. (ed.) (2010) *Reforming the International Financial System For Development*, New York: Columbia University Press.

Sutcliffe, B. (2008) 'Marxism and Development', in Dutt, A.K. and Ros, J. (eds) *International Handbook of Development Economics, Volume 1*, Cheltenham: Edward Elgar.

Svensson, J. (1999) 'Aid, Growth and Democracy', *Economics and Politics*, 11: 275–297.

Svensson, J. (2000) 'Foreign Aid and Rent-Seeking', *Journal of International Economics*, 51 (2): 437–461.

Svensson, L.E. (1994) 'The Operation and Collapse of Fixed Exchange Rate Regimes', NBER Working Paper Series No. 4971.

Swan, T.W. (1956) 'Economic Growth and Capital Accumulation', *Economic Record*, 32: 334–361.

Sweezy, P. (1972) 'The Ressurgence of Financial Control: Fact or Fancy?' *Socialist Revolution*, 8 (2): 157–192.

Szirmai, A. (2005) *The Dynamics of Socio-Economic Development: An Introduction*, Cambridge: Cambridge University Press.

Taylor, I. (2005) 'The Developmental State in Africa: The Case of Botswana', in Mbabazi, P. and Taylore, I. (eds) *The Potentiality of 'Developmental States' in Africa*, Dakar: CODESRIA.

Teboul, R. and Moustier, E. (2001) 'Foreign Aid and Economic Growth: The Case of the Countries South of the Mediterranean', *Applied Economics Letters*, 8 (3): 187–190.

Teunissen, J.J. (2006) 'Should Developing Countries Support the US Dollar? By Way of Introduction', in Teunissen, J.J. and Akkerman, A. (eds) *Global Imbalances and the US Debt Problem: Should Developing Countries Support the US Dollar?*, The Hague: FONDAD.

The Economist (1977) 'The Dutch Disease', *The Economist*, 26 November: 82–83

The Economist (2005) 'The Hidden Wealth of the Poor: A Survey of Microfinance', *The Economist*, 5 November.

The Economist (2008a) 'A Worthless Currency', *The Economist*, 17 July.

The Economist (2008b) Poor People, Rich Returns' *The Economist*, 15 May.

The Economist (2008c) 'Doing Good by Doing Very Nicely Indeed', *The Economist*, 26 June.

Thérien, J.P. (2002) 'Debating Foreign Aid: Right Versus Left', *Third World Quarterly*, 23 (3): 449–466.

Thirkell-White, B. (2005) *The IMF and the Politics of Financial Globalisation: From the Asian Crisis to a New International Financial Architecture*, Basingstoke: Palgrave Macmillan.

Thirwall, A.P. (2007) 'Keynes and Economic Development', *Economia Aplicada*, 11 (3): 447–457 .

Thirwall, A.P. (2011) *Economics Of Development: Theory and Evidence*, Basingstoke: Palgrave Macmillan.

Thomas, A. (2000) 'Meanings and Views of Development', in Allen, T. and Thomas, A. (eds) *Poverty and Development into the 21st Century*, Oxford: Oxford University Press.

Tobin, J. (1978) 'A Proposal for International Monetary Reform', *Eastern Economic Journal*, 4 (3–4): 153–159.

Tobin, J. (1996) 'Prologue', in ul Haq, M., Kaul, I. and Grunberg, I. (eds) *The Tobin Tax: Coping with Financial Volatility*, Oxford: Oxford University Press.

Tobin, J. (2001) 'An Interview with James Tobin: They are Misusing My Name', *Der Spiegel*, 2 September.

Tonelson, A. (2002) *The Race to the Bottom: Why a Worldwide Worker Surplus and Uncontrolled Free Trade are Sinking American Living Standards*, Boulder, CO: Westview Press.

Toporowski, J. (2010) *Why the World Economy Needs a Financial Crash and Other Critical Essays on Finance and Financial Economics*, Anthem Press.

Tornell, A., Westermann, F. and Martinez, L. (2004) 'The Positive Link Between Financial Liberalization, Growth and Crises', NBER Working Paper No. 10293, Washington, DC: National Bureau of Economic Research.

Toussaint, E. and Millet, D. (2010) *Debt, the IMF, and the World Bank: Sixty Questions, Sixty Answers*, New York: Monthly Review Press.

Townsend, R. (1994) 'Risk and Insurance in Village India', *Econometrica*, 62: 539–591.

Toye, J. (1987) *Dilemmas of Development: Reflections on the Counter-Revolution in Development Theory and Policy*, Oxford: Blackwell.

Tribe, M., Nixson, F. and Sumner A. (2010) *Economics and Development Studies*, London: Routledge.

Triffin, R. (1960) *Gold and the Dollar Crisis: The Future of Convertibility*, New Haven, CT: Yale University Press.

Tripathi, S. (2006) 'Microcredit Won't Make Poverty History', *The Guardian*, 17 October, available at: www.guardian.co.uk/business/2006/oct/17/businesscomment.internationalaidanddevelopment (accessed on 11 August 2010).

Truman, H.S. (1949) 'Inaugural address of Harry S. Truman', speech, presidential inauguration ceremony, Washington, DC, 20 January 1949.

Tsai, P-L. (1995) 'Foreign Direct Investment and Income Inequality: Further Evidence', *World Development*, 23 (3): 469–483.

Tucker, R.C. (ed.) (1978) *The Marx-Engels Reader*, New York: Norton.

Turner, M.S. (1989) *Joan Robinson and the Americans*, M.E. Sharpe.

Turner, M.S. (1990) 'The Cambridge Keynesians and the "Bastard Keynesians": A Comment on Economists and their Understanding of the Inflationary Aspects of Keynesian Policy', *Journal of Economic Issues*, 24 (3): 886–890.

UN (2010a) *The Millennium Development Goals Report 2010*, New York: United Nations.

UN (2010b) 'Keeping the Promise – United To Achieve The Millennium Development Goals'. Resolution adopted by the General Assembly, A/RES/65/1, 19 October, available at: www.un.org/en/mdg/summit2010/pdf/outcome_documentN1051260.pdf

UN (2010c) 'Press release: UN Summit concludes with adoption of global action plan to achieve development goals by 2015', New York: United Nations, available at: www.un.org/millenniumgoals/pdf/Closing%20press%20release%20FINAL-FINAL%20Rev3.pdf (accessed 11 August 2010).

UN Commission of Experts (2009) 'Report of the Commission of Experts of the President of the United Nations General Assembly on Reforms of the International Monetary and Financial System', New York: United Nations, available at: www.un.org/ga/econcrisissummit/docs/FinalReport_ CoE.pdf (accessed 9 May 2010).

UN DESA (2005) *World Economic and Social Survey 2005: Financing for Development*, New York: UN Department of Economic and Social Affairs.

UN Millennium Project (2005) *Investing in Development: A Practical Plan to Achieve the Millennium Development Goals: Overview*, New York: United Nations.

UN OHRLLS (2010) 'UN Office Of The High Representative for the Least Developed Countries, Landlocked Developing Countries and Small Island Developing States', available at: www.unohrlls. org/ (accessed 30 October 2010).

UNCDF (2005) 'Microfinance and the Millennium Development Goals', available at: www. yearofmicrocredit.org/docs/mdgdoc_MN.pdf (accessed 11 August 2010).

UNCTAD (1999) *World Investment Report 1999: FDI and the Challenge of Development*, New York: UNCTAD.

UNCTAD (2000) *World Investment Report 2000: Cross-Border Mergers and Acquisitions and Development*, New York: UNCTAD.

UNCTAD (2007) *World Investment Report 2007: Transnational Corporations, Extractive Industries and Development*, New York: UNCTAD.

UNCTAD (2009) *Trade and Development Report 2009: Transnational Corporations, Agricultural Production and Development*, New York: United Nations.

UNCTAD (2010) *World Investment Report 2010: Investing in a Low-Carbon Economy*, New York and Geneva: UNCTAD.

UNCTAD (2011) *World Investment Report 2011: Non-equity Modes of International Production and Development*, New York and Geneva: UNCTAD.

UNCTAD (2012) *World Investment Report 2012: Towards a New Generation of Investment Policies*, New York and Geneva: UNCTAD.

UNDP (1995) *Human Development Report 1995: Gender and Human Development*, Oxford: Oxford University Press.

UNDP (1996) *Human Development Report 1996: Economic Growth and Human Development*, Oxford: Oxford University Press.

UNDP (2006) *Human Development Report 2006: Beyond Scarcity: Power, Poverty and the Global Water Crisis*, Basingstoke: Palgrave/United Nations Development Programme.

UNDP (2009) *Human Development Report 2009: Overcoming Barriers: Human Mobility and Development*, Basingstoke: Palgrave/United Nations Development Programme.

UNDP (2010) *Human Development Report 2010: The Real Wealth of Nations: Pathways to Human Development*, Basingstoke: Palgrave/United Nations Development Programme.

UNDP (2011) *Human Development Report 2011: Sustainability and Equity: A Better Future for All*, Basingstoke: Palgrave/United Nations Development Programme.

UNGA (1970) 'International Development Strategy for the Second United Nations Development Decade', UN General Assembly Resolution 2626 (XXV), 24 October, available at: www.un.org/ documents/ga/res/25/ares25.htm (accessed 1 August 2010).

UNPD (2009) *Trends in International Migrant Stock: The 2008 Revision*, United Nations database, POP/DB/MIG/Stock/Rev.2008, available at: www.unmigration.org (accessed 2 September 2010).

USAID (2003) *Foreign Aid in the National Interest: Promoting Freedom, Security, and Opportunity*, Washington, DC: US Agency for International Development, available at www.usaid.gov/fani (accessed 1 August 2010).

USAID (2010) 'This is USAID', available at: www.usaid.gov/about_usaid/ (accessed 1 August 2010).

van der Veen, A.M. (2011) *Ideas, Interests, and Foreign Aid*, Cambridge: Cambridge University Press.

Van Heerde, J. and Hudson, D. (2010), '"The Righteous Considereth the Cause of the Poor"'? Public Attitudes towards Poverty in Developing Countries', *Political Studies*, 58: 389–409.

Vandemoortele, J. (2002) *Are the MDGs Feasible?* New York: UNDP Bureau for Development Policy.

Vernon, R. (1966) 'International Investment and International Trade in the Product Life Cycle', *The Quarterly Journal of Economics*, 80 (2): 190–207.

Vertovec, S. (2006) 'Diasporas Good? Diasporas Bad?', Working Paper No. 41, Oxford: Centre on Migration Policy and Society.

Vreeland, J.R. (2003) *The IMF and Economic Development*, Cambridge: Cambridge University Press

Vreeland, J.R. (2007) *The International Monetary Fund: Politics of Conditional Lending*, London: Routledge.

Wade, R.H. (1990) *Governing the Market*, Princeton, NJ: Princeton University Press.

Wade, R.H. (2000) 'Governing the Market: A Decade Later', *LSE Development Studies Institute Working Paper No. 00-03*, London: London School of Economcis.

Wade, R. H. (2003) 'What Strategies are Viable for Developing Countries Today? The World Trade Organization and the Shrinking of "Development Space" ', *Review of International Political Economy*, 10 (4): 621–644.

Wade, R.H. (2004) 'On the Causes of Increasing World Poverty and Inequality, or Why the Matthew Effect Prevails', *New Political Economy*, 9 (2): 163–188.

Wade, R. H. (2010) 'What Strategies are Viable for Developing Countries Today? The World Trade Organization and the Shrinking of "Development Space" ', *Review of International Political Economy*, 10 (4): 621–644.

Wade, R.H. and Veneroso, F. (1998) 'The Asian Crisis: The High Debt Model versus the Wall Street-IMF-Treasury Complex', *New Left Review*, 228: 3–24.

Wallerstein, I. (1979) *The Capitalist World-Economy,* Vol. 2, Cambridge: Cambridge University Press.

Walras, L. ([1954] 2003) *Elements of Pure Economics*, London: Routledge.

Warren, B. (1980) *Imperialism, Pioneer of Capitalism*, ed. J. Sender, London: Verso.

Warwick Commission (2009) *The Warwick Commission on International Financial Reform: In Praise of Unlevel Playing Fields*, Warwick: University of Warwick.

Washington Post (2006) 'Editorial: Counting Aid Dollars', *Washington Post*, 16 April.

Watkins, K. (2009) 'Why Dead Aid is Dead Wrong', *Prospect*, 158, 4 May.

Watson, M. (1999) 'Rethinking Capital Mobility, Re-regulating Financial Markets', *New Political Economy*, 4 (1): 55–75.

Watson, M. (2005) *Foundations of International Political Economy*, Basingstoke: Palgrave.

WDI (2012) *World Bank World Development Indicators*, available at: http://data.worldbank.org/data-catalog/world-development-indicators.

WDI (2013) *World Bank World Development Indicators*, available at: http://data.worldbank.org/data-catalog/world-development-indicators.

Weber, H. (2002) 'The Imposition of a Global Development Architecture: The Example of Microcredit', *Review of International Studies*, 28: 537–555.

Weber, H. (2004) 'The New Economy and Social Risk: Banking on the Poor', *Review of International Political Economy*, 11 (4): 356–386.

Weiss, L. (1998) *The Myth of the Powerless State: Governing the Economy in a Global Era*, Cambridge: Polity.

Wellink, N. (2002) *Central Banks as Guardians of Financial Stability*, speech by Dr Nout Wellink, president of De Nederlandsche Bank and president of the Bank for International Settlements, at the seminar 'Current issues in central banking, on the occasion of the opening of the new office building of the Central Bank of Aruba, Oranjestad', 14 November, published in the *BIS Review* 63/2002, available at: www.bis.org/review/r021120a.pdf.

White, H. (2006) 'Impact-Evaluation: The Experience of the Independent Evaluation Group of the World Bank', Washington, DC: World Bank.

WHO (2001) 'Smallpox: WHO Factsheet', available at: www.who.int/mediacentre/factsheets/smallpox/en/ (accessed 27 December 2011).

WHO (2012) 'Structural Adjustment Programmes (SAPs)', available at: www.who.int/trade/glossary/story084/en/index.html (accessed 30 March 2012).

Williams, D. (1999) 'Constructing the Economic Space: The World Bank and the Making of Homo Oeconomicus', *Millennium: Journal of International Studies*, 28 (1): 79–99.

Williamson, J. (1990) 'What Washington Means by Policy Reform', in Williamson, J. (ed.) *Latin American Adjustment: How Much Has Happened?*, Washington, DC: Institute for International Economics.

Williamson, J. (2000) 'What Should the World Bank think about the Washington Consensus?' *World Bank Research Observer*, 15 (2): 251–264.

Williamson, J. (2003) 'From Reform Agenda to Damaged Brand Name', *Finance and Development*, 40 (3): 10–13.

Williamson, O.E. (1975) *Markets and Hierarchies, Analysis and Antitrust Implications: A Study in the Economics of Internal Organization*.

Woo-Cummings, M. (1996) *Capital Ungoverned: Liberalizing Finance in Interventionist States*, Ithaca, NY: Cornell University Press.

Woo-Cummings, M. (ed.) (1999) *The Developmental State*, Ithaca, NY: Cornell University Press.

Wood, A. (2008) 'Looking Ahead Optimally in Allocating Aid', *World Development*, 36 (7): 1135–1151.

Wood, R.E. (1992) *From Marshall Plan to Debt Crisis: Foreign Aid and Development Choices in the World Economy*, Berkeley, CA: University of California Press.

Woodruff, C. and Zenteno, R.M. (2001) 'Remittances and Microenterprises in Mexico', UCSD, Graduate School of International Relations and Pacific Studies Working Paper, University of California at San Diego.

Woods, N. (2008) 'Whose Aid? Whose Influence? China, Emerging Donors and the Silent Revolution in Development Assistance', *International Affairs*, 84 (6): 1205–1221.

Woodward, D. (2001) *The Next Crisis? Direct and Equity Investment in Developing Countries*, London: Zed.

Woodward, R. (2004) 'The Organisation for Economic Cooperation and Development', *New Political Economy*, 9 (1): 113–127.

Woodward, R. (2009) *The Organisation for Economic Co-operation and Development*, London: Routledge.

World Bank (1989) *Sub-Saharan Africa: From Crisis to Sustainable Growth; A Long-Term Perspective Study*, Washington, DC: World Bank.

World Bank (1990) *World Development Report 1990: Poverty*, Oxford: Oxford University Press.

World Bank (1993) *The East Asian Miracle: Economic Growth and Public Policy*, Washington, DC: World Bank.

World Bank (1995) *Bureaucrats in Business: The Economics and Politics of Government Ownership*, New York: Oxford University Press.

World Bank (1998a) *Beyond the Washington Consensus: Institutions Matter*, Washington, DC: World Bank.

World Bank (1998b) 'Assessing Aid: What Works, What Doesn't, and Why', *World Bank Policy Research Report*, New York: Oxford University Press/IBRD.

World Bank (2001a) *Global Development Finance 2001: Building Coalitions for Effective Development Finance*, Washington, DC: World Bank.

World Bank (2001b) *World Development Report 2000/01: Attacking Poverty*, Washington, DC: World Bank.

World Bank (2002) 'Education for Dynamic Economies: Action Plan to Accelerate Progress Towards Education for All', prepared by the World Bank Staff for the Spring Development Committee Meeting, April.

World Bank (2003) *Supporting Sound Policies with Adequate and Appropriate Financing: Implementing the Monterrey Consensus at the Country Level*, Washington, DC: World Bank.

World Bank (2004) *Global Development Finance: Harnessing Cyclical Gains for Development*, Washington, DC: World Bank.

World Bank (2005a) *Global Development Finance 2005: Mobilizing Finance and Managing Vulnerability*, Washington, DC: World Bank.

World Bank (2005b) *World Development Report 2005: A Better Investment Climate for Everyone*, Washington, DC: World Bank.

World Bank (2006a) *Global Development Finance 2006: The Development Potential of Surging Capital Flows*, Washington, DC: World Bank.

World Bank (2006b) *Global Economic Prospects 2006: Economic Implications of Remittances and Migration*, Washington, DC: World Bank.

World Bank (2006c) *The Development Impact of Workers' Remittances in Latin America*, Report No. 37026, Washington, DC: World Bank.

World Bank (2008) *Migration and Remittances Factbook 2008*, Washington, DC: World Bank.

World Bank (2010a) *Global Development Finance 2010: External Debt of Developing Countries*, Washington, DC: World Bank.

World Bank (2010b) 'An Analysis of Trends in the Average Total Cost of Migrant Remittance Services', Payment Systems Development Group Policy Note, Version 1, 23 April, available at: http://remittanceprices.worldbank.org/documents/remittancepriceworldwide-Analysis.pdf (accessed 3 September 2010).

World Bank (2010c) 'About Remittance Prices Worldwide', available at: http://remittanceprices.worldbank.org/About/ (accessed 3 September 2010).

World Bank (2010d) *Remittance Price Comparison Databases Minimum Requirements and Overall Policy Strategy*, Global Remittances Working Group, Washington, DC: World Bank, available at: http://remittanceprices.worldbank.org/~/media/FPDKM/Remittances/Documents/StandardsNationalDatabases.pdf (accessed 28 February 2013).

World Bank (2011) *Migration and Remittances Factbook 2011*, 2nd edn, Washington, DC: World Bank.

Worrell *et al.* (2000) *The Political Economy of Exchange Rate Policy in the Caribbean*, produced for the Inter-American Development Bank, available at: www.iadb.org/res/publications/pubfiles/pubr-401.pdf.

Wray, L.R. (1998) *Understanding Modern Money: The Key to Full Employment and Price Stability*, Cheltenham: Edward Elgar.

Wray, L.R. (2000) 'Modern Money', in Smithin, J. (ed.) *What is Money?* London: Routledge.

Wright, G. and Mutesasira, L. (2001) 'The Relative Risks of Savings to Poor People', MicroSave Briefing Note #06, September.

Wriston, W. (1992) 'Decline of the Central Bankers,' *New York Times*, 20 September.

WTO (2010) 'The Challenges of Trade Financing', available at: www.wto.org/english/thewto_e/coher_e/challenges_e.htm (accessed 31 October 2010).

WTO (2011) *International Trade Statistics 2011*, Geneva: World Trade Organisation.

WTO/OECD (2009), *Aid for Trade at a Glance 2009*, Paris: OECD.

Younger, S.D. (1992), 'Aid and the Dutch Disease: Macroeconomic Management When Everybody Loves You', *World Development*, 20: 1587–1597.

Yunus, M. (1999) *Banker to the Poor: Micro-lending and the Battle Against World Poverty*, New York: Public Affairs.

Yunus, M. (2006) Nobel lecture, Norway, Oslo, 10 December, available at: http://nobelprize.org/nobel_prizes/peace/laureates/2006/yunus-lecture-en.html (accessed 11 August 2010).

Zedillo, E., Al-Hamad, A.Y., Bryer, D., Chinery-Hesse, M., Delors, J., Grynspan, R., Livshit, A.Y., Osman, A.M., Rubin, R., Singh, M. and Son, M. (2001) 'Recommendations of the High-Level Panel on Financing for Development', UN General Assembly document A/55/1000, New York: United Nations.

Zelizer, V. (1994) *The Social Meaning of Money: Pin Money, Paychecks, Poor Relief and Other Currencies*, Princeton, NJ: Princeton University Press.

Zhou, X. (2009) 'Reform the International Monetary System', Beijing: People's Bank of China, available at: www.pbc.gov.cn/english/detail.asp?col=6500&id=178 (accessed 9 May 2010).

Index

Locators shown in *italics* refer to tables, figures, and boxes.